Lecture Notes of
the Unione Matematica Italiana

Luc Tartar

From Hyperbolic Systems to Kinetic Theory

A Personalized Quest

 Springer

Luc Tartar
Department of Mathematical Sciences
Carnegie Mellon University
Pittsburgh, PA 15213-3890
USA
tartar@andrew.cmu.edu

ISBN 978-3-540-77561-4   ISBN 978-3-540-77562-1 (eBook)

DOI 10.1007/978-3-540-77562-1

Lecture Notes of the Unione Matematica Italiana ISSN print edition: 1862-9113
                                        ISSN electronic edition: 1862-9121

Library of Congress Control Number: 2007942545

Mathematics Subject Classification (2000): 35K05, 35L45, 35L60, 35L65, 35L67, 35Q30, 70F45, 76A02,
                            76N15, 76P05, 82C22, 82C40

Cover design: WMXDesign GmbH

Printed on acid-free paper

9 8 7 6 5 4 3 2 1

springer.com

Dedicated to Robert DAUTRAY

He helped me at a critical time, when I could no longer bear the rejection in the academic world (partly for having refused the current methods of falsifications, and partly because I was too interested in science for a mathematician), and he also guided me in my readings while I worked at Commissariat à l'Énergie Atomique, so that I did not get lost like many other mathematicians in the jungle of models which physicists have generated, and I could understand what mathematical tools should be developed for helping understand in a better way how nature works.

to Peter LAX

He gave an example of how a good mathematician can work, by putting some order in a corner of the physical world where the preceding knowledge was made up of a few examples and too many guesses. Why have there been so few mathematicians who wanted to follow his example?

to Lucia
to my children, Laure, Michaël, André, Marta
and to my grandson, Lilian

Preface

After publishing *An Introduction to Navier–Stokes Equation and Oceanography* [20],[1,2] and *An Introduction to Sobolev Spaces and Interpolation Spaces* [21],[3] the revised versions of my lecture notes for graduate courses that I had taught in the spring of 1999 and in the spring of 2000, I want to follow with another set of lecture notes for a graduate course that I had taught in the fall of 2001, with the title "Introduction to kinetic theory". For this one, there had been no version available on the Internet, and I had not even written the notes for the last four lectures, and after a few years, I find it useful to make the text available to a larger audience by publishing a revised and completed version, but I had to change the title in a significant way.

In [21], I had written that my reasons for publishing lecture notes is to tell the readers some of what I have understood, the technical mathematical aspects of the course, the scientific questions behind the theories, and more, and I shall have succeeded if many become aware, and go forward on the path of discovery, not mistaking research and development, knowing when and why they do one or the other, and keeping a higher goal in mind when for practical reasons they decide to obey the motto of the age for a while, *publish or perish*.

In the fall of 2001, I had done precisely that, and I had taught the mathematical results that I had proven during my quest for understanding about

---

[1] Claude Louis Marie Henri NAVIER, French mathematician, 1785–1836. He had worked in Paris, France. He introduced the equation now known as the *Navier–Stokes equation* in 1821, although he did not understand about shear stress.

[2] Sir George Gabriel STOKES, Irish-born mathematician, 1819–1903. He had worked in London, and in Cambridge, England, holding the Lucasian chair (1849–1903).

[3] Sergei L'vovich SOBOLEV, Russian mathematician, 1908–1989. He had worked in Leningrad, in Moscow, and in Novosibirsk, Russia. There is now a Sobolev Institute of Mathematics of the Siberian branch of the Russian Academy of Sciences, Novosibirsk, Russia. I first met Sergei SOBOLEV when I was a student, in Paris in 1969, and conversed with him in French, which he spoke perfectly (all educated Europeans at the beginning of the 20th century learned French).

kinetic theory, which I had started in the early 1970s, but I had also taught about what is wrong with kinetic theory, which I had started to understand in the early 1980s, and I had tried to teach a little about continuum mechanics and physics with the critical mind of a mathematician, so that the students could understand what were the results of my detective work on this particular question of kinetic theory, and understand how to attack other questions of continuum mechanics or physics by themselves later (having in mind the defects that have already been found on each question, by me or by others).

In [21], I had suggested to the readers who already know something about continuum mechanics or physics to look at my lecture notes, to read about the defects which I know about in classical models, because other authors rarely mention these defects even though they have heard about them. This set of lecture notes, written with a concern towards kinetic theory, is of this type. I had suggested to the readers who do not yet know much about continuum mechanics or physics, to start with more classical descriptions about the problems, for example by consulting the books which have been prepared under the direction of Robert DAUTRAY,[4] and of Jacques-Louis LIONS,[5] whom he had convinced to help him, [5]–[13].

I have mentioned that my personal point of view, which is that one should not follow the path of the majority when reason clearly points to a different direction, probably owes a lot to having been raised as the son of a (Calvinist) Protestant minister,[6] but I had lost the faith when I was twelve or thirteen years old, and I may not have explained well why I later found myself forced to practice the art of the detective in deciding what had to be discarded from what I could reasonably trust until some new information became available. Becoming a mathematician had been one of the reasons, because mathematicians must know what is proven and what is only conjectured, and when later I became interested in understanding continuum mechanics and physics from a mathematical point of view, I found that the analysis that must be done in organizing the information, as well as the misinformation that "scientists" transmit about the real world, is quite similar to the analysis that must be done in organizing the information and misinformation that various religious

---

[4] Ignace Robert DAUTRAY (KOUCHELEVITZ), French physicist, born in 1928.

[5] Jacques-Louis LIONS, French mathematician, 1928–2001. He received the Japan Prize in 1991. He had worked in Nancy and in Paris, France; he held a chair (analyse mathématique des systèmes et de leur contrôle, 1973–1998) at Collège de France, Paris. The laboratory dedicated to functional analysis and numerical analysis which he initiated, funded by CNRS (Centre National de la Recherche Scientifique) and Université Paris VI (Pierre et Marie Curie), is now named after him, the Laboratoire Jacques-Louis Lions. I first had Jacques-Louis LIONS as a teacher at École Polytechnique in Paris in 1966–1967, and I did research under his direction, until my thesis in 1971.

[6] Jean CALVIN (CAUVIN), French-born theologian, 1509–1564. He had worked in Paris and in Strasbourg, France, in Basel and in Genève (Geneva), Switzerland.

traditions transmit, and in both these approaches, one can observe the perverse influence of political factors.

The particular difficulty that I had encountered myself around 1980 was related to the political perversion of the French academic system itself, because I found myself facing an unimaginable situation of forgeries, organized by a "mathematician" and continued by a "physicist", which turned into a nightmare when I was repeatedly confronted with the racist behaviour of those who insisted that it was normal that I should not have the same rights as others.[7]

Fortunately, Robert DAUTRAY provided me with a new job outside this strange "academic" world,[8] and I was extremely grateful to him for that, as it contrasted a lot with the rejection that I was feeling in the mathematical world, including the strange opposition of my mentors, Laurent SCHWARTZ and Jacques-Louis LIONS,[9] who had chosen the side of the forgers against me, probably because they had some different, wrong information. However, I am even more grateful to Robert DAUTRAY for something that very few people could have provided me, as my understanding of physics could not have improved in the way it did without his help, which was mostly through telling me what to read, and it is natural that I should dedicate this set of lecture notes to him, although he may not agree entirely with my personal analysis on the subject of kinetic theory.

My new job, or more precisely what I had understood about what I had to do, had been both simple and impossible, to understand physics in a better way, through a mathematical approach, of course. I felt that Robert DAUTRAY understood that physics had reached a few dead ends, where physicists were hitting some walls which had been created before them, by other physicists who had invented the wrong games for understanding how nature works. It should not have been too critical, as it is natural that guessing produces a few answers that are not completely right, although they may not be completely wrong, and using the art of the engineer one can make things work even though one does not have the correct equations for describing the processes that one wants to tame, but this approach in science has its limitations. In order to go forward, one needs to apply a scientific approach, and practice the art of the detective to discover what has been done wrong, and then one needs to do it in a better way, ideally in the right way, if that is possible. I thought that Robert DAUTRAY was not only aware of that, but that he saw that some of

---

[7] This happened in one of the campuses of University Paris XI (Paris Sud), Orsay, France, from 1979 to 1982.

[8] I worked at CEA (Commissariat à l'Énergie Atomique) in Limeil, France, from 1982 to 1987.

[9] Laurent SCHWARTZ, French mathematician, 1915–2002. He received the Fields Medal in 1950. He had worked in Nancy, in Paris, France, at École Polytechnique, which was first in Paris (when I had him as a teacher in 1965–1966), and then in Palaiseau, and at Université Paris 7 (Denis Diderot), Paris.

this work of providing more order must be done by mathematicians, at least well-trained mathematicians.

The job of a detective is certainly made quite difficult if he/she is forbidden to ask questions to important witnesses, or if he/she realizes that there is a wall of silence and that there is information that could be useful for his/her search which some powerful group does not want him/her to discover. That type of difficulty exists in physics, as well as in other sciences, including mathematics. At the beginning, some guessed rule had been successful in one situation, and although it was dangerous to apply a similar guess indiscriminately for all kinds of problems, it had been done, but what made this practice quite unfortunate was then to create a dogma, and to teach it to new generations of students. Because no hints were given that some of these rules could be slightly wrong, or even completely misleading, these physicists were not really trained as scientists, and it is not surprising that many of them ended up working like engineers, mistaking physics and technology, and not caring much for the fact that some of the currently taught "laws of physics" are obviously wrong: they are simply the laws that physicists have guessed in their quest about the laws that nature follows, and it would have been surprising that their first guess had been right.

Before 1982, I had mostly thought about questions concerning continuum mechanics, developing homogenization and the compensated compactness method, partly with François MURAT,[10] but I had also understood a question of the appearance of nonlocal effects by homogenization of some hyperbolic equations, and I thought that this was a more rational explanation than the strange games of spontaneous absorption and emission that physicists had invented, so that their probabilistic games were just one possible approach to describing the correct effective equations, confirming what I had already discovered before, that probabilities are introduced by physicists when they face a situation that they do not understand, so that it should be pointed out how crucial it is to introduce probabilities as late as possible in the analysis of a problem, ideally not at all if possible, but certainly further and further away from one generation to the next. However, up to 1982, I did not see how to include quantum mechanics and statistical mechanics in my approach to the partial differential equations of continuum mechanics and physics.

After 1982, the first step was relatively easy, and in reading what Robert DAUTRAY had told me I identified a few points which are certainly wrong in the laws that physicists use; however, making them right seemed to require the development of new mathematical tools. The tool of H-measures [18], which I started describing at the end of 1986, was something that I had already guessed two years before, but its extension to semi-linear hyperbolic systems

---

[10] François MURAT, French mathematician, born in 1947. He works at CNRS (Centre National de la Recherche Scientifique) and Université Paris VI (Pierre et Marie Curie), Paris, France.

has eluded me since, and I see that extension as necessary to explain some of the strange rules about quantum mechanics, and then derive better rules than those of statistical mechanics.

At the end of 1983, a year before the first hint about new mathematical tools, I already "knew" what is wrong with kinetic theory, which is the subject of this set of lecture notes, as a consequence of having "understood" what is wrong with quantum mechanics. As I am a mathematician, I use quotes because I want to emphasize that it was not yet mathematical knowledge, and it was not about a precise conjecture either because I could not formulate one at the time, but I had acquired the certitude that some aspects of what the physicists say will not appear in the new mathematical framework that I was searching for.

The main mistake of physicists had been to stick to 18th century ideas of classical mechanics, instead of observing that if the 19th century ideas about continuum mechanics are inadequate for explaining what is observed at a microscopic level, it is because one needs new mathematical tools for 20th century mechanics/physics (turbulence, plasticity, atomic physics), which have no probability in them, of course, as the use of probabilities is the sign that one does not understand what is going on. It had been a mistake to concentrate too much effort on problems of partial differential equations which show finite-dimensional effects, for which 18th century mechanics is adapted, instead of observing that the more interesting problems of partial differential equations all show infinite-dimensional effects, which cannot be grasped with 18th/19th century ideas; actually, my subject of research since the early 1970s had been precisely focused on studying the effect of microstructures in partial differential equations, a subject which I have decided to describe as *beyond partial differential equations*. The certitude that mathematics brings is that there are absolutely no particles at atomic level, there are only waves, so that there cannot be any particles interacting in the way that had been assumed by MAXWELL,[11] and by BOLTZMANN.[12]

Nevertheless, one should be careful not to disparage MAXWELL and BOLTZMANN for the fact that their pioneering work in kinetic theory has some defects, because they had shown a good physical intuition for the way to correct an important defect of continuum mechanics, which is that the constitutive relations used are wrong, because they result from the inexact postulate that the relations valid at equilibrium are true at all times.

That there are no particles and that they are waves could have been understood earlier, as a consequence of an observation of POINCARÉ in his study

---

[11] James CLERK MAXWELL, Scottish physicist, 1831–1879. He had worked in Aberdeen, Scotland, in London and in Cambridge, England, holding the first Cavendish professorship of physics (1871–1879).

[12] Ludwig BOLTZMANN, Austrian physicist, 1844–1906. He had worked in Graz and Vienna, Austria, in Leipzig, Germany, and then again in Vienna.

of relativity,[13] that instantaneous forces at a distance do not make any sense, which EINSTEIN after him had probably not understood so well,[14] and that "particles" feel a field that transmits the interactions as waves, but POINCARÉ had died many years before the wave nature of "particles" was confirmed by an observation of L. DE BROGLIE in his study of "electrons",[15] that they are waves. Unfortunately, the idea that there are only waves and no particles was then completely messed up in the following development of quantum mechanics, which led to that strange dogmatic discipline where "nonexistent particles" are assumed to play "esoteric probabilistic games".

At the end of 1983, I had then "understood" that there are absolutely no particles at a microscopic level, so that *real gases are not made of particles*, and I understood it in a mathematical way in the late 1980s, by introducing H-measures [18], which are related to oscillations and concentration effects in weakly converging sequences, and then by proving transport equations for them when one considers sequences of solutions of particular linear hyperbolic systems. Better mathematical results are still needed in order to understand the case of semi-linear hyperbolic systems, which I believe is the mathematical problem to study to explain all the strange effects which are observed at a microscopic level.

Although MAXWELL and BOLTZMANN had done quite a good job in postulating their equations for kinetic theory, because it is not yet clear a century and a half after them how to write the equations correctly, it is useful to describe some defects in their work to show some limitations of kinetic theory, in the same way that one shows the limitations of classical mechanics by pointing out that NEWTON's work was unchallenged for two centuries,[16] until relativity was introduced by POINCARÉ, and then EINSTEIN, so that one

---

[13] Jules Henri POINCARÉ, French mathematician, 1854–1912. He had worked in Paris, France. There is now an Institut Henri Poincaré (IHP), dedicated to mathematics and theoretical physics, part of Université Paris VI (Pierre et Marie Curie), Paris.

[14] Albert EINSTEIN, German-born physicist, 1879–1955. He received the Nobel Prize in Physics in 1921, for his services to theoretical physics, and especially for his discovery of the law of the photoelectric effect. He had worked in Bern, in Zürich, Switzerland, in Prague, now capital of the Czech Republic, at ETH (Eidgenössische Technische Hochschule), Zürich, Switzerland, in Berlin, Germany, and at IAS (Institute for Advanced Study), Princeton, NJ. The Max Planck Institute for Gravitational Physics in Potsdam, Germany, is named after him, the Albert Einstein Institute.

[15] Prince Louis Victor Pierre Raymond DE BROGLIE, 7th Duc de Broglie, French physicist, 1892–1987. He received the Nobel Prize in Physics in 1929, for his discovery of the wave nature of electrons. He had worked in Paris, France.

[16] Sir Isaac NEWTON, English mathematician, 1643–1727. He had worked in Cambridge, England, holding the Lucasian chair (1669–1701). The Isaac Newton Institute for Mathematical Sciences in Cambridge, England, is named after him.

knows now that one needs relativistic corrections when the velocities involved can be compared with the speed of light c.

Some have thought that what I had understood with H-measures was well known, but it is exactly as if one says that Laurent SCHWARTZ's theory of distributions had been introduced by DIRAC,[17] and the authors of such remarks only show that they cannot recognize mathematics when they see it. However, such deceptive statements were also made by good mathematicians, and in that case it shows something else: in each religion, there is a fundamentalist party who is interested in enforcing dogmas, not always because all these people believe in them, but often because some prefer to slow down the advance of knowledge (usually for keeping the power they have over the naive who believe in these dogmas), and in the case that I consider it means slowing down the evolution of science in general, and physics in particular, and it is not too difficult to understand the political motivation of those who behave in this way, and they often associate with people who do not hide that their work is political, but insist in brainwashing the naive that it is correct.

Although I advocate using reason for criticizing without concessions the points of view that are taught in order to find better "truths", one should observe that this approach is more suited to mathematicians than to physicists or engineers, but not all mathematicians have been trained well enough for following that path, and that might explain why some people initially trained as mathematicians write inexact statements, which they often do not change after being told about their mistakes, which others repeat then without knowing that they propagate errors; if their goal had not been to mislead others, a better strategy would have been to point out that some statements were only conjectures.

Of course, although a few problems of continuum mechanics or physics have led to some of the mathematical questions described in this course, I have added some results for the usual reason that mathematicians are supposed to discover general structures hidden behind particular results, and describe something more general after having done a systematic study, akin to a cleaning process.

I had not consciously been following the path that Peter LAX had opened,[18] of developing mathematics for a better understanding of continuum mechanics and physics. I first heard him talk at the Lions–Schwartz seminar at IHP

---

[17] Paul Adrien Maurice DIRAC, English physicist, 1902–1984. He received the Nobel Prize in Physics in 1933, jointly with Erwin SCHRÖDINGER, for the discovery of new productive forms of atomic theory. He had worked in Cambridge, England, holding the Lucasian chair (1932–1969).

[18] Peter David LAX, Hungarian-born mathematician, born in 1926. He received the Wolf Prize in 1987, for his outstanding contributions to many areas of analysis and applied mathematics, jointly with Kiyoshi ITO. He received the Abel Prize in 2005. He works at NYU (New York University), New York, NY.

(Institut Henri Poincaré) in Paris, in the late 1960s, about N-waves for the Burgers equation,[19] to show that there are two invariants for integrable data (whose sum is the classical invariant $\int_{\mathbb{R}} u(\cdot, t)\, dx$), and about the Korteweg–de Vries equation (not yet popularized as the KdV equation),[20,21] to discuss its infinite list of invariants. Then, I heard him talk in 1971 in Madison, WI, during my first visit to United States, at a meeting of MRC (Mathematics Research Center) in Madison, WI, organized by Eduardo ZARANTONELLO,[22] and Peter LAX talked about "entropies" for systems, but I did not know enough about hyperbolic systems of conservation laws at the time to appreciate the importance of the results that he was presenting. Actually, I knew almost nothing of that subject, which was not really known among mathematicians in France in the early 1970s, and I may have helped to make it better known by teaching a few courses on the subject in the late 1970s, but I had first heard about the details in a course by Joel SMOLLER in Orsay in 1973,[23] then in discussions with Ron DIPERNA,[24] and with Constantine DAFERMOS,[25] and then in a course by Takaaki NISHIDA in Orsay in the late 1970s,[26] before I started teaching it myself.

Although I had understood early that Laurent SCHWARTZ was not interested in continuum mechanics or physics, I had taken some time to make the same observation concerning Jacques-Louis LIONS, but in the late 1970s, once that I was explaining the point of view that one should try to understand more about the physical meaning of the equations that one is studying, I had been surprised to hear Jacques-Louis LIONS defend the opposite position, that in his opinion this was not strictly necessary, so after that I had no

---

[19] Johannes Martinus BURGERS, Dutch-born mathematician, 1895–1981. He had worked at University of Maryland, College Park, MD.

[20] Diederik Johannes KORTEWEG, Dutch mathematician, 1848–1941. He had worked in Amsterdam, The Netherlands.

[21] Gustav DE VRIES, Dutch mathematician, 1866–1934. He had worked in Breda, in Alkmaar, and then as a high school teacher in Haarlem, The Netherlands.

[22] Eduardo H. ZARANTONELLO, Argentinian mathematician, born in 1918. He has worked in La Plata, in Córdoba, in San Juan, and in San Luis y Cuyo, Argentina, but when I first met him in 1971, during my first trip to the US, he was working at MRC (Mathematics Research Center) in Madison, WI; ten years ago he was still working, in Mendoza, Argentina.

[23] Joel Alan SMOLLER, American mathematician. He works at University of Michigan, Ann Arbor, MI.

[24] Ronald John DI PERNA, American mathematician, 1947–1989. He had worked at Brown University, Providence, RI, at University of Michigan, Ann Arbor, MI, at University of Wisconsin, Madison, WI, at Duke University, Durham, NC, and at UCB (University of California at Berkeley), Berkeley, CA.

[25] Constantine M. DAFERMOS, Greek-born mathematician, born in 1941. He has worked at Cornell University, Ithaca, NY, and at Brown University, Providence, RI.

[26] Takaaki NISHIDA, Japanese mathematician, born in 1942. He works at Kyoto University, Kyoto, Japan.

more doubts about his interests, and our paths separated. A few years ago, Peter LAX recalled a discussion from the 1950s where Jacques-Louis LIONS had been criticized by British applied mathematicians for focusing too much on functional analysis and for caring very little about continuum mechanics, and that Jacques-Louis LIONS found nothing better than replying with a joke,[27] which showed that he was already against understanding more about continuum mechanics.

In the early 1970s, after working with François MURAT on an extension of the work of Sergio SPAGNOLO on G-convergence,[28] before I borrowed the term homogenization from Ivo BABUŠKA for designing it and François MURAT chose to call our approach H-convergence,[29] it had been the work of Évariste SANCHEZ-PALENCIA that helped me understand the connection of our work with continuum mechanics,[30] and after that I insisted more and more about the usefulness of understanding about the possible physical meanings of the equations that one studies. The main features which I tried then to develop in my research work now look to me very similar to those which Peter LAX had chosen for himself, to learn about results in continuum mechanics and physics and, after developing an intuition for a particular field, to select a good subject and to put some order in it by creating an adapted mathematical framework, and eventually introduce new mathematical tools for studying it.

In some way, the qualities that Peter LAX has shown are not so common among mathematicians, even those who have been in contact with him. When I first met Ralph PHILLIPS in the spring of 1983,[31] in Stanford,[32] CA, I asked him a question about a remark of Leonardo DA VINCI,[33] which I thought must be classical for specialists of scattering,[34] but I was surprised to discover

---

[27] Jacques-Louis LIONS's answer was that the British could not be trusted, since the time they had burnt Jeanne D'ARC (Joan of ARC).

[28] Sergio SPAGNOLO, Italian mathematician, born in 1941. He works at Universitá di Pisa, Pisa, Italy.

[29] Ivo M. BABUŠKA, Czech-born mathematician, born in 1926. He worked at Charles University, Prague, Czech Republic, at University of Maryland, College Park, MD, and at University of Texas, Austin, TX.

[30] Enrique Évariste SANCHEZ-PALENCIA, Spanish-born mathematician, born in 1941. He works at CNRS (Centre National de la Recherche Scientifique) and Université Paris VI (Pierre et Marie Curie), Paris, France. I have always known him under the French form of his first name, Henri, but he now uses his second name, Évariste.

[31] Ralph Saul PHILLIPS, American mathematician, 1913–1998. He had worked at USC (University of Southern California), Los Angeles, CA, and at Stanford University, Stanford, CA.

[32] Leland STANFORD, American businessman, 1824–1893. Stanford University, and the city of Stanford where it is located, are named after him.

[33] Leonardo DA VINCI, Italian artist, engineer and scientist, 1452–1519. He had worked in Milano (Milan) and in Firenze (Florence), Italy.

[34] In the beginning of 1982, while I was visiting the Scuola Normale Superiore in Pisa, Italy, I was told to take the train to Firenze (Florence) to see an exhibition

that Ralph PHILLIPS had no physical intuition at all, and that for him scattering theory was just a chapter of functional analysis, so that he had not thought of using his collaboration with Peter LAX on the subject for learning about the physical phenomena which could be covered by their mathematical theory. Some have worked on a subject that Peter LAX had initialized, like that of hyperbolic systems of conservation laws, and many have pushed their work in directions totally disconnected from reality, despite a warning from Constantine DAFERMOS that *the umbilical cord that joins the theory of systems of conservation laws with continuum physics is still vital for the proper development of the subject and it should not be severed.*

For example, why are there people who play with models where there are shocks which do not satisfy some of the conditions that Peter LAX had introduced, and who forget to point out that the models that they use have been postulated by engineers, and why is it that they do not see that they are obviously incompatible with classical ideas in thermodynamics? Of course, I have been teaching for many years that thermodynamics is flawed and should be improved, but that does not mean that any model which is incompatible with classical thermodynamics can be considered a good model of physical reality!

When Peter LAX introduced "entropy conditions" for systems,[35] he was generalizing the work for the scalar case of Eberhard HOPF,[36] and of KRUZHKOV,[37] who had found an intrinsic way for expressing a condition introduced by Olga OLEINIK,[38] and he had observed that if a sequence of approximations like that created by the method of artificial viscosity converges almost everywhere, then an "entropy condition" holds, but he knew how difficult it was to obtain enough estimates for proving that desired strong convergence.[39] Some

---

of a manuscript of Leonardo DA VINCI. To explain the fact that the surface of the moon reflects the light from the sun in every direction, Leonardo had assumed that there were oceans on the moon and that because of waves the light could be reflected in various directions. We know now that there are no oceans on the moon, so that he was wrong, but I had admired Leonardo's inventiveness, and I had thought that he had not been too far from guessing why a rough surface can reflect light in every direction.

[35] Constantine DAFERMOS prefers to call them E-conditions, as these notions are not always linked to thermodynamic entropy, and I had chosen myself to write "mathematical entropies" in making the distinction.

[36] Eberhard Frederich Ferdinand HOPF, Austrian-born mathematician, 1902–1983. He had worked at MIT (Massachusetts Institute of Technology), Cambridge, MA, in Leipzig and in München (Munich), Germany, and at Indiana University, Bloomington, IN, where I met him in 1980.

[37] Stanislav Nikolaevich KRUZHKOV, Russian mathematician, 1936–1997. He had worked in Moscow, Russia.

[38] Olga Arsen'evna OLEINIK, Ukrainian-born mathematician, 1925–2001. She had worked in Moscow, Russia. I do not remember when I first met her, before 1976.

[39] In the mid 1960s, James GLIMM had found a way to estimate the total variation of the solution of some systems, for initial data having a small variation. It was only a few years after that approach of Peter LAX that I introduced a different

authors do not seem to have understood that they just repeat Peter LAX's argument when they write articles with statements that if something converges strongly, then the Hilbert expansion is true,[40] without pointing out the known defects of that conjecture of HILBERT that letting the "mean free path between collisions" tend to 0 in the Boltzmann equation gives the Euler equation for an ideal gas,[41] that the Boltzmann equation has been derived by assuming that a gas is rarefied and that, apart from having also postulated irreversibility by introducing probabilities, it does not make any sense to apply it to a dense gas by making a "mean free path between collisions" tend to 0, and that as real gases are not ideal, it means that either the Boltzmann equation does not apply to real gases or that the Hilbert expansion is false.

As in my preceding lecture notes, [20] and [21], I have given information in footnotes about the people who have participated in the creation of the knowledge related to the subject of the course, and I refer to the prefaces of those lecture notes in explaining my motivation, and I just want to repeat the motto of Hugo of Saint Victor,[42] *Learn everything, and you will see afterward that nothing is useless*, as it corresponds to what I have understood in my quest about how creation of knowledge occurs.

I have often heard people say about famous scientists from the past, that luck played an important role in their discoveries, but the truth must be that they would have missed the importance of the new hints that had occurred if they had not known beforehand all the aspects of their problems. Those who present chance as an important factor in discovery probably wish that every esoteric subject that they like be considered important and funded, but that is not at all what the quoted motto is about.

I hope that the many pieces of the puzzle that I describe in this course will help a few mathematicians to understand a way to follow the path of Peter LAX, by doing mathematics on problems which have been selected with care, so that in the end they help clarify a piece of that important puzzle, understanding physics in a better way.

I would not have been able to complete the publication of my first two lecture notes and to think about revising and completing this third set of lecture notes without the support of Lucia OSTONI, and I want to thank her for that and for much more, having given me the stability that I had lacked

---

method, based on the results of compensated compactness that I had introduced with François MURAT, but it appeared difficult to apply for systems, and the first to succeed was Ron DiPERNA.

[40] David HILBERT, German mathematician, 1862–1943. He had worked in Königsberg (then in Germany, now Kaliningrad, Russia) and in Göttingen, Germany.

[41] Leonhard EULER, Swiss-born mathematician, 1707–1783. He had worked in St Petersburg, Russia, in Berlin, Germany, and then again in St Petersburg.

[42] Hugo VON BLANKENBURG, German-born theologian, 1096–1141. He had worked at the monastery of Saint Victor in Paris, France.

so much in the last twenty-five years, so that I could feel safer in resuming my research of giving a sounder mathematical foundation to 20th century continuum mechanics and physics.

I want to thank my good friends Carlo SBORDONE and Franco BREZZI for having proposed to publish my lecture notes in a series of Unione Matematica Italiana. I also want to thank the referee for the improvements that he has suggested.

Milano, July 2007   *Luc TARTAR*
Correspondant de l'Académie des Sciences, Paris
Membro Straniero dell'Istituto Lombardo Accademia di Scienze e Lettere,
Milano
University Professor of Mathematics
Department of Mathematical Sciences
Carnegie Mellon University
Pittsburgh, PA 15213-3890
United States of America

Notes on names cited in footnotes for the Preface, LUCAS,[43] P. CURIE and M. SKŁODOWSKA-CURIE,[44] FIELDS,[45] DIDEROT,[46] CAVENDISH,[47] PLANCK,[48]

---

[43] Reverend Henry LUCAS, English clergyman and philanthropist, 1610–1663.

[44] Pierre CURIE, French physicist, 1859–1906. He and his wife, Marie SKŁODOWSKA-CURIE, Polish-born physicist, 1867–1934, received the Nobel Prize in Physics in 1903, in recognition of the extraordinary services they have rendered by their joint research on the radiation phenomena discovered by Professor Henri BECQUEREL, jointly with Antoine Henri BECQUEREL. Marie SKŁODOWSKA-CURIE also received the Nobel Prize in Chemistry in 1911, in recognition of her services to the advancement of chemistry by the discovery of the elements radium and polonium, by the isolation of radium and the study of the nature and compounds of this remarkable element. They had worked in Paris, France. University Paris VI in Paris, France, is named after them, Université Pierre et Marie Curie.

[45] John Charles FIELDS, Canadian mathematician, 1863–1932. He had worked in Meadville, PA, and in Toronto, Ontario.

[46] Denis DIDEROT, French philosopher and writer, 1713–1784. He had worked in Paris, France, and he was the editor-in-chief of the Encyclopédie. Université Paris 7, Paris, France, is named after him.

[47] Henry CAVENDISH, English physicist and chemist (born in Nice, not yet in France then), 1731–1810. He was wealthy and lived in London, England.

[48] Max Karl Ernst Ludwig PLANCK, German physicist, 1858–1947. He received the Nobel Prize in Physics in 1918, in recognition of the services he rendered to the advancement of physics by his discovery of energy quanta. He had worked in Kiel

SCHRÖDINGER,[49] WOLF,[50] ITO[51] ABEL,[52] BROWN,[53] DUKE,[54] CORNELL,[55] CHARLES IV,[56] BERKELEY,[57] Jeanne D'ARC,[58] and for the preceding footnotes, NOBEL,[59] BECQUEREL,[60] James GLIMM.[61]

---

and in Berlin, Germany. There is a Max Planck Society for the Advancement of the Sciences, which promotes research in many institutes, mostly in Germany (I spent my sabbatical year 1997–1998 at the Max Planck Institute for Mathematics in the Sciences in Leipzig, Germany).

[49] Erwin Rudolf Josef Alexander SCHRÖDINGER, Austrian-born physicist, 1887–1961. He received the Nobel Prize in Physics in 1933, jointly with Paul Adrien Maurice DIRAC, for the discovery of new productive forms of atomic theory. He had worked in Vienna, Austria, in Jena and in Stuttgart, Germany, in Breslau (then in Germany, now Wrocław, Poland), in Zürich, Switzerland, in Berlin, Germany, in Oxford, England, in Graz, Austria, and in Dublin, Ireland.

[50] Ricardo WOLF, German-born inventor, diplomat and philanthropist, 1887–1981. He emigrated to Cuba before World War I; from 1961 to 1973 he was Cuban Ambassador to Israel, where he stayed afterwards. The Wolf Foundation was established in 1976 with his wife, Francisca SUBIRANA-WOLF, 1900–1981, to promote science and art for the benefit of mankind.

[51] Kiyosi ITO, Japanese mathematician, born in 1915. He received the Wolf Prize in 1987, for his fundamental contributions to pure and applied probability theory, especially the creation of the stochastic differential and integral calculus, jointly with Peter LAX. He worked in Kyoto, Japan, although he worked at some time at Aarhus University, Aarhus, Denmark (1966–1969) and at Cornell University, Ithaca, NY (1969–1975).

[52] Niels Henrik ABEL, Norwegian mathematician, 1802–1829.

[53] Nicholas BROWN Jr., American merchant, 1769–1841. Brown University, Providence, RI, is named after him.

[54] Washington DUKE, American industrialist, 1820–1905. Duke University, Durham, NC, is named after him.

[55] Ezra CORNELL, American philanthropist, 1807–1874. Cornell University, Ithaca, NY, is named after him.

[56] CHARLES IV of Luxembourg, 1316–1378. German king and King of Bohemia (in 1346) and Holy Roman Emperor (in 1355) as Karl IV. Charles University, which he founded in Prague in 1348, is named after him.

[57] George BERKELEY, Irish-born philosopher and Anglican Bishop, 1685–1753. The city of Berkeley, CA, is named after him.

[58] Jeanne D'ARC, French national heroine, and saint, 1412–1431. She was beatified in 1909, and canonized in 1920.

[59] Alfred NOBEL, Swedish industrialist and philanthropist, 1833–1896. He created a fund to be used as awards for people whose work most benefited humanity.

[60] Antoine Henri BECQUEREL, French physicist, 1852–1908. He received the Nobel Prize in Physics in 1903, in recognition of the extraordinary services he has rendered by his discovery of spontaneous radioactivity, jointly with Pierre CURIE and Marie SKŁODOWSKA-CURIE. He had worked in Paris, France.

[61] James G. GLIMM, American mathematician, born in 1934. He worked at MIT (Massachusetts Institute of Technology), Cambridge, MA, at NYU (New York University), New York, NY, and at SUNY (State University of New York), Stony Brook, NY.

Detailed Description of Lectures

a.b: refers to definition, lemma or theorem # b in lecture # a, while (a.b) refers to equation # b in lecture # a.

<u>Lecture 1</u>: Historical Perspective.

Conservation laws (1.1)–(1.2), linearized wave equation (1.3)–(1.5), quasi-linear wave equation (1.6), gas dynamics (1.7)–(1.9), Burgers equation (1.10)–(1.13).

<u>Lecture 2</u>: Hyperbolic Systems: Riemann Invariants, Rarefaction Waves.

Linear system (2.1), 2.1: linear hyperbolic or strictly hyperbolic system, eigenvalues and eigenvectors (2.2)–(2.4), solution of linear hyperbolic system (2.5)–(2.7), 2.2: quasi-linear hyperbolic or strictly hyperbolic system (2.8), gas dynamics (2.9)–(2.28), 2.3: Riemann problem (2.29), solution of linear case (2.30), 2.4: Riemann invariants (2.31), integral curves (2.32), Riemann invariants for gas dynamics (2.33)–(2.35), 2.5: simple waves, equations for Riemann problem (2.36)–(2.39), 2.6: linearly degenerate of genuinely nonlinear fields (2.40)–(2.41), the case of gas dynamics (2.42)–(2.44).

<u>Lecture 3</u>: Hyperbolic Systems: Contact Discontinuities, Shocks.

Contact discontinuities (3.1)–(3.2), conservation forms (3.3)–(3.4), 3.1: weak solutions (3.5)–(3.7), 3.2: Rankine–Hugoniot conditions (3.8)–(3.10), the case of gas dynamics (3.11)–(3.12), 3.3: shocks (3.13)–(3.17), 3.4: entropy and entropy flux (3.18).

<u>Lecture 4</u>: The Burgers Equation and the 1-D Scalar Case.

Burgers equation (4.1), Burgers–Hopf equation and Hopf–Cole transform (4.2)–(4.5), one sided inequality for u_x implying uniqueness (4.6), Lax–Friedrichs scheme (4.7)–(4.8), CFL condition (4.9), order-preserving property (4.10)–(4.11), 4.1: Crandall–Tartar lemma, application to Lax–Friedrichs scheme (4.12)–(4.13).

<u>Lecture 5</u>: The 1-D Scalar Case: the E-Conditions of Lax and of Oleinik.

Galilean transformation (5.1), nonuniqueness (5.2)–(5.3), 5.1: Oleinik E-condition (5.4)–(5.5), 5.2: Lax E-condition (5.6)–(5.7), rarefaction wave (5.8)–(5.11), shock (5.12).

<u>Lecture 6</u>: Hopf's Formulation of the E-Condition of Oleinik.

Hopf's entropy condition (6.1)–(6.2), a family of entropy giving Oleinik E-condition (6.3), Lax generalization to systems (6.4)–(6.7), Lax–Friedrichs scheme (6.8), viscous shock profile (6.9)–(6.14).

<u>Lecture 7</u>: The Burgers Equation: Special Solutions.

One-sided inequality for u_x implying uniqueness (7.1), perturbation of a constant (7.2)–(7.8), perturbation of Riemann data (7.9)–(7.12), various scalings (7.13)–(7.19), perturbation of a rarefaction wave (7.20)–(7.27).

<u>Lecture 8</u>: The Burgers Equation: Small Perturbations; the Heat Equation.

The danger of linearization (8.1)–(8.8), heat equation (8.9), Fokker–Planck equation (8.10), elementary solution of heat equation (8.11)–(8.12), difference scheme for 1-D heat equation (8.13)–(8.16).

<u>Lecture 9</u>: The Fourier Transform; the Asymptotic Behaviour for the Heat Equation.

Fourier transform of integrable functions (9.1)–(9.3), derivation and multiplication (9.4)–(9.5), Fourier transform on $S(\mathbb{R}^N)$ and $S'(\mathbb{R}^N)$ (9.6)–(9.8), Plancherel formula (9.9), inverse Fourier transform (9.10), affine change of variable (9.11), Fourier transform of a convolution product (9.12)–(9.13), Fourier transform for the heat equation (9.14)–(9.22), semi-group (9.23)–(9.24), scaling and decay (9.25)–(9.28), 9.1: relation with moments and decay (9.29)–(9.30), matrix of inertia and anisotropic Gaussians (9.31)–(9.33), solving a diffusion equation with anisotropic Gaussians (9.34)–(9.42).

<u>Lecture 10</u>: Radon Measures; the Law of Large Numbers.

Radon measures (10.1), Fourier transform of a Radon measure (10.2)–(10.4), centre of mass and convolution (10.5)–(10.7), 10.1: law of large numbers (10.8), matrix of inertia and convolution (10.9)–(10.11), 10.2: strong law of large numbers (10.12).

<u>Lecture 11</u>: A 1-D Model with Characteristic Speed $\frac{1}{\varepsilon}$.

Explicit difference schemes (11.1), 1-D model with velocities $\pm\frac{1}{\varepsilon}$ (11.2), 11.1: limit as $\varepsilon \to 0$ (11.3)–(11.4),

<u>Lecture 12</u>: A 2-D Generalization; the Perron–Frobenius Theory.

2-D model with velocities $\pm\frac{1}{\varepsilon}$ along axes (12.1)–(12.4), 12.1: reducible matrices, 12.2: a condition for irreducibility, 12.3: $\rho(A)$ is a simple eigenvalue with positive eigenvector, 12.4: the case of other eigenvalues of modulus $\rho(A)$, 12.5: primitive or imprimitive irreducible matrices, 12.6: a criterion using the length of loops, 12.7: asymptotic behaviour of $A^n w$ as $n \to \infty$.

<u>Lecture 13</u>: A General Finite-Dimensional Model with Characteristic Speed $\frac{1}{\varepsilon}$.

The model (13.1)–(13.5), 13.1: $M e = 0$ and e positive, L^∞ estimate (13.6), 13.2: coerciveness on $e^\perp$ (13.7), estimates (13.8)–(13.10), convergence (13.11)–(13.17).

<u>Lecture 14</u>: Discrete Velocity Models.

Conservations in a collision (14.1), probabilities (14.2), general model (14.3), properties of coefficients (14.4)–(14.8), entropy (14.9)–(14.10), conservations and decay of entropy (14.11)–(14.13), four velocities Maxwell model (14.14)–(14.15), general semi-linear case (14.16)–(14.17), 14.1: local existence (14.18), 1-D four velocities model and Broadwell model (14.19)–(14.20), 14.2: finite propagation speed, 14.3: condition for positivity (14.21), 14.4: forward invariant sets for ordinary differential equations, 14.5: characterization of forward invariant sets (14.22), 14.6: forward invariant sets for a semi-linear system, characterization (14.23)–(14.24), a model with a bounded forward invariant set (14.25)–(14.27), Carleman model (14.28)–(14.29), formal (Hilbert) expansion for Broadwell model (14.30)–(14.37), restriction of convolution product on circle (14.38).

Lecture 15: The Mimura–Nishida and the Crandall–Tartar Existence Theorems.

15.1: Mimura–Nishida existence theorem (15.1)–(15.2) and (15.8)–(15.12), 15.2: Crandall–Tartar existence theorem (15.3)–(15.5), 15.3: use of bounds on entropy (15.6)–(15.7).

Lecture 16: Systems Satisfying My Condition (S).

Condition (S) (16.1)–(16.2), 16.1: spaces V_c and W_c, 16.2: product on $W_{c_1} \times W_{c_2}$ with $c_1 \neq c_2$ (16.3), 16.3: global existence ($t \in \mathbb{R}$) for small data in L^1 (16.4)–(16.13), a case of necessity for small data (16.14), 16.4: local existence for data in L^1 (16.15), asymptotic behaviour (16.16), 16.5: a Mimura–Nishida type estimate (16.17).

Lecture 17: Asymptotic Estimates for the Broadwell and the Carleman Models.

Asymptotic behaviour for the Broadwell model (17.1)–(17.4), 2-D four velocities model (17.5)–(17.7), Illner–Reed estimate for Carleman model (17.8), self-similar solutions of Carleman model (17.9)–(17.12).

Lecture 18: Oscillating Solutions; the 2-D Broadwell Model.

Oscillating solutions of Carleman model (18.1)–(18.2), 18.1: div-curl lemma (18.3)–(18.5), 18.2: application (18.6), systems stable by weak convergence (18.7)–(18.9), Gagliardo–Nirenberg estimate (18.10), application to 2-D four velocities model (18.11)–(18.17).

Lecture 19: Oscillating Solutions: the Carleman Model.

Rescaling of a solution (19.1), bounded sequences of solutions (19.2)–(19.4), general system of two equations (19.5), extracting converging subsequences (19.6)–(19.9), an infinite system (19.10)–(19.11), uniqueness (19.12)–(19.14), strength of oscillations and differential inequalities (19.15)–(19.18).

Lecture 20: The Carleman Model: Asymptotic Behaviour.

Integrable nonnegative data and rescaling (20.1)–(20.4), 20.1: strong convergence in $|x| > t + \varepsilon$ (20.5)–(20.7), 20.2: a subsequence converges to a solution of Carleman with support in $|x| \leq t$, formal (Hilbert) limit of the Broadwell model (20.8)–(20.12), 20.3: the case of Carleman model (20.13)–(20.14), Kurtz scaling (20.15)–(20.18), oscillating solutions for Broadwell model (20.19)–(20.23).

Lecture 21: Oscillating Solutions: the Broadwell Model.

Properties of weak limits and the weak limit X_{111} of $u_n v_n w_n$ (21.1)–(21.8), 21.1: estimate for X_{111} (21.9), inequality for σ_w (21.10), periodically modulated case (21.11)–(21.12), 21.2: the Carleman model (21.13)–(21.20), 21.3: the Broadwell model (21.21)–(21.31), a system for Fourier coefficients (21.32).

(31.9)–(31.10), 31.4: discrete analogue of half derivatives (31.11)–(31.17), 31.5: compactness by integration (31.18)–(31.23).

Lecture 32: Wave Front Sets; H-Measures.
   First-order equations and bicharacteristic rays (32.1)–(32.2), Wigner transform (32.3)–(32.5), H-measures (32.6)–(32.9), localization principle (32.10).

Lecture 33: H-Measures and "Idealized Particles".
   H-measures for the wave equations (33.1)–(33.5), internal energy and equipartition of energy (33.6)–(33.7).

Lecture 34: Variants of H-Measures.
   Geometrical optics (34.1)–(34.5), my proposal for introducing a characteristic length (34.6), Gérard's proposal of semi-classical measures (34.7), P.-L. Lions & Paul's proposal to define them with Wigner transform (34.8)–(34.9), an observation of Wigner (34.10), 34.1: k-point correlation measures (34.11)–(34.12), 34.2: properties of correlation measures (34.13)–(34.14). Conclusion.

35: Biographical Information.
   Basic biographical information for people whose name is associated with something mentioned in the lecture notes.

36: Abbreviations and Mathematical Notation.

References.

Index.

Contents

1

Historical Perspective

The goal of these lectures is to study partial differential equations related to questions of kinetic theory, and to elucidate some of the questions of *continuum mechanics* or *physics* which lie behind these problems.

One may arrive at these questions from different ways and many interesting mathematical questions arise in the various approaches.

From a *classical mechanics* point of view, one imagines a collection of rigid bodies moving under some set of forces, for example gravitational attraction between them, and one wants to study the evolution of such a system. Of course, one should also consider electromagnetic effects, and ALFVÉN has explained by electromagnetic effects some of the features observed in galaxies,[1] which astrophysicists pretend to explain by gravitational effects only, and I have read that there are anomalies in the movement of the planet Jupiter, which might be related to its important magnetic properties, because of *Lorentz forces*,[2] but then one could not just play with ordinary differential equations as is usual in classical mechanics, and one would have to add what is known as the *Maxwell equation*, which is a system of partial differential equations, and it becomes the realm of continuum mechanics, but in these lecture

---

[1] Hannes Olof Gösta ALFVÉN, Swedish-born physicist, 1908–1995. He received the Nobel Prize in Physics in 1970, for fundamental work and discoveries in magneto-hydrodynamics with fruitful applications in different parts of plasma physics, jointly with Louis NÉEL. He had worked in Uppsala and Stockholm, Sweden, in UCSD (University of California at San Diego), La Jolla, CA, and USC (University of Southern California), Los Angeles, CA.

[2] Hendrik Antoon LORENTZ, Dutch physicist, 1853–1928. He received the Nobel Prize in Physics in 1902, jointly with Pieter ZEEMAN, in recognition of the extraordinary service they rendered by their research into the influence of magnetism upon radiation phenomena. He had worked in Leiden, The Netherlands. The Institute for Theoretical Physics in Leiden, The Netherlands, is named after him, the Lorentz Institute.

notes I shall call it the *Maxwell–Heaviside equation*,[3] because if MAXWELL had unified the previous results on electricity and on magnetism obtained by AMPÈRE,[4] GAUSS,[5] BIOT and SAVART,[6,7] and FARADAY,[8] it is to HEAVISIDE that one owes the simplified version of the Maxwell–Heaviside equation using vector calculus.[9]

In considering rigid bodies which are submitted to forces acting at a distance, one uses the point of view of NEWTON, which he developed for gravitation. As explained by FEYNMAN in taped lectures given at Cornell University,[10] the difficulty that NEWTON had overcome was that in his day, it was explained that planets turn around the sun (or the moon around the earth) because angels were pulling them, and he did not question the existence of angels, but the fact that these angels were believed to pull the planets in a tangential way, and NEWTON's first contribution was to observe that the force applied by the angels was towards the sun for a planet, or towards the earth for the moon. Then, he realized that the force pulling the moon must be the same as the force drawing apples towards the ground, so that he had discovered the name of these angels, gravitation. NEWTON added a curious argument for having the gravitational force decay in *distance*$^{-2}$, while he could

---

[3] Oliver HEAVISIDE, English engineer, 1850–1925. He had worked as a telegrapher, in Denmark, in Newcastle upon Tyne, England, and then did research on his own, living in the south of England.

[4] André Marie AMPÈRE, French mathematician, 1775–1836. He had worked in Bourg, in Lyon, and in Paris, France.

[5] Johann Carl Friedrich GAUSS, German mathematician, 1777–1855. He had worked in Göttingen, Germany.

[6] Jean-Baptiste BIOT, French mathematician and physicist, 1774–1862. He had worked in Beauvais, and in Paris, France, holding a chair (physique mathématique, 1801–1862) at Collège de France, Paris.

[7] Félix SAVART, French physicist, 1791–1841. He had worked at Collège de France, Paris, France (physique générale et expérimentale, 1836–1841).

[8] Michael FARADAY, English chemist and physicist, 1791–1867. He had worked in London, England, as Fullerian professor of chemistry at the Royal Institution of Great Britain.

[9] MAXWELL had imagined mechanical devices for transmitting the electric field and the magnetic field, and I read that HEAVISIDE replaced a set of 20 equations in 20 variables that MAXWELL had written by a set of 4 equations in 2 variables. HEAVISIDE had also developed an *operational calculus*, which was given a mathematical explanation by Laurent SCHWARTZ, using his theory of distributions.

[10] Richard Phillips FEYNMAN, American physicist, 1918–1988. He received the Nobel Prize in Physics in 1965, jointly with Sin-Itiro TOMONAGA and Julian SCHWINGER, for their fundamental work in quantum electrodynamics, with deep-ploughing consequences for the physics of elementary particles. He had worked at Cornell University, Ithaca, NY, and at Caltech (California Institute of Technology), Pasadena, CA.

have deduced that from one of Kepler's laws.[11] Forces at a distance pertain to classical mechanics, and they only involve ordinary differential equations, but a general existence theory for ordinary differential equations was not known until CAUCHY for the analytic case,[12] and LIPSCHITZ for more general cases.[13] Again, NEWTON's point of view is classical mechanics, but there is something wrong about forces acting at a distance in an instantaneous way; one difficulty is about a force acting at a distance (and about what a force is anyway, but that goes beyond continuum mechanics too), but another difficulty is about action being instantaneous, and if one tries to give a precise meaning to instantaneity, one is bound to find the question of relativity, which was first studied by POINCARÉ, and he observed that the Maxwell–Heaviside equation is invariant by the Lorentz group, but it was EINSTEIN who really understood the physical meaning of the question. However, it seems to be POINCARÉ's understanding of relativity that a particle feels the action of the field and tells the field that it is there, so that the field transmits (at the velocity of light c) the information between particles: a particle does not store any information about the positions of the other particles, and mathematically it leads to the study of *semi-linear hyperbolic systems* having only the velocity of light c as characteristic velocity, but again there is a problem with the precise notion of a particle, which goes beyond continuum mechanics.

In the case of a universe made up of a finite number of classical particles, LAGRANGE had an interesting thought,[14] that if one was given the initial position of all the particles, then the whole future of the universe could be described, but he overlooked a few problems; of course, there were already some hints at his time that the world is not described by ordinary differential equations, like the *Euler equation* for *ideal fluids*, but if it had been as he imagined, he had not actually proven a global existence theorem for ordinary differential equations because of possible *collisions*; another difficulty is that one may need infinite accuracy on the initial data because of possible chaotic effects, as was first observed by POINCARÉ (although the term chaos was coined much later, and is used now by people who usually forget to say how much they owe to POINCARÉ for the tools that they use, not always in an accurate way if one considers the reactions provoked by those who had the idea

---

[11] Johannes KEPLER, German-born mathematician, 1571–1630. He had worked in Graz, Austria, in Prague, now capital of the Czech republic, and in Linz, Austria.

[12] Augustin Louis CAUCHY, French mathematician, 1789–1857. He was made Baron by CHARLES X. He had worked in Paris, France, went into exile after the 1830 revolution and worked in Torino (Turin), Italy, returned from exile after the 1848 revolution, and worked in Paris again.

[13] Rudolf Otto Sigismund LIPSHITZ, German mathematician, 1832–1903. He had worked in Breslau (then in Germany, now Wrocław, Poland) and in Bonn, Germany.

[14] Giuseppe Lodovico LAGRANGIA (Joseph Louis LAGRANGE), Italian-born mathematician, 1736–1813. He had worked in Torino (Turin) Italy, in Berlin, Germany, and in Paris, France. He was made Count in 1808 by NAPOLÉON I.

of explaining instability by saying that it is as if the movement of a butterfly in Brazil could create a storm in New York; predictably, if one considers the low level of scientific knowledge nowadays, it was misunderstood that butterflies in Brazil have an effect on the weather in New York. Some people have been upset enough to take the time to show that this kind of effect is precluded by some models of hydrodynamics (but they have probably not explained that lots of terms had been thrown out of the models that they use, just because they were believed to be small, although the derivative of something small is not always small, and the long-term effect of those terms had never been ascertained). Both these reactions were a little silly, because all that had been said was that a small cause (like the movement of a butterfly in Brazil, or any other thing that one likes to think of as very small) might create a large effect (like a storm in New York, or any other thing that one likes to think of as very large); however, one should observe that most people do not even understand the difference between the quantifiers $\exists$ and $\forall$, and one should have explained that for some systems of ordinary differential equations a very small perturbation at some point may become very large later on (at a different point), but not every small perturbation at a point has this property (because perturbing in the direction of the flow is just a translation in time, which remains under control), and that for a given system this is not valid for all points, and that this effect does not happen for all systems. Anyway, the world is not described by ordinary differential equations, and those who believe that partial differential equations always behave like ordinary differential equations should start by learning about which terms have been neglected in arriving at the model that they use, and they should then prove, and not postulate, that these terms can really be neglected because their later effects will always remain very small, unlike the chaotic behaviour that they pretend to specialize upon.

Using rigid bodies is also an approximation, and one could think of considering elastic bodies, but that would also force us to use partial differential equations instead of ordinary differential equations, and a particular difficulty would actually arise because of questions of *finite elasticity* which are not yet well understood.[15]

---

[15] Finite is not opposite to infinite but to infinitesimal: if a point x in an initial configuration is moved to a point $u(x)$, an hypothesis of infinitesimal deformation consists in assuming that $\nabla u(x)$ is near I, and this leads to linearized elasticity, while in finite elasticity one only assumes that $\nabla u(x)$ is near a rotation (but it may be far from rotations for materials like rubber), and that leads to problems which are not so well understood from a mathematical point of view, because one should look at the evolution problem, of course, and one cannot use the simplistic view that elastic materials minimize their potential energy, which is a fake continuum mechanics point of view, which has been pushed forward by some adepts of the calculus of variations, probably because it is irrelevant from a physical point of view.

Knowing all these limitations is just a way to know in advance that some questions are not too physical, like the asymptotic behaviour of an approximate system of ordinary differential equations for example, and a lot of what is said is hardly relevant from a realistic point of view, because one assumes that the models used are exact, and there are always a few things which have been neglected so that the model can be accurate for a large time, but not for an infinite time. For example, fluids are not incompressible, and one can measure a *finite speed of progation* of sound (about 300 meters per second for air and 1,500 meters per second for water) while it is infinite for an incompressible fluid; discussing the asymptotic behaviour of a truncated system is usually not relevant, in particular for turbulence, which is not about letting time go to infinity anyway, except possibly in infinite domains when one may do a rescaling of space.

Assuming that one works with rigid bodies, one must compute the resultant of the forces applied to the body, and consider that it is applied to the centre of gravity of the body, which will move according to the classical law of motion, resulting from the work of NEWTON, after some initial thoughts by Galileo,[16] that force is mass × acceleration, and the resultant torque which will make the body rotate, and for which one needs to know the matrix of inertia of the body. In most treatments of kinetic theory, torque effects are neglected as if the body were points, but in the case of colliding spheres anyone having played billiards knows about the importance of spin for the result of a collision, and such questions should be addressed.

We are not interested in asymptotic behaviour but in the finite-time existence in the case of a large number of particles;[17] of course, if one lets the number of particles go to infinity, the mass of each particle must be scaled accordingly. An important problem is to study possible collisions or near collisions.

---

[16] Galileo GALILEI, Italian mathematician, 1564–1642. He had worked in Siena, in Pisa, in Padova (Padua), Italy, and again in Pisa.

[17] Celestial mechanics is interested in a small number of particles (planets), and apart from proving that no collisions will occur, one wants to know if a solution stays globally bounded (once one moves with the centre of gravity), and that means analysing if a planet can escape to infinity; this cannot happen if there is not enough energy in the system, as the total energy (kinetic energy plus potential energy) is conserved, and in the case of two bodies the escape velocity can be easily computed. The escape velocity from the attraction of the earth is around 11.2 km s^{-1}; it corresponds to the kinetic energy $\frac{v^2}{2}$ being able to compensate for the difference in potential between the surface of the earth and infinity, equal to $\frac{GM}{R}$, and $\frac{GM}{R^2}$ is the acceleration of gravity, around 9.81 m s^{-2}, and the radius of the earth R is around 6,378 km (the gravitational constant G has been measured as around 6.67 ×10^{-11} N m^2 kg^{-2}, where N stands for newton, the unit of force, so that the mass of the earth M is around 5.98 ×10^{24} kg).

A puzzling fact is that, although the system is Hamiltonian,[18] a notion already introduced by LAGRANGE, so that it conserves energy, one finds numerical evidence that the energy decreases when one computes solutions with a large number of particles. The same effect would be observed by two mathematical observers, one using time in the usual way and the other reversing time (and velocities), so that one observes some kind of irreversibility, which only occurs because the number of particles gets very large; of course, for a given number of particles, one could make the numerical methods precise enough for avoiding the loss of energy, but if the number of particles tends to infinity and the masses of the particles are rescaled it becomes a different problem to ascertain what the limit is. The observation that some energy is "lost" is not in contradiction with other approaches, where one uses *internal energy*, and where irreversibility also occurs, so that one seems to be losing energy but the "lost" part is just hidden and it can be followed as *heat*, and a technical word will be associated with this effect, *entropy*, and one will have to understand what it means, but the main mathematical difficulty will be that the actual postulates concerning the second principle in *thermodynamics* are inadequate and should be improved, but in ways which have not been understood yet.

From the continuum mechanics point of view, partial differential equations were used for describing the movement of a gas or a liquid, and in these equations various *thermodynamical quantities* appeared like the density ϱ and the pressure p, as in the Euler equation for ideal (inviscid) fluids

$$\frac{\partial \varrho}{\partial t} + \sum_{i=1}^{3} \frac{\partial(\varrho\, u_i)}{\partial x_i} = 0, \tag{1.1}$$

expressing the *conservation of mass*, and

$$\frac{\partial(\varrho\, u_j)}{\partial t} + \sum_{i=1}^{3} \frac{\partial(\varrho\, u_i\, u_j)}{\partial x_i} + \frac{\partial p}{\partial x_j} = 0 \text{ for } j = 1, 2, 3, \tag{1.2}$$

expressing the balance of linear momentum, and later other quantities were added, the absolute temperature θ (or T), the internal energy (per unit of mass) e, the entropy (per unit of mass) s, and so on. Early in the study of gases, it had been found by BOYLE in 1662,[19] and by MARIOTTE in 1676,[20] that the product of pressure by volume is constant, and for a long time it was implicitly assumed that one worked at constant temperature, and it is worth recalling how the notion of temperature had evolved.

---

[18] Sir William Rowan HAMILTON, Irish mathematician, 1805–1865. He had worked in Dublin, Ireland.

[19] Robert BOYLE, Irish-born physicist, 1627–1691. He had worked in Oxford, and in London, England.

[20] Edme MARIOTTE, French physicist and priest, 1620–1684. He had been prior of Saint Martin sous Beaune, near Dijon, France.

The first thermoscope was invented by Galileo in 1593, the first thermometer using air by SANTORIO,[21] and the first thermometer using liquid by REY,[22] but the first sealed thermometer that used liquid (alcohol) was invented in 1654 by FERDINAND II,[23] and called a Florentine thermometer. In 1661, BOYLE was shown such a Florentine thermometer by R. SOUTHWELL.[24] Mercury was first substituted for alcohol in Florence, at Accademia del Cimento (academy of experiment), founded in 1657 by FERDINAND II and his brother Leopold,[25] but it was in 1714 that FAHRENHEIT found a way to avoid mercury clinging to the glass,[26] and introduced his scale of temperature, with a mixture of water, ice, and cooking salt at 0°, a mixture of water and ice at 32° and boiling water at 212°. RÉAUMUR had a scale in 1730 where water froze at 0° and boiled at 80°.[27] CELSIUS had a scale in 1742 with water freezing at 100° and boiling at 0° in 1741,[28] and a few people are credited for inverting the scale as it is used today (and named after CELSIUS since 1948, as it was called degrees centigrade before): J.-P. CRISTIN (in 1743),[29] EKSTRÖM,[30] LINNÉ (in 1745),[31] and STRÖMER.[32] An absolute temperature scale was introduced in 1862, by THOMSON,[33] later to become Lord Kelvin, and JOULE,[34] and is now named after Lord Kelvin, where the temperature is obtained by adding 273.15 to the temperature in degrees Celsius.

---

[21] Santorio SANTORIO (SANCTORIUS of Padua), Italian physician, 1561–1636. He had worked in Padova (Padua), Italy.

[22] Jean REY, French physician and chemist, 1583–1645.

[23] Ferdinando DÉ MEDICI, Italian statesman, 1610–1670. In 1621 he became Grand Duke of Tuscany as FERDINAND II. He had lived in Firenze (Florence), Italy.

[24] Sir Robert SOUTHWELL, Irish-born diplomat, 1635–1702.

[25] Leopoldo DÉ MEDICI, Italian noble, 1617–1675. He was named cardinal in 1667. He had lived in Firenze (Florence), Italy.

[26] Gabriel Daniel FAHRENHEIT, German-born physicist, 1686–1736. He had worked in Amsterdam, The Netherlands.

[27] René Antoine FERCHAULT DE RÉAUMUR, French scientist, 1683–1757. He had worked in Paris, France.

[28] Anders CELSIUS, Swedish astronomer, 1701–1744. He had worked in Uppsala, Sweden.

[29] Jean-Pierre CRISTIN, French scientist, 1683–1755.

[30] Daniel EKSTRÖM, Swedish instrument maker, 1711–1755. He had worked in Uppsala, Sweden.

[31] Carl LINNAEUS (Carl VON LINNÉ), Swedish naturalist, 1707–1778. He had worked in Uppsala, Sweden.

[32] Mårten STRÖMER, Swedish astronomer, 1707–1770. He had worked in Uppsala, Sweden.

[33] William THOMSON, Irish-born physicist, 1824–1907. In 1892 he was made Baron Kelvin of Largs, and thereafter known as Lord Kelvin. He had worked in Glasgow, Scotland.

[34] James Prescot JOULE, English scientist, 1818–1889. He had lived in Manchester, England, being a brewer with an interest in science.

Thinking that the temperature could be considered constant, it is natural to consider that in the Euler equation the pressure p is a smooth function of ϱ (barotropic model) with $\frac{dp}{d\varrho} > 0$. If one considers small and smooth perturbations around a constant solution $u = u_0$ and $\varrho = \varrho_0$, one may use *Galilean invariance* and assume that $u_0 = 0$,[35] and the linearized problem around $(0, \varrho_0)$ is then

$$\frac{\partial \varrho}{\partial t} + \sum_{i=1}^{3} \varrho_0 \frac{\partial u_i}{\partial x_i} = 0 \tag{1.3}$$

from conservation of mass, and

$$\varrho_0 \frac{\partial u_j}{\partial t} + \frac{dp}{d\varrho}(\varrho_0) \frac{\partial \varrho}{\partial x_i} = 0 \text{ for } j = 1, 2, 3, \tag{1.4}$$

from balance of linear momentum, so that one has

$$\frac{\partial^2 \varrho}{\partial t^2} - \frac{dp}{d\varrho}(\varrho_0) \Delta \varrho = 0, \tag{1.5}$$

a *wave equation* where perturbations propagate at the velocity $\sqrt{\frac{dp}{d\varrho}(\varrho_0)}$. However, if one used the Boyle–Mariotte law $p = A\,\varrho$, one found that $\sqrt{A}$ is rather different than the measured velocity at which perturbations propagate, the speed of sound, first estimated by NEWTON.

Improving the Boyle–Mariotte law by measuring the effects of temperature was done by GAY-LUSSAC in 1802,[36] whose law states that at fixed volume the pressure is proportional to the absolute temperature (although the notion was not defined yet), and he mentions a law found in 1787 (but not published) by CHARLES,[37] that at constant pressure the volume is proportional to the absolute temperature. In 1811, AVOGADRO stated his law,[38] that equal volumes of any two different gases at the same temperature and pressure contain an equal number of molecules,[39] a number called the Avogadro number

---

[35] If a new frame moves at constant velocity a with respect to an initial frame, then one uses $\widetilde{x} = x - t\,a$ in the new frame, so the new velocity is $\widetilde{u}(\widetilde{x}, t) = u(x, t) - a$, but the change for any thermodynamical quantity f is $\widetilde{f}(\widetilde{x}, t) = f(x, t)$; the Euler equation is invariant by such transformations (which form a group). Some authors *mistakenly* use the term Galilean invariance for other groups of transformations, like the group of rotations in x space, in which case the correct qualifier is isotropic, and the Euler equation describes an isotropic fluid.

[36] Joseph Louis GAY-LUSSAC, French physicist, 1778–1850. He had worked in Paris, France.

[37] Jacques Alexandre César CHARLES, French physicist, 1746–1823. He had worked in Paris, France.

[38] Lorenzo Romano Amedeo Carlo AVOGADRO, Count of Quaregna and Cerreto, Italian physicist, 1776–1856. He had worked in Torino (Turin), Italy.

[39] That law is not true at high pressure, where gases may liquefy.

$(6.0221367 \times 10^{23})$ by PERRIN,[40] who had measured it in relation with Brownian motion,[41] and this had led to the law $PV = nRT$ for perfect gases, where n is the number of moles, and R is the perfect gas constant (8.314 joules per mole per kelvin).

In 1807, POISSON used a law $p = C \varrho^{\gamma}$,[42] which may have been suggested by LAPLACE,[43] and there is a value of γ which gives the measured value of the speed of sound, but I doubt that LAPLACE or POISSON knew the explanation that one teaches now in thermodynamics, related to adiabatic transformations.[44] Working in a one-dimensional situation (the barrel of a gun), POISSON used the equation

$$\frac{\partial^2 w}{\partial t^2} - \frac{\partial}{\partial x}\left(f\left(\frac{\partial w}{\partial x}\right)\right) = 0, \tag{1.6}$$

with $f(z) = C z^{\gamma}$, which is quasi-linear,[45] and he studied special solutions (*rarefaction waves*), which he left in an implicit form.

This equation is related to the *Lagrangian point of view* (already introduced by EULER), where one follows material points; from an initial position y one considers $x = \Phi(y,t)$ the solution of $\frac{dx}{dt} = u(x(t),t)$ with $x(0) = y$ (where u is the velocity field, supposed to be smooth enough), and while in the (physical) *Eulerian point of view* one considers functions of x and t, the

---

[40] Jean Baptiste PERRIN, French physicist, 1870–1942. He received the Nobel Prize in Physics in 1926, for his work on the discontinuous structure of matter, and especially for his discovery of sedimentation equilibrium. He had worked in Paris, France.

[41] Robert BROWN, Scottish-born botanist, 1773–1858. He had collected specimens in Australia, and then worked in London, England.

[42] Siméon Denis POISSON, French mathematician, 1781–1840. He had worked in Paris, France.

[43] Pierre-Simon LAPLACE, French mathematician, 1749–1827. He had been made Count in 1806 by NAPOLÉON I and Marquis in 1817 by LOUIS XVIII. He had worked in Paris, France. NAPOLÉON I wrote in his memoir, written on St Helena, that he had removed LAPLACE from the office of minister of the interior, which he held in 1799, after only six weeks, "because he brought the spirit of the infinitely small into the government".

[44] Intuitively, a wave propagates too fast for an equilibrium in temperature to take place, so the process is not isothermal (i.e. at constant temperature), and the Boyle–Mariotte law does not apply; as there is no time for heat exchange, the process is called adiabatic (i.e. without heat transfer), a term equivalent to isentropic (as the second law of thermodynamics is $\delta Q = \theta\, ds$, and $\theta > 0$, $\delta Q = 0$ is equivalent to $ds = 0$).

[45] A semi-linear equation is linear in the highest-order derivatives with coefficients independent of lower-order derivatives, while a quasi-linear equation is linear in the highest-order derivatives but with coefficients which may depend upon lower-order derivatives. For example, $w_{tt} - c^2 \Delta w = F(w, w_t, w_{x_1}, \ldots, w_{x_N})$ is semi-linear, while $w_{tt} - A(w, w_t, w_{x_1}, \ldots, w_{x_N}) \Delta w = F(w, w_t, w_{x_1}, \ldots, w_{x_N})$ is quasi-linear.

(mathematical) Lagrangian point of view expresses them as functions of y and t. In all dimensions N one has $\frac{\partial f}{\partial t}\big|_y = \frac{\partial f}{\partial t} + \sum_{i=1}^{N} u_i \frac{\partial f}{\partial x_i}$; in one dimension one has $\frac{1}{\varrho(y,0)} \frac{\partial f}{\partial y}\big|_t = \frac{1}{\varrho(x,t)} \frac{\partial f}{\partial x}$. I do not find the Lagrangian point of view so useful for fluids in more than one dimension, because of turbulence effects,[46] and the Lagrangian point of view is more often used for solids. The Lagrangian point of view requires us to use the mathematical *Piola stress tensor*,[47] also introduced by KIRCHHOFF,[48] and called the *Piola–Kirchhoff stress tensor*, usually not symmetric, instead of the physical *Cauchy stress tensor*, always symmetric, which appears in the physical Eulerian point of view.[49] In Lagrangian coordinates, the Euler equation becomes $\frac{\partial \varrho}{\partial t} + \frac{\varrho^2}{\varrho_0} \frac{\partial u}{\partial y} = 0$ and $\frac{\partial u}{\partial t} + \frac{1}{\varrho_0} \frac{\partial p}{\partial y}$, which imply $\frac{\partial}{\partial t}\left(\frac{\varrho_0}{\varrho^2} \frac{\partial \varrho}{\partial t}\right) - \frac{\partial}{\partial y}\left(\frac{1}{\varrho_0} \frac{\partial p}{\partial y}\right) = 0$, and $w = \int^y \frac{1}{\varrho}$ satisfies an equation of the type considered by POISSON in the case where ϱ_0 is constant.

In 1848, CHALLIS noticed that there must be something wrong with the formula derived by POISSON in the case of periodic initial data,[50] and STOKES explained that the profile of a solution was getting steeper and steeper until it approached a discontinuous solution; he was then the first to derive the correct *jump conditions* for discontinuous solutions, as a consequence of conservation of mass and the *balance of momentum*. Jump conditions were rediscovered later by RIEMANN in his thesis in 1860,[51] where he used conservation of mass, balance of momentum and conservation of entropy,[52] instead of conservation of mass, balance of momentum and *conservation of energy*. It is important to notice that Peter LAX has generalized some notions from *gas dynamics* to other quasi-linear systems of *conservation laws*, and he has given a new meaning to the term *Riemann invariants*, but also, extending the work done for a scalar equation by Olga OLEINIK, and then Eberhard HOPF, he has

---

[46] Although turbulent flows are only said to occur in three dimensions, there are effects of a similar type in two dimensions for fluids when one uses a more realistic physical description (and exactly two-dimensional or one-dimensional flows are only a mathematical approximation, of course), but turbulence is certainly not about letting t tend to infinity (except possibly in an infinite region unchanged by rescaling in space).

[47] Gabrio PIOLA, Italian mathematician, 1794–1850. He had worked in Milano (Milan), Italy.

[48] Gustav Robert KIRCHHOFF, German physicist, 1824–1887. He had worked in Breslau (then in Germany, now Wrocław, Poland).

[49] The appearance of plasticity or turbulence renders the Lagrangian point of view problematic, and numerical analysts tend to use a mixture or Eulerian and Lagrangian points of view for this reason.

[50] James CHALLIS, English astronomer, 1803–1882. He had worked in Cambridge, England.

[51] Georg Friedrich Bernhard RIEMANN, German mathematician, 1826–1866. He had worked in Göttingen, Germany.

[52] The term (thermodynamic) entropy was only coined by CLAUSIUS in 1865, although the idea may go back to CARNOT, and RIEMANN must have used a function of (thermodynamic) entropy without using the name entropy.

used the term "entropy" in designing other functions not directly linked to thermodynamical entropy; the details of this important question will be explained later, and meanwhile I shall add the qualifier thermodynamical when referring to the usual physical quantity appearing in the second law of thermodynamics. Nowadays, the jump conditions are called the *Rankine–Hugoniot conditions*,[53,54] probably because STOKES did not reproduce his derivation of 1848 of the jump conditions when he edited his complete works in 1880, apologizing for his "mistake", because he had been (wrongly) convinced by Lord Rayleigh,[55] and by THOMSON (not yet Lord Kelvin) that his discontinuous solutions were not physical, because they did not conserve energy. This shows that none of them understood at that time that the missing energy had been transformed into heat, but if one has learnt thermodynamics, one should not disparage these great scientists of the 19th century for their curious mistake, and one should recognize that there are things which take time to understand. Actually, some mathematicians should pay more attention to what thermodynamics says, and by publishing too much on questions that they have not studied enough, it tends to make engineers and physicists believe that mathematicians do not know what they are talking about, and they should also observe that thermodynamics is not a good name, as it is not about dynamics but about equilibria. By describing the various pieces of the puzzle that I have studied, and by pointing out the limitations that I know, an important one being that the laws discovered experimentally by looking at equilibria are used all the time, even out of equilibrium, I want to convince the reader that one should try to go beyond the actual version of thermodynamics, and that one should create a good theory for questions out of equilibrium; that is usually treated by kinetic theory, the subject of these lectures, which has other defects which I shall point out.

Because the mathematical model used has no temperature variable, the part of the energy transformed into heat is apparently "lost", and the first way to correct this defect is to use a model which takes into account the first law of thermodynamics, expressing the conservation of *total* energy by the introduction of an internal energy (per unit mass) e (and $de = -p\,d\left(\frac{1}{\varrho}\right) + \delta Q$ for a gas, where δQ is the heat received, which the second law of thermodynamics relates to thermodynamical entropy as $\delta Q = \theta\,ds$, a form which may be due to DUHEM,[56] leading to the system of gas dynamics,

---

[53] William John Macquorn RANKINE, Scottish engineer, 1820–1872. He had worked in Glasgow, Scotland.

[54] Pierre Henri HUGONIOT, French engineer, 1851–1887.

[55] John William STRUTT, third Baron Rayleigh (known as Lord Rayleigh), English physicist, 1842–1919. He received the Nobel Prize in Physics in 1904, for his investigations of the densities of the most important gases and for his discovery of argon in connection with these studies. He had worked in Cambridge, England, holding the Cavendish professorship (1879–1884), after MAXWELL.

[56] Pierre Maurice Marie DUHEM, French mathematician, 1861–1916. He had worked in Lille and in Bordeaux, France.

$$\frac{\partial \varrho}{\partial t} + \sum_{i=1}^{3} \frac{\partial (\varrho\, u_i)}{\partial x_i} = 0, \tag{1.7}$$

for conservation of mass,

$$\frac{\partial (\varrho\, u_j)}{\partial t} + \sum_{i=1}^{3} \frac{\partial (\varrho\, u_i\, u_j)}{\partial x_i} + \frac{\partial p}{\partial x_j} = 0 \text{ for } j = 1, 2, 3, \tag{1.8}$$

for the balance of linear momentum, and

$$\frac{\partial}{\partial t}\left(\frac{\varrho\, |u|^2}{2} + \varrho\, e\right) + \sum_{i=1}^{3} \frac{\partial}{\partial x_i}\left(\frac{\varrho\, |u|^2 u_i}{2} + \varrho\, e\, u_i + p\, u_i\right) = 0, \tag{1.9}$$

for the balance of energy. The unknowns are the velocity u and some thermodynamical quantities, the density ϱ, the pressure p and the internal energy per unit of mass e (in the absence of a force field, the total energy per unit of mass is $E = \frac{\varrho\, |u|^2}{2} + e$, but in a force field deriving from a potential V one must add V to the preceding quantity); of course, there are not enough equations, but there is a relation between ϱ, p and e, given by the *equation of state*, which results from measurements of equilibria (and interpolation between the measured values, of course). The model does not take into account the effects of viscosity (which would appear in the three equations describing the balance of momentum) and heat conductivity (which would appear in the last equation describing the balance of energy). Energy cannot disappear in this model, because internal energy is supposed to take into account all the energy transformed into heat and stored inside the body (at a mesoscopic level), but the analysis of this model will show that something else disappears, and that will involve thermodynamical entropy; for this question I shall show some of the general principles (i.e. valid for many other systems, but restricted to one space dimension), which Peter LAX initiated in 1957.[57]

The question of appearance of discontinuities can be better described in a simpler model, the *Burgers equation*,

$$\frac{\partial u}{\partial t} + u\, \frac{\partial u}{\partial x} = 0 \text{ for } x \in \mathbb{R}, t > 0; \; u(x,0) = u_0(x) \text{ for } x \in \mathbb{R}, \tag{1.10}$$

---

[57] COURANT and FRIEDRICHS had written in 1948 a book on questions of *shocks*, which summarized many technical reports on questions which had been of importance during World War II. Peter LAX once told me that he had once to lecture on this subject, and instead of going through all the particular examples which were treated in the book, he prefered to start by developing a general mathematical framework encompassing all of them.

where u has the dimension of a velocity.[58] Using the method of characteristic curves,[59] the equation of a characteristic curve is $\frac{dx}{dt} = u(x(t), t)$ with $x(0) = y$, and then along this curve one has $\frac{d(u(x(t),t))}{dt} = 0$, so that the characteristic curves are lines and the solution is given by

$$u(x(t), t) = u_0(y) \text{ and } x(t) = y + t\, u_0(y), \tag{1.11}$$

as long as it makes sense; based on the remark that $y = x - t\, u$, the implicit solution found by POISSON was similar to writing a solution of the Burgers equation in the implicit form

$$u(x, t) = u_0\big(x - t\, u(x, t)\big). \tag{1.12}$$

Of course, the solution u is assumed to be smooth in this computation, and (1.7) shows that, apart from the case where u_0 is nondecreasing, there cannot exist a smooth solution for all $t > 0$, but one can also deduce such a property from the implicit equation (1.8).[60] What CHALLIS had noticed is similar to observing that if $u_0(x) = \sin x$ then the implicit equation cannot have a unique solution for all t; indeed $u = 0$ and $x - t\, u = j\,\pi$, i.e. $x = j\,\pi$, gives a few solutions, and $u = 1$ and $x - t\, u(x, t) = 2k\,\pi + \frac{\pi}{2}$, i.e. $x = t + 2k\,\pi + \frac{\pi}{2}$ gives a few solutions, but one has trouble deciding between $u = 0$ and $u = 1$ if one has $j\,\pi = t + 2k\,\pi + \frac{\pi}{2}$ for integers j, k, and this indeed happens (for all $x \in \mathbb{R}$) for $t = \frac{\pi}{2}$. It is simpler to use (1.7) and observe that if $y_1 < y_2$ but $u_0(y_1) > u_0(y_2)$, the characteristic lines through y_1 and y_2 intersect at a positive time $-\frac{y_2 - y_1}{u_0(y_2) - u_0(y_1)}$ and that it is impossible to have a smooth solution until that time as both $u_0(y_1)$ and $u_0(y_2)$ (which are different) compete to be the value of $u(x, t)$ for the point of intersection of the two characteristic lines; one deduces that the time of existence of a smooth solution is exactly

$$T_c = \frac{1}{\beta} \text{ if } \inf_{x \in \mathbb{R}} \frac{du_0}{dx}(x) = -\beta < 0, \tag{1.13}$$

---

[58] Some people prefer to write the equation $\frac{\partial v}{\partial t} + c\, v\, \frac{\partial v}{\partial x} = 0$ for a characteristic velocity c, with v having no dimension.

[59] For solving a first-order partial differential equation $\frac{\partial u}{\partial t} + \sum_{i=1}^{N} a_i(x, t) \frac{\partial u}{\partial x_i} = f(x, t, u)$, with initial data u_0, one first computes for every $y \in \mathbb{R}^N$ the characteristic curve going through y, defined by the system of ordinary differential equations $\frac{dx_i}{dt} = a_i(x(t), t)$ for $i = 1, \ldots, N$, and $x(0) = y$; then $v(t) = u(x(t), t)$ is the solution of a scalar differential equation $\frac{dv}{dt} = f(x(t), t, v)$ with $v(0) = u_0(y)$. I do not know who developed the method in a precise way, perhaps CAUCHY who had the first abstract theory of differential equations, but POISSON must have understood that, because he had a good physical intuition, according to what I heard about his work on the three-dimensional wave equation in the lectures of Laurent SCHWARTZ, when I was a student at École Polytechnique.

[60] For example, if $\frac{du_0}{dx} > -\alpha$ with $\alpha \geq 0$ then the implicit equation has a unique solution as long as $0 \leq t < \frac{1}{\alpha}$, because the function $v \mapsto v - u_0(x - t\, v)$ has a derivative $\geq 1 - t\alpha > 0$; in particular if u_0 is nondecreasing the solution exists for all $t > 0$.

because for $t < \frac{1}{\beta}$ only one characteristic line goes through (x, t) whatever x is, but if $t > \frac{1}{\beta}$ (and u_0 is bounded) there exists x with two different chracteristic lines going through the point (x, t).

What should one do after the appearance of the first singularity? STOKES's proposal is to accept discontinuous solutions, and that is related to considering solutions in the sense of distributions, but for that one will have to use equations in conservative form, because $u\, u_x$ does not make sense when u is discontinuous, because u_x has a Dirac mass at this point, and one can only multiply a Dirac mass by a function which is continuous at that point,[61] and u is not, but one can define the derivative of $\frac{u^2}{2}$ in the sense of distributions. A way to see the difficulty is to consider a function v equal to -1 for $x < 0$ and $+1$ for $x > 0$ (i.e. $u = -1 + 2H$, where H is the Heaviside function, whose derivative in the sense of distribution is the Dirac mass at 0), so that the derivative of u is $u_x = 2\delta_0$; as $u^2 = 1$ one has $3u^2 u_x = 6\delta_0$, but as $u^3 = u$ the derivative of u^3 is $2\delta_0$, one deduces $(u^3)_x \neq 3u^2 u_x$; one should be careful then that some nonlinear calculus rules are not allowed for discontinuous solutions.

[Taught on Monday August 27, 2001.]

Notes on names cited in footnotes for Chapter 1, NÉEL,[62] ZEEMAN,[63] FULLER,[64]

---

[61] That is not exactly true because one can multiply a Dirac mass by any Borel function, and Borel functions are defined at every point (and are not equivalence classes, as locally integrable functions are). However, in order to solve partial differential equations one uses the theory of distributions, where Lebesgue-measurable functions are identified if they coincide almost everywhere (a function is measurable if and only if it coincides with a Borel function outside a set of Lebesgue measure 0). If one thinks in terms of mathematics, one may well like to use Borel functions, but the theory of distributions is adapted to the laws of continuum mechanics and physics (at least for their linear equations), in particular because of the point of view that I developed in the early 1970s, that weak convergence is a good model for explaining the relations between different levels, microscopic/mesoscopic/macroscopic.

[62] Louis Eugène Félix NÉEL, French physicist, 1904–2000. He received the Nobel Prize in Physics in 1970, for fundamental work and discoveries concerning antiferromagnetism and ferrimagnetism which have led to important applications in solid state physics, jointly with Hannes ALFVÉN. He had worked in Strasbourg, and in Grenoble, France.

[63] Pieter ZEEMAN, Dutch physicist, 1865–1943. He received the Nobel Prize in Physics in 1902, jointly with Hendrik LORENTZ, in recognition of the extraordinary service they rendered by their research into the influence of magnetism upon radiation phenomena. He had worked in Leiden, and in Amsterdam, The Netherlands.

[64] John FULLER, English politician and philanthropist, 1757–1834.

TOMONAGA,[65] SCHWINGER,[66] CHARLES X,[67] BONAPARTE/NAPOLÉON I,[68] CLAUSIUS,[69] CARNOT,[70] COURANT,[71] FRIEDRICHS,[72] BOREL,[73] LEBESGUE,[74] and for the preceding footnotes, PURDUE,[75] HARVARD.[76]

---

[65] Sin-Itiro TOMONAGA, Japanese-born physicist, 1906–1979. He received the Nobel Prize in Physics in 1965, jointly with Julian SCHWINGER and Richard FEYNMAN, for their fundamental work in quantum electrodynamics, with deep-ploughing consequences for the physics of elementary particles. He had worked in Tokyo, Japan, in Leipzig, Germany, in Tsukuba, Japan, and at IAS (Institute for Advanced Study), Princeton, NJ.

[66] Julian Seymour SCHWINGER, American physicist, 1918–1994. He received the Nobel Prize in Physics in 1965, jointly with Sin-Itiro TOMONAGA and Richard FEYNMAN, for their fundamental work in quantum electrodynamics, with deep-ploughing consequences for the physics of elementary particles. He had worked at UCB (University of California at Berkeley), Berkeley, CA, at Purdue University, West Lafayette, IN, and at Harvard University, Cambridge, MA.

[67] Charles-Philippe de France, 1757–1836, comte d'Artois, duc d'Angoulême, pair de France, was King of France from 1824 to 1830 under the name CHARLES X.

[68] Napoléon BONAPARTE, French general, 1769–1821. He became Premier Consul after his coup d'état in 1799, was elected Consul à vie in 1802, and he proclaimed himself emperor in 1804, under the name NAPOLÉON I (1804–1814, and 100 days in 1815).

[69] Rudolf Julius Emmanuel CLAUSIUS, German physicist, 1822–1888. He had worked in Berlin, Germany, in Zürich, Switzerland, in Würzburg and in Bonn, Germany.

[70] Sadi Nicolas Léonard CARNOT, French engineer, 1796–1832. He had worked in Paris, France.

[71] Richard COURANT, German-born mathematician, 1888–1972. He had worked in Göttingen, Germany, and at NYU (New York University), New York, NY. The department of mathematics of NYU is named after him, the Courant Institute of Mathematical Sciences.

[72] Kurt Otto FRIEDRICHS, German-born mathematician, 1901–1982. He had worked in Aachen and in Braunschweig, Germany, and at NYU (New York University), New York, NY.

[73] Félix Edouard Justin Emile BOREL, French mathematician, 1871–1956. He had worked in Lille and in Paris, France.

[74] Henri Léon LEBESGUE, French mathematician, 1875–1941. He had worked in Rennes, in Poitiers, and in Paris, France, holding a chair (mathématiques, 1921–1941) at Collège de France, Paris.

[75] John PURDUE, American industrialist, 1802–1876. Purdue University, West Lafayette, IN, is named after him.

[76] John HARVARD, English clergyman, 1607–1638. Harvard University, Cambridge, MA, is named after him.

2

Hyperbolic Systems: Riemann Invariants, Rarefaction Waves

The book of COURANT and FRIEDRICHS [3] helped mathematicians entering an important domain of continuum mechanics by collecting a lot of information scattered in the engineering literature, but the analysis done by Peter LAX in extracting a general mathematical framework out of it was crucial. When I looked at this book in the late 1970s for lectures that I was teaching, I noticed an historical section, and it had some influence on my interest in the history of ideas in mathematics, or in science in general, and when I lectured on *quasi-linear hyperbolic systems* of conservation laws in the spring of 1991 I tried to read some of the earlier texts that were mentioned there. Cathleen MORAWETZ,[1] who had been asked to edit the book by her advisor (FRIEDRICHS) when she was a graduate student, told me a few years ago that the historical section was initially much larger, but had to be trimmed because the book was too long.

Questions of shock waves had been important in industrial or military applications, and while COURANT and FRIEDRICHS had been involved on the American side, with Peter LAX working out some of the mathematical questions, similar work had been done in USSR by Sergei GODUNOV and Olga OLEINIK,[2] and there was some work in England and in France, where the mathematical community was not really aware of these questions. After some initial mathematical work in the early 1950s on a scalar equation, which will be described later, it was time in the late 1950s to try to handle systems in a mathematical way, and this is the trend that Peter LAX started; however,

---

[1] Cathleen SYNGE-MORAWETZ, Canadian-born mathematician, 1923. She works at NYU (New York University), New York, NY. Her father, John SYNGE had been the head of the mathematics department at Carnegie Tech (Carnegie Institute of Technology), now CMU (Carnegie Mellon University), Pittsburgh, PA, from 1946 to 1948.

[2] Sergei Konstantinovich GODUNOV, Russian mathematician, born in 1929. He works at the Sobolev institute of mathematics of the Siberian branch of the Russian Academy of Sciences, Novosibirsk, Russia.

one should be aware that numerical codes have been written since the early 1950s on questions for which the mathematical understanding is missing, and by using experimental information and conjectures these codes perform quite well, but Peter LAX tried to attack the problem from a mathematical point of view, working on equations where the physical intuition might not exist.

There was a different group of mathematicians working on questions related to *viscous fluids*, which give rise to partial differential equations of parabolic type (because one cheats with physics by pretending that the fluids are incompressible), the prototype being the *Navier–Stokes equation*,[3] and the first mathematical work was done by Jean LERAY in the 1930s,[4] using his work with SCHAUDER of extending the Brouwer topological degree to an infinite-dimensional setting;[5,6] the work was continued by Eberhard HOPF, Olga LADYZHENSKAYA,[7] and others in the 1960s, like Ciprian FOIAS,[8] Jacques-Louis LIONS, and James SERRIN.[9]

Although real problems from continuum mechanics occur in three space dimensions, the framework for quasi-linear hyperbolic systems mostly deals with problems in one space dimension; even for linear hyperbolic systems, the multidimensional situation is much more difficult than the one-dimensional case. After understanding how to define and solve linear hyperbolic systems (in one space dimension), the elements of the general theory will be presented (*Riemann problem*, Riemann invariants, shocks and *contact discontinuities*, "entropies"), with the system of gas dynamics as an example.

One considers a linear system with constant coefficients

---

[3] NAVIER had introduced the equation in 1821 by a molecular approach, and it was rederived more mathematically in 1843 by SAINT-VENANT and in 1845 by STOKES.

[4] Jean LERAY, French mathematician, 1906–1998. He received the Wolf Prize in 1979, for pioneering work on the development and application of topological methods to the study of differential equations, jointly with André WEIL. He had worked in Nancy, France, in a prisoner of war camp in Austria (1940–1945), in Paris, France, holding a chair (théorie des équations différentielles et fonctionnelles, 1947–1978) at Collège de France, Paris.

[5] Juliusz Pawel SCHAUDER, Polish mathematician, 1899–1943. He had worked in Lvov (then in Poland, now in Ukraine).

[6] Luitzen Egbertus Jan BROUWER, Dutch mathematician, 1881–1966. He had worked in Amsterdam, The Netherlands.

[7] Olga Aleksandrovna LADYZHENSKAYA, Russian mathematician, 1922–2004. She had worked at the Steklov Mathematical Institute, in Leningrad, USSR, then St Petersburg, Russia. I first met her in 1991 in Bath, England.

[8] Ciprian Ilie FOIAS, Romanian-born mathematician, born in 1933. He worked in Bucharest, Romania, at Université Paris-Sud, Orsay, France (where he was my colleague in 1978–1979), at Indiana University, Bloomington, IN, and at Texas A&M, College Station, TX.

[9] James B. SERRIN Jr., American mathematician, born in 1926. He works at University of Minnesota, Minneapolis, MN.

$$\frac{\partial U}{\partial t} + A\,\frac{\partial U}{\partial x} = 0 \text{ for } x \in \mathbb{R}, t > 0; \ U(\cdot, 0) = U_0 \text{ in } \mathbb{R}, \tag{2.1}$$

where $U(x,t)$ is a vector with p components, and A is a $p \times p$ matrix independent of x, t or U. Every partial differential equation with constant coefficients can be rewritten as such a system but I am only interested here in linear hyperbolic systems, which must exhibit an effect of *finite speed of propagation*. Another definition of hyperbolicity (in a given direction, which is time in the physical examples) is that the *Cauchy problem* should be well posed. Linear hyperbolic equations have been studied by Lars GÅRDING,[10] Lars HÖRMANDER,[11] Peter LAX, and Jean LERAY.

Definition 2.1. *One says that the system is* hyperbolic *if A has only real eigenvalues and is diagonalizable; the system is said to be* strictly hyperbolic *if A has only distinct real eigenvalues (so that it is diagonalizable).*

One orders the eigenvalues in increasing order

$$\lambda_1 \le \ldots \le \lambda_p, \tag{2.2}$$

and one chooses a basis of eigenvectors $r_j, j = 1, \ldots, p$, i.e.

$$A\,r_j = \lambda_j r_j \text{ for } j = 1, \ldots, p, \tag{2.3}$$

and also uses the dual basis $l_j, j = 1, \ldots, p$, i.e. $l_j(r_k) = \delta_{j,k}$ the Kronecker symbol,[12] for $j, k = 1, \ldots, p$, so that

$$A^T \ell_k = \lambda_k \ell_k \text{ for } k = 1, \ldots, p. \tag{2.4}$$

Peter LAX calls the r_j right eigenvectors and the ℓ_k left eigenvectors, and one may think of r_j as a column vector and of ℓ_k as a row vector; of course, this is related to the fact that the r_j belong to a vector space $E = \mathbb{R}^p$, and the ℓ_k belong to its dual E', and that no Euclidean structure on $\mathbb{R}^p$ is necessary,[13,14] so that it should not be identified with its dual. If A is

---

[10] Lars GÅRDING, Swedish mathematician, born in 1919. He worked at Lund University, Lund, Sweden.

[11] Lars HÖRMANDER, Swedish mathematician, born in 1931. He received the Fields Medal in 1962, and the Wolf Prize in 1988, for fundamental work in modern analysis, in particular, the application of pseudo-differential and Fourier integral operators to linear partial differential equations, jointly with Friedrich HIRZEBRUCH. He worked in Stockholm, Sweden, at Stanford University, Stanford, CA, at IAS (Institute for Advanced Study), Princeton, NJ, and in Lund, Sweden.

[12] Leopold KRONECKER, German mathematician, 1823–1891. He had worked in Berlin, Germany.

[13] EUCLID of Alexandria, "Egyptian" mathematician, about 325 BCE–265 BCE. It is not known where he was born, but he had worked in Alexandria, Egypt, shortly after it was founded by ALEXANDER the Great, in 331 BCE.

[14] BCE = Before Common Era, CE = Common Era.

hyperbolic, the explicit solution of the Cauchy problem is easily obtained by decomposing the unknown vector $U(x,t)$ on the basis of eigenvalues $r_j, j = 1, \ldots, p$,

$$U(x,t) = \sum_{j=1}^{p} u_j(x,t) r_j, \qquad (2.5)$$

and applying ℓ_k to the equation; as $\langle \ell_k, U \rangle = u_k$ and $\langle \ell_k, A\,U \rangle = \langle A^T \ell_k, U \rangle = \lambda_k u_k$, one finds that

$$\frac{\partial u_k}{\partial t} + \lambda_k \frac{\partial u_k}{\partial x} = 0, k = 1, \ldots, p, \qquad (2.6)$$

giving $u_k(x,t) = u_k(x - \lambda_k t, 0) = \langle \ell_k, U_0(x - \lambda_k t) \rangle$, and one deduces that the solution is given by the formula

$$U(x,t) = \sum_{j=1}^{p} \langle \ell_j, U_0(x - \lambda_j t) \rangle r_j, x, t \in \mathbb{R}. \qquad (2.7)$$

This formula shows that the eigenvalues of the matrix A are velocities, called *characteristic velocities*, of propagation of some particular modes corresponding to the eigenvectors r_j (the dual basis is only a technical tool). For a quasi-linear system, one starts with a similar definition.

Definition 2.2. *The quasi-linear system (shown here with an initial datum)*

$$\frac{\partial U}{\partial t} + A(U) \frac{\partial U}{\partial x} = 0 \text{ for } x \in \mathbb{R}, t > 0; \ U(\cdot, 0) = U_0 \text{ in } \mathbb{R} \qquad (2.8)$$

is hyperbolic *if for* U *in a domain* $D \subset \mathbb{R}^p$,[15] *the matrix* $A(U)$ *has real eigenvalues and is diagonalizable for every* $U \in D$, *and* strictly hyperbolic *if* $A(U)$ *has real distinct eigenvalues for every* $U \in D$ *(which one orders* $\lambda_1(U) < \ldots < \lambda_p(U)$*).*

Assuming that A has distinct eigenvalues and is a smooth function in D, then the eigenvalues are smooth functions and one may define a basis of right eigenvectors $r_j(U), j = 1, \ldots, p$ which are smooth functions, and the dual basis of left eigenvectors is then also smooth.

---

[15] In gas dynamics, the density ϱ is nonnegative, and the case where $\varrho = 0$ is related to cavitation and creates mathematical difficulties; the internal energy e is nonnegative, and it designates a part of the energy which is hidden at mesoscopic level; the pressure is nonnegative, and it is interpreted in kinetic theory as resulting from particles bouncing on the boundary of the container and exchanging momentum with it (negative pressures like suction involve viscosity).

Is then the system of gas dynamics strictly hyperbolic? One considers the simple case where $u_2 = u_3 = 0$,[16] which is a reasonable hypothesis when the gas moves in a small tube at slow velocities, so that the system is

$$
\begin{aligned}
&\varrho_t + (\varrho\, u)_x = 0, \\
&(\varrho\, u)_t + (\varrho\, u^2 + p)_x = 0, \\
&\left(\frac{\varrho\, u^2}{2} + \varrho\, e\right)_t + \left(\frac{\varrho\, u^3}{2} + \varrho\, e\, u + p\, u\right)_x = 0,
\end{aligned}
\tag{2.9}
$$

and if one assumes that the solution (ϱ, u, e) is smooth in (x, t) and that the equation of state gives p as a smooth function of (ϱ, e), one can rewrite the system by noticing that

$$
\big((\varrho\, u)_t + (\varrho\, u^2 + p)_x\big) - u\,\big(\varrho_t + (\varrho\, u)_x\big) = \varrho\, u_t + \varrho\, u\, u_x + p_x,
\tag{2.10}
$$

and

$$
\begin{aligned}
&\left(\left(\frac{\varrho\, u^2}{2} + \varrho\, e\right)_t + \left(\frac{\varrho\, u^3}{2} + \varrho\, e\, u + p\, u\right)\right) + \left(\frac{u^2}{2} - e\right)\left(\varrho_t + (\varrho\, u)_x\right) \\
&\quad - u\left((\varrho\, u)_t + (\varrho\, u^2 + p)_x\right) = \varrho\, e_t + \varrho\, u\, e_x + p\, u_x,
\end{aligned}
\tag{2.11}
$$

so that, as long as $\varrho > 0$, one finds the system

$$
\begin{aligned}
&\varrho_t + u\, \varrho_x + \varrho\, u_x = 0, \\
&u_t + u\, u_x + \frac{1}{\varrho}\, p_x = 0, \\
&e_t + u\, e_x + \frac{p}{\varrho}\, u_x = 0,
\end{aligned}
\tag{2.12}
$$

which for $U^e = \begin{pmatrix} \varrho \\ u \\ e \end{pmatrix}$ belonging to the domain $D^e = (0, \infty) \times \mathbb{R} \times (0, \infty)$

corresponds to $A(U^e)$ given by

$$
A(U^e) = \begin{pmatrix} u & \varrho & 0 \\ \frac{1}{\varrho}\frac{\partial p}{\partial \varrho}\big|_e & u & \frac{1}{\varrho}\frac{\partial p}{\partial e}\big|_\varrho \\ 0 & \frac{p}{\varrho} & u \end{pmatrix} = u\, I + \begin{pmatrix} 0 & \varrho & 0 \\ \frac{1}{\varrho}\frac{\partial p}{\partial \varrho}\big|_e & 0 & \frac{1}{\varrho}\frac{\partial p}{\partial e}\big|_\varrho \\ 0 & \frac{p}{\varrho} & 0 \end{pmatrix},
\tag{2.13}
$$

and the fact that $A(U^e) - u\, I$ only depends upon the thermodynamical variables (ϱ, e) is related to the Galilean invariance of the system of gas dynamics. The characteristic polynomial of $A(U^e) - u\, I$ is $-\lambda^3 + \lambda\left(\frac{\partial p}{\partial \varrho}\big|_e + \frac{p}{\varrho^2}\frac{\partial p}{\partial e}\big|_\varrho\right) = 0$, so the gas dynamics system is strictly hyperbolic if and only if one has

$$
c^2 = \frac{\partial p}{\partial \varrho}\Big|_e + \frac{p}{\varrho^2}\frac{\partial p}{\partial e}\Big|_\varrho > 0,
\tag{2.14}
$$

---

[16] Without this condition, the system is hyperbolic but not strictly hyperbolic. The two added components of velocity, u_2, u_3 solve the equations $(u_j)_t + u_1(u_j)_x = 0$ for $j = 2, 3$, so that the eigenvalue u_1 has multiplicity 3. The equation of balance of energy is $\left(\frac{\varrho |u|^2}{2} + \varrho\, e\right)_t + \left(\frac{\varrho |u|^2 u_1}{2} + \varrho\, e\, u_1 + p\, u_1\right)_x = 0$, and it gives $\varrho\, e_t + \varrho\, u_1\, e_x + p\, (u_1)_x = 0$.

and $c > 0$ is the local speed of sound; the eigenvalues are then

$$\lambda_1(U^e) = u - c; \ \lambda_2(U^e) = u; \ \lambda_3(U^e) = u + c, \tag{2.15}$$

and one may choose as right eigenvectors

$$r_1(U^e) = \begin{pmatrix} -\varrho^2 \\ \varrho c \\ -p \end{pmatrix}; \ r_2(U^e) = \begin{pmatrix} \frac{\partial p}{\partial e}\big|_\varrho \\ 0 \\ -\frac{\partial p}{\partial \varrho}\big|_e \end{pmatrix}; \ r_3(U^e) = \begin{pmatrix} +\varrho^2 \\ \varrho c \\ +p \end{pmatrix}. \tag{2.16}$$

The computations are made simpler if one uses the first and second law of thermodynamics, so that

$$de = -p\, d\left(\frac{1}{\varrho}\right) + \theta\, ds. \tag{2.17}$$

Multiplying the equation in ϱ by $\frac{p}{\varrho^2}$ and adding to the equation in u gives $\theta(s_t + u\, s_x) = 0$, and as $\theta > 0$, one deduces that

$$s_t + u\, s_x = 0. \tag{2.18}$$

If one assumes now that the equation of state gives p as a smooth function of (ϱ, s), one considers the system

$$\begin{aligned} \varrho_t + u\,\varrho_x + \varrho\,u_x &= 0, \\ u_t + u\,u_x + \frac{1}{\varrho}\,p_x &= 0, \\ s_t + u\, s_x &= 0, \end{aligned} \tag{2.19}$$

which for $U^s = \begin{pmatrix} \varrho \\ u \\ s \end{pmatrix}$ belonging to the domain $D^s = (0, \infty) \times \mathbb{R} \times \mathbb{R}$ corresponds to $A(U^s)$ given by

$$A(U^s) = \begin{pmatrix} u & \varrho & 0 \\ \frac{1}{\varrho}\frac{\partial p}{\partial \varrho}\big|_s & u & \frac{1}{\varrho}\frac{\partial p}{\partial s}\big|_\varrho \\ 0 & 0 & u \end{pmatrix} = u\,I + \begin{pmatrix} 0 & \varrho & 0 \\ \frac{1}{\varrho}\frac{\partial p}{\partial \varrho}\big|_s & 0 & \frac{1}{\varrho}\frac{\partial p}{\partial s}\big|_\varrho \\ 0 & 0 & 0 \end{pmatrix}, \tag{2.20}$$

and again the fact that $A(U^s) - u\,I$ only depends upon the thermodynamical variables (ϱ, s) is related to Galilean invariance. As the velocity of propagation should not depend upon the basis used for the space U, one finds the same eigenvalues

$$\lambda_1(U^s) = u - c; \ \lambda_2(U^s) = u; \ \lambda_3(U^s) = u + c, \tag{2.21}$$

but the local velocity of sound c is given by the simpler formula (giving the same value than before)

$$c^2 = \frac{\partial p}{\partial \varrho}\Big|_s, \tag{2.22}$$

and one may choose the right eigenvectors as

$$
r_1(U^s) = \begin{pmatrix} -\varrho \\ c \\ 0 \end{pmatrix} ; \quad r_2(U^s) = \begin{pmatrix} -\frac{\partial p}{\partial s}\big|_\varrho \\ 0 \\ \frac{\partial p}{\partial \varrho}\big|_s \end{pmatrix} ; \quad r_3(U^s) = \begin{pmatrix} +\varrho \\ c \\ 0 \end{pmatrix} .
\tag{2.23}
$$

One should notice that this computation has not relied on what θ is but only on $\theta \neq 0$, so that if one replaces s by $\varphi(s)$ and θ by $\frac{\theta}{\varphi'(s)}$ with a function φ such that $\varphi' > 0$ for example, one obtains the same result; actually the equation for s implies the conservation law

$$
\big(\varrho f(s)\big)_t + \big(\varrho u f(s)\big)_x = 0,
\tag{2.24}
$$

for all smooth functions f, and it is not clear at this point why the thermodynamical entropy s cannot be replaced by $\varphi(s)$.[17] It is this property of thermodynamical entropy, that it corresponds to new conserved quantities for smooth solutions, that led Peter LAX to call any conserved quantity an "entropy" (which I often qualify as mathematical, so that the uninformed reader will not be mistaken); one should notice that looking for "entropies" does not require a system to be hyperbolic, and classical results in this direction are often related to a theorem of A. NOETHER.[18]

The laws of thermodynamics for a gas have given $de = -p\,d\big(\frac{1}{\varrho}\big) + \theta\,ds$, which implies that

$$
\frac{\partial s}{\partial \varrho}\bigg|_e = -\frac{p}{\theta\,\varrho^2}; \quad \frac{\partial e}{\partial s}\bigg|_\varrho = \frac{1}{\theta},
\tag{2.25}
$$

and

$$
\frac{\partial e}{\partial \varrho}\bigg|_s = \frac{p}{\varrho^2}; \quad \frac{\partial e}{\partial s}\bigg|_\varrho = \theta;
\tag{2.26}
$$

writing a function f as $f\big(\varrho, e(\varrho, s)\big)$, one has

---

[17] The experimental facts have shown that when two separate bodies are put in contact, heat flows from one body to the other if they do not have the same temperature (and heat flows from the hotter one to the colder one), and equilibrium occurs when the temperatures of the two bodies coincide, and not when their values of $\frac{\theta}{\varphi'(s)}$ coincide; however, this requires some heat conductivity, which is missing in the model considered here. It is for this kind of reason that one should be careful about the properties of a system of equations that one uses for describing physical reality, and one has to prove that the system has a property which is observed, and if it is lacking some real property it does not mean that the model is useless, but it points out in what situations the model should be used and in what other situations the model should not be used.

[18] Amalie (Emmy) NOETHER, German-born mathematician, 1882–1935. She had worked in Göttingen, Germany, and then in Bryn Mawr, PA.

$$\left.\frac{\partial f}{\partial \varrho}\right|_s = \left.\frac{\partial f}{\partial \varrho}\right|_e + \frac{p}{\varrho^2}\left.\frac{\partial f}{\partial e}\right|_\varrho$$
$$\left.\frac{\partial f}{\partial s}\right|_\varrho = \theta \left.\frac{\partial f}{\partial e}\right|_\varrho$$

$$(2.27)$$

and writing a function f as $f(\varrho, s(\varrho, e))$, one has

$$\left.\frac{\partial f}{\partial \varrho}\right|_e = \left.\frac{\partial f}{\partial \varrho}\right|_s - \frac{p}{\theta \varrho^2}\left.\frac{\partial f}{\partial s}\right|_\varrho$$
$$\left.\frac{\partial f}{\partial e}\right|_\varrho = \frac{1}{\theta}\left.\frac{\partial f}{\partial s}\right|_\varrho.$$

$$(2.28)$$

Definition 2.3. *The* Riemann problem *is a particular case of the Cauchy problem, where the initial datum U_0 has the form*

$$U_0(x) = \begin{cases} U_- & \text{if } x < 0 \\ U_+ & \text{if } x > 0. \end{cases} \tag{2.29}$$

Although the computations done previously on the linear case seem to have assumed some smoothness of the solution, they are actually true in the sense of distributions and one may take for U_0 any measurable and locally integrable function (or any distribution, if one likes); in the linear case the solution of the Riemann problem is then piecewise constant, of the form

$$U(x,t) = \begin{cases} a_0 = U_- & \text{for } x < \lambda_1 t \\ a_j & \text{for } \lambda_j t < x < \lambda_{j+1} t \text{ and } 1 \le j \le p-1 \\ a_p = U_+ & \text{for } \lambda_p t < x \end{cases} \tag{2.30}$$

showing that the initial discontinuity splits in general into p discontinuities, propagating at one of the characteristic velocities, and there is such a discontinuity propagating at velocity λ_j if and only if $\langle \ell_j, U_+ - U_- \rangle \ne 0$.

The solution of the Riemann problem is slightly different for the quasi-linear situation. There are sectors of the (x,t) plane where the solution is constant, but these sectors do not cover the whole plane in general, and in describing the solution one may need some sectors where the solution changes continuously, the centred rarefaction waves, whose study involves the Riemann invariants, and one may also need some discontinuities; two types of discontinuities may occur, contact discontinuities as in the linear case, and shocks. Which shocks are acceptable is a difficult question, related to explaining how to deal with irreversible phenomena, and understanding the terms "entropies" and *entropy conditions* will be a part of the answer.

To describe the regular parts in the solution of the Riemann problem, Peter LAX introduced the notion of Riemann invariants, which generalize what RIEMANN had done on a particular example, of course.

Definition 2.4. *A function w, defined on the domain $D \subset \mathbb{R}^p$, is called a j-Riemann invariant (for some $j \in \{1,\ldots,p\}$) if it satisfies*

$$\langle \nabla w(U), r_j(U) \rangle = 0 \text{ in } D. \tag{2.31}$$

The equation for a j-Riemann invariant is a differential equation, for the vector field r_j, and one can apply the method of characteristic curves to find a local solution if the value of w is given on a *noncharacteristic surface* S, i.e. a surface S such that for every $U \in S$ the vector $r_j(U)$ is not tangent to S. The characteristic curves are obtained by solving the differential system

$$\frac{dV}{d\tau} = r_j(V(\tau)), \tag{2.32}$$

and along each of these curves (which are defined locally because r_j is smooth) a j-Riemann invariant must be constant. If $w_1, w_2, \ldots, w_k$ are j-Riemann invariants, then for every smooth function h of k variables, $h(w_1, w_2, \ldots, w_k)$ is also a j-Riemann invariant, and one can describe locally all the j-Riemann invariants if one knows $p - 1$ functions on a noncharacteristic hypersurface S whose differentials are linearly independent, and the independence stays true along the characteristic curve for the corresponding j-Riemann invariants that they define, a classical result for linear ordinary differential equations; indeed, if $\sum_{k=1}^{p}(r_j)_k \frac{\partial w}{\partial x_k} = 0$, then deriving in x_ℓ gives $\sum_{k=1}^{p}(r_j)_k \frac{\partial^2 w}{\partial x_k \partial x_\ell} + \sum_{k=1}^{p} \frac{\partial (r_j)_k}{\partial x_\ell} \frac{\partial w}{\partial x_k} = 0$, i.e. $M(\tau) = \nabla w(V(\tau))$ satisfies a linear differential equation $\frac{dM}{d\tau} + B(\tau)M(\tau) = 0$, where the matrix B has entries $B_{\ell,k} = \frac{\partial (r_j)_k}{\partial x_\ell}$, and one deduces that if M vanishes at some value of τ it must vanish for all values of τ.

Which are the Riemann invariants for the system of gas dynamics? Using the variables (ϱ, u, s), the equation for a 1-Riemann invariant is $-\varrho \frac{\partial w}{\partial \varrho} + c \frac{\partial w}{\partial u} = 0$, which has two independent solutions, one being $w = s$ and another one being $w = u + g(\varrho, s)$, where g satisfies $\frac{\partial g}{\partial \varrho} = \frac{c}{\varrho}$ (so g is defined modulo a function of s, and as functions of s are 1-Riemann invariants it does not matter much), and then

the general 1-Riemann invariant is $h(u + g(\varrho, s), s)$ with $\frac{\partial g}{\partial \varrho} = \frac{c}{\varrho}$
for an arbitrary smooth function h; $\qquad$ (2.33)

the equation for a 2-Riemann invariant is $-\frac{\partial p}{\partial s}\big|_\varrho \frac{\partial w}{\partial \varrho} + \frac{\partial p}{\partial \varrho}\big|_s \frac{\partial w}{\partial s} = 0$, which has two independent solutions, one being $w = u$ and another one being $w = p$, so

the general 2-Riemann invariant is $h(u, p)$
for an arbitrary smooth function h; $\qquad$ (2.34)

and, either by repeating the same computation, or by using the fact that a change of orientation of the x axis exchanges the order of eigenvalues (as it changes their sign) and changes u into $-u$, the formulas for 1-Riemann invariants give 3-Riemann invariants by changing u into $-u$, so

the general 3-Riemann invariant is $h(u - g(\varrho, s), s)$ with $\frac{\partial g}{\partial \varrho} = \frac{c}{\varrho}$
for an arbitrary smooth function h. $\qquad$ (2.35)

Again, one sees an interesting property of functions of s, that they are both 1-Riemann invariants and 3-Riemann invariants, so that the surfaces $s = constant$ are well defined without having to invoke thermodynamics, but s could be replaced by an arbitrary function of s (so one cannot discover the special role of temperature from the equations); this also shows that the system of gas dynamics is special, because the vector fields r_1 and r_3 satisfy an integrability condition,[19] of a type studied by CLEBSCH,[20] and by FROBENIUS.[21]

Definition 2.5. *A regular solution U of the system in an open set Ω of the (x,t) plane is called a j-simple wave (or a j-wave) if, for every j-Riemann invariant w, $w\big(U(x,t)\big)$ is constant in Ω.*

It means that the values taken by U in the open set Ω are all taken from one integral curve of the vector field r_j, so if one has a parametrization $V(\tau)$ of the integral curve, it just means that τ is a function of (x,t), so that

$$U(x,t) = V\big(\tau(x,t)\big), \text{ with } \frac{dV}{d\tau} = r_j\big(V(\tau)\big) \text{ and } A\,r_j = \lambda_j r_j, \qquad (2.36)$$

so that τ satisfies

$$\frac{\partial \tau}{\partial t} + \lambda_j\big(V(\tau)\big)\frac{\partial \tau}{\partial x} = 0, \qquad (2.37)$$

an equation already considered in the special case of the Burgers equation; the same computation as before shows that the characteristic curves are straight lines, where τ is constant.

The solution of the Riemann problem has some constant sectors and may have some sectors where the solution is smooth and not constant, and it is then a j-wave where all the characteristic lines go through 0, a *centred wave* (of course, the solution may have different sectors, corresponding to j-waves with different values of j).

The reason that one looks for centred waves is a question of invariance. Because the equation is invariant by the group of transformations $(x,t) \mapsto$

---

[19] The commutator of the operators A_1 and A_3 of derivation in the directions r_1 and r_3 should be a linear combination of A_1 and A_3. If a nonzero vector field v in $\mathbb{R}^N$ is given and one wants to find a hypersurface which is perpendicular to v at each point, it means that one looks for a smooth function f such that $v = c\,grad\,f$ with a function $c \neq 0$, from which one deduces that $\frac{\partial v_i}{\partial x_j} - \frac{\partial v_j}{\partial x_i} = \frac{\partial c}{\partial x_j}\frac{\partial f}{\partial x_i} - \frac{\partial c}{\partial x_i}\frac{\partial f}{\partial x_j} = \frac{\partial d}{\partial x_j}v_i - \frac{\partial d}{\partial x_i}v_j$ for all i,j, where $d = \log c$, and conversely, $\frac{\partial v_i}{\partial x_j} - \frac{\partial v_j}{\partial x_i} = \frac{\partial d}{\partial x_j}v_i - \frac{\partial d}{\partial x_i}v_j$ for all i,j implies $\frac{\partial(e^{-d}v_i)}{\partial x_j} - \frac{\partial(e^{-d}v_j)}{\partial x_i} = 0$ for all i,j, so that $e^{-d}v_i = \frac{\partial w}{\partial x_i}$ for all i, locally.

[20] Rudolf Friedrich Alfred CLEBSCH, German mathematician, 1833–1872. He had worked in Berlin, in Karlsruhe, in Giessen and in Göttingen, Germany.

[21] Ferdinand Georg FROBENIUS, German mathematician, 1849–1917. He had worked in Zürich, Switzerland, and in Berlin, Germany.

$(k\,x, k\,t)$ and the initial datum U_0 of the Riemann problem is also invariant by the same transformations, one first looks for a solution which is invariant by these transformations,[22] i.e. a solution of the form

$$U(x,t) = W\left(\frac{x}{t}\right). \tag{2.38}$$

Denoting $\sigma = \frac{x}{t}$ one finds that the equation says

$$A\big(W(\sigma)\big)W'(\sigma) = \sigma\,W'(\sigma), \tag{2.39}$$

so that when $W'(\sigma) \neq 0$ it should be an eigenvector and σ a corresponding eigenvalue; as the system is strictly hyperbolic one must have $\sigma = \lambda_j\big(W(\sigma)\big)$ for a fixed value of j around the point, and one finds then that $W'(\sigma)$ must be proportional to $r_j\big(W(\sigma)\big)$, so that the values taken by W are on an integral curve of the vector field r_j, and U is then a j-simple wave; however, there is more to it, because one must move along this curve in such a way that $\sigma = \lambda_j\big(W(\sigma)\big)$ is satisfied, and that is not always possible. In order to move along the integral curve and satisfy $\sigma = \lambda_j\big(V(\sigma)\big)$, one must look at the way λ_j varies along the curve, and this question led Peter LAX to the following definition.

Definition 2.6. *The jth characteristic field is said to be* linearly degenerate *in D if one has*

$$\langle \nabla\,\lambda_j(U), r_j(U)\rangle = 0 \ \text{for all } U \in D. \tag{2.40}$$

The jth characteristic field is said to be genuinely nonlinear *in D if one has*

$$\langle \nabla\,\lambda_j(U), r_j(U)\rangle \neq 0 \ \text{for all } U \in D, \tag{2.41}$$

and one may assume that $\langle \nabla\,\lambda_j(U), r_j(U)\rangle = 1$ in D, by multiplying r_j by a nonzero function.[23]

Of course, there are intermediate cases, and one should not think of having understood the general case by treating only these two extreme possibilities,

---

[22] There are cases of equations invariant by a group, with solutions only invariant by a subgroup (or not invariant at all if the subgroup is restricted to identity); for example, for the eigenvalue problem $-\Delta u = \lambda u$ with Dirichlet boundary condition on all the boundary (or with Neumann boundary condition on all the boundary) of a disc, where the problem and the boundary data are invariant by rotation, but for particular values of λ there are solutions of the form $f_n(r)\cos n\theta$, with $n \neq 0$ (and f_n is related to Bessel functions). In the case of an evolution problem like here, the question is different because one expects existence and uniqueness of a solution (if one finds the right way to define what kind of solution one is looking for), and the solution must then inherit the invariance.

[23] If one changes the normalization of the eigenvectors r_j, the j-Riemann invariants stay the same, as well as the integral curves of the vector field r_j, and only their parametrization changes.

but the theory is much simpler if each field is either genuinely nonlinear or linearly degenerate, and it was natural that Peter LAX had first considered this case; actually, if it was not the case that for the system of gas dynamics the second field is linearly degenerate, one could have thought that such a condition was purely artificial and only occurred for linear systems.

For the system of gas dynamics, using the description in (ϱ, u, s), the quantity for the first characteristic field is $\left(-\frac{\partial c}{\partial \varrho}\right)(-\varrho) + 1\,c$, while for the third characteristic field it is $\left(+\frac{\partial c}{\partial \varrho}\right)(+\varrho) + 1\,c$, and for the second it is 0, as r_2 has its second component 0. One deduces that the first and third characteristic fields are genuinely nonlinear if and only if

$$\frac{\partial(\varrho c)}{\partial \varrho} = \varrho\,\frac{\partial c}{\partial \varrho} + c \neq 0, \text{ or equivalently } \frac{\partial^2(\varrho p)}{\partial \varrho^2} \neq 0, \tag{2.42}$$

because $\varrho\left(\frac{\partial^2(\varrho p)}{\partial \varrho^2}\right) = \varrho^2\frac{\partial^2 p}{\partial \varrho^2} + 2\varrho\,\frac{\partial p}{\partial \varrho} = \frac{\partial\left(\varrho^2\frac{\partial p}{\partial \varrho}\right)}{\partial \varrho} = \frac{\partial(\varrho^2 c^2)}{\partial \varrho}$.

Assuming that the jth characteristic field is genuinely nonlinear and that one has normalized $r_j(U)$ by $\langle \nabla\,\lambda_j(U), r_j(U)\rangle = 1$, so that if $\frac{dV}{d\tau} = r_j\big(V(\tau)\big)$ one deduces that $\frac{d(\lambda_j(V(\tau)))}{d\tau} = 1$, and the constraint $\lambda_j\big(V(\tau)\big) = \sigma$ is then easy to implement; moving along an integral curve of the normalized vector field r_j from τ_1 to τ_2 with $\tau_1 < \tau_2$ corresponds to the following centred j-wave solution:

$$U(x,t) = V(\tau_1) \text{ for } x < \lambda_j\big(V(\tau_1)\big)t$$
$$U(x,t) = V\left(\tfrac{x}{t} - \lambda_j\big(V(\tau_1)\big) + \tau_1\right) \text{ for } \lambda_j\big(V(\tau_1)\big)t < x < \lambda_j\big(V(\tau_2)\big)t \tag{2.43}$$
$$U(x,t) = V(\tau_2) \text{ for } x > \lambda_j\big(V(\tau_2)\big)t,$$

which is a smooth (locally Lipschitz continuous) solution of the Riemann problem with $U_- = V(\tau_1)$ and $U_+ = V(\tau_2)$. The preceding solution is also called a *rarefaction wave*, as when t increases the nonconstant part of the solution spreads over larger and larger intervals in x.

By using the invariance of the equation by scaling (with $k = -1$) and translation in time, one sees that if $T_0 > 0$ and U is a solution of the equation $\frac{\partial U}{\partial t} + A(U)\frac{\partial U}{\partial x} = 0$ for $t > 0$, then $\tilde{U}$ defined by

$$\tilde{U}(x,t) = U(-x, T_0 - t) \text{ for } x \in \mathbb{R}, 0 < t < T_0, \tag{2.44}$$

is a solution; if one applies this procedure to the rarefaction wave found above, one finds a *compression wave*, which starts from a smooth (Lipschitz continuous) datum at $t = 0$, but converges at $t \to T_0$ to the datum of a translated Riemann problem with $\tilde{U}(x, T_0) = V(\tau_2)$ for $x < 0$ and $\tilde{U}(x, T_0) = V(\tau_1)$ for $x > 0$.

One has seen then that using the j-waves permits one to move in the space $\mathbb{R}^p$ along special curves but only in some directions, so that this does not solve the general case of the Riemann problem; in other cases one must use discontinuities, either *contact discontinuities* or *shocks*, which will be discussed next.

[Taught on Wednesday August 29, 2001.]

Notes on names cited in footnotes for Chapter 2, SYNGE,[24] CARNEGIE,[25] MELLON,[26] DE SAINT-VENANT,[27] WEIL,[28] STEKLOV,[29] FOURIER,[30] HIRZE-BRUCH,[31] ALEXANDER the Great,[32] DIRICHLET,[33] F.E. NEUMANN,[34] BESSEL.[35]

---

[24] John Lighton SYNGE, Irish mathematician, 1897–1995. He had worked in Toronto (Ontario), at OSU (Ohio State University), Columbus, Ohio, and at Carnegie Tech (Carnegie Institute of Technology), now CMU (Carnegie Mellon University), Pittsburgh, PA, where he had been the head of the mathematics department from 1946 to 1948, and in Dublin, Ireland.

[25] Andrew CARNEGIE, Scottish-born businessman and philanthropist, 1835–1919. Besides endowing the school that became Carnegie Institute of Technology and later Carnegie Mellon University when it merged with the Mellon Institute of Industrial Research, he funded about three thousand public libraries, named Carnegie libraries in United States.

[26] Andrew William MELLON, American financier and philanthropist, 1855–1937. He had founded the Mellon Institute of Industrial Research in Pittsburgh, PA, which merged in 1967 with the Carnegie Institute of Technology to form Carnegie Mellon University.

[27] Adhémar Jean Claude BARRÉ DE SAINT-VENANT, French mathematician, 1797–1886. He had worked in Paris, France.

[28] André WEIL, French-born mathematician, 1906–1998. He received the Wolf Prize in 1979, for his inspired introduction of algebro-geometry methods to the theory of numbers, jointly with Jean LERAY. He had worked in Aligarh, India, in Haverford, PA, in Swarthmore, PA, in São Paulo, Brazil, in Chicago, IL, and at IAS (Institute for Advanced Study), Princeton, NJ.

[29] Vladimir Andreevich STEKLOV, Russian mathematician, 1864–1926. He had worked in Kharkov and in St Petersburg (then Petrograd), Russia. The Steklov Mathematical Institute in St Petersburg, Russia, is named after him.

[30] Jean-Baptiste Joseph FOURIER, French mathematician, 1768–1830. He had worked in Auxerre, in Paris, France, accompanied BONAPARTE in Egypt, was prefect in Grenoble, France, until the fall of NAPOLÉON I, and worked in Paris again. The first of three universities in Grenoble, Université de Grenoble I, is named after him, and the Institut Fourier is its department of mathematics.

[31] Friedrich HIRZEBRUCH, German mathematician, born in 1927. He received the Wolf Prize in 1988, for outstanding work combining topology, algebraic and differential geometry, and algebraic number theory; and for his stimulation of mathematical cooperation and research, jointly with Lars HÖRMANDER. He worked in Erlangen, Germany, at Princeton University, Princeton, NJ, and in Bonn, Germany.

(*footnotes 32 to 35 on next page*)

---

[32] Alexandros Philippou Makedonon, 356–323 BCE, was King of Macedon as ALEXANDER III, and is referred to as ALEXANDER the Great, in relation to the large empire that he conquered.

[33] Johann Peter Gustav LEJEUNE DIRICHLET, German mathematician, 1805–1859. He had worked in Breslau (then in Germany, now Wrocław, Poland), in Berlin and in Göttingen, Germany.

[34] Franz Ernst NEUMANN, German mathematician, 1798–1895. He had worked in Königsberg (then in Germany, now Kaliningrad, Russia).

[35] Friedrich Wilhelm BESSEL, German mathematician, 1784–1846. He had worked in Königsberg (then in Germany, now Kaliningrad, Russia).

3

Hyperbolic Systems: Contact Discontinuities, Shocks

If the jth characteristic field is linearly degenerate, i.e. $\langle \nabla \lambda_j(U), r_j(U) \rangle = 0$ for $U \in D$, then on every integral curve of $\frac{dV}{d\tau} = r_j(V(\tau))$ the eigenvalue λ_j is constant, and one cannot construct (nonconstant) centred wave solutions, i.e. of the form $U\left(\frac{x}{t}\right)$ and taking values in one integral curve.

However, if λ_j^* is the constant value of λ_j on a given integral curve which is parametrized by $V(\tau)$ (for $\tau_1 \leq \tau \leq \tau_2$), one can construct solutions of the form

$$U(x,t) = V\left(f(x - \lambda_j^* t)\right) \text{ for } \lambda_j^* t + z_1 < x < \lambda_j^* t + z_2 \text{ and} \atop f \text{ smooth in } (z_1, z_2) \text{ with } \tau_1 \leq f(\cdot) \leq \tau_2. \qquad (3.1)$$

Indeed, $U_x = f'(x - \lambda_j^* t)\frac{dV}{d\tau}$ and $U_t = -\lambda_j^* f'(x - \lambda_j^* t)\frac{dV}{d\tau}$, and $\frac{dV}{d\tau}$ is an eigenvector of $A\left(V(\tau)\right)$ for the eigenvalue λ_j^*, showing that U is a solution. One can then construct a sequence of smooth functions f_n converging to a discontinuous function f_∞ which takes the value τ_1 for $z < 0$ and τ_2 for $z > 0$ (for example $f_n(z) = \tau_1$ for $z < 0$, $f_n(z) = (1 - n\,z)\tau_1 + n\,z\,\tau_2$ for $0 < z < \frac{1}{n}$ and $f_n(z) = \tau_2$ for $z > \frac{1}{n}$), and the corresponding U_n is a sequence of smooth (Lipschitz continuous) solutions of the equation, which converges to a discontinuous function U_∞ defined by

$$U_\infty(x,t) = V(\tau_1) \text{ if } x < \lambda_j^* t; \; U_\infty(x,t) = V(\tau_2) \text{ if } x > \lambda_j^* t. \qquad (3.2)$$

If the system is in *conservative form*, i.e.

$$A(U) = \nabla F(U), \text{ so that } \frac{\partial U}{\partial t} + A(U)\frac{\partial U}{\partial x} = 0 \text{ is written as } \frac{\partial U}{\partial t} + \frac{\partial F(U)}{\partial x} = 0,$$
$$(3.3)$$

then $U_n \to U_\infty$ and $F(U_n) \to F(U_\infty)$ strongly (in L_{loc}^p strong for every $p \in [1, \infty)$ for example), and passing to the limit in the sense of distributions gives

$$\frac{\partial U_\infty}{\partial t} + \frac{\partial\left(F(U_\infty)\right)}{\partial x} = 0, \qquad (3.4)$$

but the writing as $A(U_\infty)\frac{\partial U_\infty}{\partial x}$ does not make sense. However, although the meaning of a discontinuous solution is not clear, one tends to accept U_∞ as a good discontinuous solution, because it is a strong limit of smooth solutions; notice that one then accepts the discontinuous function jumping from $V(\tau_1)$ for $x < \lambda_j^* t$ to $V(\tau_2)$ for $x > \lambda_j^* t$, as well as the discontinuous function jumping from $V(\tau_2)$ for $x < \lambda_j^* t$ to $V(\tau_1)$ for $x > \lambda_j^* t$. In the conservative case, where the notion of a discontinuous solution is clearer, these discontinuous functions are *weak solutions*, which satisfy then the *Rankine–Hugoniot condition*, and they are particular cases of *j-contact discontinuities*, according to the following definitions and properties.

Definition 3.1. *A function U defined in an open set Ω of the (x,t) plane is a* weak solution *of the system in conservative form* $U_t + \big(F(U)\big)_x = 0$ *if U and $F(U)$ are Lebesgue-measurable and locally integrable in Ω and satisfy the equation in the sense of distributions in Ω, i.e.*

$$\int_\Omega \left(U\, \frac{\partial \varphi}{\partial t} + F(U)\, \frac{\partial \varphi}{\partial x} \right) dx\, dt = 0 \text{ for all } \varphi \in C_c^\infty(\Omega), \qquad (3.5)$$

where $C_c^\infty(\Omega)$ is the space of infinitely smooth functions with compact support in Ω;[1] U is a weak solution *of the Cauchy problem*

$$U_t + \big(F(U)\big)_x = f \text{ for } x \in \mathbb{R}, t > 0, \text{ and } U(\cdot, 0) = U_0 \text{ in } \mathbb{R}, \qquad (3.6)$$

if

$$\int_{\mathbb{R}\times(0,\infty)} \left(-U\, \frac{\partial \varphi}{\partial t} - F(U)\, \frac{\partial \varphi}{\partial x} \right) dx\, dt = \int_{\mathbb{R}\times(0,\infty)} f\, \varphi\, dx\, dt + \int_\mathbb{R} U_0 \varphi(\cdot, 0)\, dx$$
$$\text{for all } \varphi \in C_c^\infty(\mathbb{R}^2),$$
$$(3.7)$$

where f is given, locally integrable in $\mathbb{R}^2$, and U_0 is given, locally integrable in $\mathbb{R}$.

Smooth solutions of $U_t + \big(F(U)\big)_x = 0$ are weak solutions, of course, as is seen by multiplying the equation by φ and integrating by parts (notice that φ is scalar, but as U and $F(U)$ are vectors, the equation is an equality between vectors, and it could be written separately for each component); a precise

---

[1] A few mathematicians are afraid of this space, probably because they remember that Laurent SCHWARTZ had described a precise topology, which makes the dual be the space of distributions; none of this nonelementary theory is needed in general, and here it is just a set of test functions. One difficult question for quasi-linear hyperbolic systems concerns shocks, and one must understand enough about distributions to realize that $F(U)$ is defined because U is a function, but that it has no meaning for general distributions, and that once $F(U)$ is locally integrable its derivative makes sense as a distribution, which is the linear mapping $\varphi \mapsto -\int F(U)\varphi_x\, dx$, but that $A(U)U_x$ does not make sense as a distribution in general.

regularity condition needed for this integration by parts to be valid is that the components of U and those of $F(U)$ belong to the Sobolev space $W^{1,1}_{loc}$, i.e. functions who partial derivatives (in the sense of distributions) are locally integrable.

In the case where the solution is piecewise smooth, with a curve of discontinuity, the concept of a weak solution makes the Rankine–Hugoniot condition appear, a condition which could as well have been called after STOKES and RIEMANN, as they had proven such conditions before RANKINE or HUGONIOT were involved in this question. In [3], COURANT and FRIEDRICHS also mentioned the work of EARNSHAW,[2] who extended in 1860 the computations of STOKES to a more general equation of state; EARNSHAW pointed to a book by PARRY,[3] and he must have understood that a supersonic effect had unknowingly been observed.[4]

Proposition 3.2. *Let Ω be an open subset of the (x,t) plane, cut by a smooth curve $x = g(t)$, defining two nonempty open subsets $\Omega_- = \{(x,t) \in \Omega \mid x < g(t)\}$ and $\Omega_+ = \{(x,t) \in \Omega \mid x > g(t)\}$. Assume that U is a smooth solution of $U_t + \big(F(U)\big)_x = 0$ in Ω_- which extends into a continuous function on $\overline{\Omega_-}$ with limits $U(g(t)_-,t)$ on the curve, and that U is also a smooth solution of $U_t + \big(F(U)\big)_x = 0$ in Ω_+ which extends into a continuous function on $\overline{\Omega_+}$ with limits $U(g(t)_+,t)$ on the curve; then U is a weak solution of $U_t + \big(F(U)\big)_x = 0$*

---

[2] Samuel EARNSHAW, English mathematician and clergyman, 1805–1888. He had worked in Cambridge and in Sheffield, England.

[3] Sir William Edward PARRY, English rear admiral and arctic explorer, 1790–1855.

[4] Thanks to the interlibrary loan system, I obtained a microfilm of the book, and I read the corresponding appendix. It was during the first of three voyages of PARRY to find the north-west passage, in 1819–1820, but that there would be three tentatives was not written in the narrative of the first voyage, of course, and I learnt that information later, from the Internet. During the winter, PARRY had to perform a few scientific experiments, and one of them was to measure the velocity of sound at low temperature; he mentioned that someone had measured the velocity of sound in Calcutta, India, with a temperature above 100° Fahrenheit. While his boat (*Hecla*) was stuck in the ice, he walked away with the physician, and had his lieutenant order a sailor to fire one of the ten guns, and they both measured the time between the flash of the detonation and the sound of the detonation, with chronometers precise to one fifth of a second. PARRY reported that one day, just after hearing the detonation, they had heard distinctly the order *fire*, which had preceded the detonation; PARRY had wondered what could have happened and what was special on that day, and he mentioned that the barometer was very low. EARNSHAW must have understood that the sound of the detonation had started supersonically and had overtaken the preceding sound, without erasing it. What was obviously not understood at the time is that the velocity of sound depends upon temperature and pressure (and the percentage of humidity in the air, which must have been important in the Calcutta measurement), so all these measurements of velocity had been done with a reading of the temperature, but without a reading of the pressure!

in Ω if and only if the following Rankine–Hugoniot condition is satisfied:

$$F\big(U(g(t)_+,t)\big) - F\big(U(g(t)_-,t)\big) = g'(t)\big(U(g(t)_+,t) - U(g(t)_-,t)\big) \quad (3.8)$$
almost everywhere along the curve.

Proof: The usual notation is to write $[f]$ for the jump of any quantity f along the curve, i.e. $f(g(t)_+,t) - f(g(t)_-,t)$, and denote by s the velocity g' at which the discontinuity moves (except when the thermodynamical quantity s is also involved), and then write the Rankine–Hugoniot condition as

$$[F(U)] = s\,[U], \quad (3.9)$$

which is an equality between vectors, of course.

For $\varphi \in C_c^\infty(\Omega)$, one decomposes $\int_\Omega (U\,\frac{\partial\varphi}{\partial t} + F(U)\,\frac{\partial\varphi}{\partial x})\,dx\,dt$ into a term $\int_{\Omega_-}$ and a term $\int_{\Omega_+}$, and each term is then integrated by parts; for example, $\int_{\Omega_-}(U\,\frac{\partial\varphi}{\partial t} + F(U)\,\frac{\partial\varphi}{\partial x})\,dx\,dt = -\int_{\Omega_-}(\frac{\partial U}{\partial t} + \frac{\partial F(U)}{\partial x})\varphi\,dx\,dt + \int_{\partial\Omega_-}(U\,n_t + F(U)n_x)\varphi\,d\ell = \int_{\partial\Omega_-}(U\,n_t + F(U)n_x)\varphi\,d\ell$, where n is the exterior normal to Ω_- (used only on the part of the curve intersecting the support of φ); similarly, an integration by parts is performed for Ω_+, with the important remark that on the curve the exterior normal to Ω_+ is exactly the opposite of the exterior normal to Ω_-, so that by adding the two terms one sees that U is a weak solution in Ω if and only if $\int_{\partial\Omega_-}([U]\,n_t + [F(U)]n_x)\varphi\,d\ell = 0$ for all $\varphi \in C_c^\infty(\Omega)$; this means that $[U]\,n_t + [F(U)]n_x = 0$ almost everywhere along the curve, and the normal to Ω_- on the curve is given by

$$\begin{pmatrix} n_x \\ n_t \end{pmatrix} = \frac{1}{\sqrt{1+(g')^2}}\begin{pmatrix} 1 \\ -g' \end{pmatrix}, \quad (3.10)$$

from which the Rankine–Hugoniot condition follows.   □

The proof actually shows that the statement is true if the curve is Lipschitz continuous, if both U and $F(U)$ belong to $W^{1,1}(\Omega_-) \cap W^{1,1}(\Omega_+)$, and if they satisfy the equation $U_t + (F(U))_x = 0$ in Ω_- and in Ω_+ (in the sense of distributions), because the trace theorem asserts the existence of some notion of limits on the curve and the integration by parts formula holds (notice that this argument is valid for F continuous, in which case g' could be infinite at some points).

In the case of the Burgers equation, where U is scalar and $F(U) = \frac{U^2}{2}$, the discontinuity must propagate at the speed $\frac{U_-+U_+}{2}$, and it will be seen later that only the discontinuities with $U_- > U_+$ are "physical". In the general one-dimensional scalar case, for $u_t + (f(u))_x = 0$ the condition is that $s = \frac{f(u_+)-f(u_-)}{u_+-u_-}$.

KRUZHKOV has extended some properties known in one dimension to the multidimensional scalar equation $\frac{\partial u}{\partial t} + \sum_{i=1}^N \frac{\partial(f_i(u))}{\partial x_i} = 0$, but the case $N \geq 2$

is not very physical, as it implies a strong anisotropy for space. I had heard of this fact long before the work of KRUZHKOV, and the remark was attributed to René THOM,[5] who argued that one needs a vector unknown, and he was actually thinking that equations of Hamilton–Jacobi type were a natural generalization,[6] while another generalization is obviously a system like gas dynamics, where there is a velocity field u and the space is isotropic but a derivative in the direction of u is natural; however, such an anisotropy exists also in the case of effects arising at the interface between two materials, but the domain of validity to be considered is limited to a small region in space, of course, moving eventually with a (shock) wave.

In the case of the system of gas dynamics, studied by HUGONIOT after the preliminary work of STOKES, EARNSHAW and RIEMANN,[7] the Rankine–Hugoniot conditions are

$$
\begin{aligned}
\left[\varrho\, u_1\right] &= D\left[\varrho\right] \\
\left[\varrho\,(u_1)^2 + p\right] &= D\left[\varrho\, u_1\right] \\
\left[\varrho\, u_1 u_2\right] &= D\left[\varrho\, u_2\right] \text{ and } \left[\varrho\, u_1 u_3\right] = D\left[\varrho\, u_3\right] \\
\left[\frac{\varrho\,|u|^2 u_1}{2} + \varrho\, e\, u_1 + p\, u_1\right] &= D\left[\frac{\varrho\,|u|^2}{2} + \varrho\, e\right]
\end{aligned}
\tag{3.11}
$$

where one writes $[f]$ for the jump of any quantity f, and one uses D for the velocity of the discontinuity, in order to avoid any confusion with the thermodynamical entropy s (used for entropy per unit of mass); using Galilean invariance, i.e. moving at velocity D, one may consider the simpler (but equivalent) question where $D = 0$, so that

moving at the velocity of the discontinuity, so that it appears stationary, $\varrho_+ u_{1+} = \varrho_- u_{1-} = Q$, flux of mass through the discontinuity,
$Q\, u_{1+} + p_+ = Q\, u_{1-} + p_-$,
if $Q \neq 0, u_{2+} = u_{2-}$ and $u_{3+} = u_{3-}$,
$Q\, \frac{|u_{1+}|^2}{2} + Q\, e_+ + p_+ u_{1+} = Q\, \frac{|u_{1-}|^2}{2} + Q\, e_- + p_- u_{1-}.$

$$\tag{3.12}$$

The case $Q = 0$ is easily checked to give u_1 and p continuous through the discontinuity, and this corresponds to a contact discontinuity, as u_1 and p are the corresponding Riemann invariants; all such contact discontinuities are accepted as physical, because, as was seen in the general case, such discontinuous weak solutions are strong limits of smooth solutions.

---

[5] René THOM, French mathematician, 1923–2002. He received the Fields Medal in 1958. He had worked in Grenoble, in Strasbourg, and at IHES (Institut des Hautes Études Scientifique) at Bures-sur-Yvette, France.

[6] Carl Gustav Jacob JACOBI, German mathematician, 1804–1851. He had worked in Königsberg (then in Germany, now Kaliningrad, Russia) and in Berlin, Germany.

[7] In 1848, STOKES did not use the equation of balance of energy, and EARNSHAW was following him in 1860, the same year in which RIEMANN did his independent work, but it seems that RIEMANN had worked with the wrong system, with entropy being conserved instead of energy; HUGONIOT's work dates from 1889.

The case $Q \neq 0$ is more technical to analyse, and some discontinuities are rejected, but the reasons for rejection are complex, and they are primarily related to the laws of thermodynamics, but if they make sense for the particular system of gas dynamics that I am considering, Peter LAX had to interpret what could be more general rules, valid for all quasi-linear hyperbolic systems, and he certainly took in consideration the question of stability of shocks, and the understanding of the scalar case.

In the scalar case, there is a complete theory, and I shall describe later the conditions imposed by Peter LAX, which are necessary conditions for admissibility of shocks, and the conditions imposed by Olga OLEINIK, which are necessary and sufficient, and the equivalent formulation by Eberhard HOPF, but one should observe that for the general scalar equation $u_t + \big(f(u)\big)_x = 0$, Galilean invariance only occurs for $f(v) = \frac{v^2}{2} + a\,v + b$, so that only the Burgers equation is a physically relevant model. For systems, a good complete theory is missing, and some conditions are named after Peter LAX, some after Constantine DAFERMOS, and some after Tai-Ping LIU.[8]

The general definition that Peter LAX introduced, j-shocks and j-contact discontinuities, is as follows.

Definition 3.3. *A discontinuous solution of the form*

$$U(x,t) = \begin{cases} U_- \text{ for } x < x_0 + s\,t, \\ U_+ \text{ for } x > x_0 + s\,t, \end{cases} \tag{3.13}$$

satisfying the Rankine–Hugoniot condition

$$F(U_+) - F(U_-) = s(U_+ - U_-), \tag{3.14}$$

is a j-shock satisfying the Lax *condition if*

$$\lambda_j(U_-) \geq s \geq \lambda_j(U_+) \tag{3.15}$$

and

$$\lambda_{j-1}(U_-) < s < \lambda_{j+1}(U_+), \tag{3.16}$$

forgetting the corresponding inequality for indices 0 or $p+1$, so that a j-shock cannot be a k-shock for $k \neq j$. It is a j-contact discontinuity if

$$\lambda_j(U_-) = s = \lambda_j(U_+), \tag{3.17}$$

but one talks of a left j-contact discontinuity if $\lambda_j(U_-) = s$, and of a right j-contact discontinuity if $s = \lambda_j(U_+)$.

---

[8] Tai-Ping LIU, Chinese-born mathematician. He has worked at University of Maryland, College Park, MD, at NYU (New York University), New York, NY, and at Stanford University, Stanford, CA.

In the last few years, shocks which do not satisfy (3.16) have been studied, and called overcompressive shocks, but that seems to be a specialty of people who do not have much interest in continuum mechanics, because they do not seem to care that the model used by engineers which they mention contradicts thermodynamics, not in a way that seems reasonable, and some of them seem to play too much with the scalar one-dimensional equation in more than one variable, without ever saying that it has no physical grounding. It seems that one must be a good mathematician, like Peter LAX, to identify a mathematical generalization of a problem from continuum mechanics or physics which is both interesting from a mathematical point of view and a good testing ground for understanding more about continuum mechanics or physics.

The reason for being so interested in shocks is that they are related to irreversibility, and irreversibility is often connected to the laws of thermodynamics, which it would be important to understand mathematically in a better way. The verb understand has a different meaning in mathematics than in other branches of science (as mathematics is a part of science) or engineering; in the present context, it is not about learning the rules of thermodynamics and applying them correctly (which one can do with a reasonable amount of effort, as for many other games invented by physicists or engineers), it is about discussing which part of the rules can be deduced from more basic principles of physics, and how to replace the other parts by more precise mathematical rules, and go beyond the rules of equilibrium thermodynamics.

Of course, many consider the *Boltzmann equation* as an answer, and the mathematical properties of the Boltzmann equation are a part of the subject of these lectures, because it is a classical model of kinetic theory, but the Boltzmann equation has a similar defect than thermodynamics, because it is postulated and irreversibility has already been put by force into the model, so that it cannot be used for studying how irreversibility occurs.

It is important for mathematicians to understand the limitations of the models used, and to avoid pretending that one has proven something which has actually been postulated, a process usually referred to as a vicious circle.

In order to study how irreversibility occurs it is important to start from conservative models which are reversible, and to explain why in some limit something seems to have been lost, and to study if something has really been lost, or if a better account can be given of what other models declare lost. Comparing various approaches to irreversibility seems then a good first step, and quasi-linear hyperbolic systems of conservation laws are particularly suited for that purpose, because an important example is the system of gas dynamics which deals precisely with the type of questions that were used for guessing the laws of thermodynamics in the first place.

One has postulated that the equation of state is always valid, but the equation of state of a gas had been discovered by looking only at various equilibria, and it is a very questionable hypothesis to assume that it also applies to the evolution problem. It is important to understand if there is a natural definition of what a solution can mean, and the concept of weak solutions can be considered basic in order to define what discontinuous solutions mean, but it will be seen that too many weak solutions exist, and a choice must be made, and up to now the choice has been purely local, accepting some shocks and rejecting others by looking only at the limiting states on both sides of the discontinuity. Because one does not know enough (at the moment, and from a mathematical point of view) about quasi-linear hyperbolic systems of conservation laws, the description will not be conclusive, but knowing about a few pieces of the puzzle is probably useful for the future solution of this difficult question, and a good way to learn more about the classical aspects is to consult the recent book [4] of Constantine DAFERMOS. After one more definition for systems, I shall turn my attention towards the case of a scalar equation (in one-dimensional space), a question which had been completed in the 1970s.

Definition 3.4. *An "entropy" φ, together with an entropy flux ψ, for the system $U_t + A(U)\,U_x = 0$, is a pair of smooth scalar functions (defined in a domain D where A is defined) satisfying*

$$\nabla\,\varphi(V)\,A(V) = \nabla\,\psi(V) \text{ for } V \in D. \qquad (3.18)$$

The important property is that any smooth solution U of $U_t + A(U)\,U_x = 0$ automatically satisfies $\big(\varphi(U)\big)_t + \big(\psi(U)\big)_t = 0$.

As was mentioned before, the choice of the term entropy in this definition of Peter LAX may be a little confusing, which is why I often add the qualifier mathematical; for example, a system of conservation laws $U_t + \big(F(U)\big)_x = 0$ always has the (trivial) entropies U_j, for $j = 1, \ldots, p$, with flux $F_j(U)$, so for the system of gas dynamics, one has the trivial entropies ϱ, $\varrho\,u_j$ for $j = 1, 2, 3$, and $\frac{\varrho |u|^2}{2} + \varrho\,e$ (so mass, momentum, and energy are mathematical entropies), but there are some nontrivial entropies like $\varrho\,f(s)$ for an arbitrary smooth function f, where s is the thermodynamical entropy per unit of mass.

[Taught on Friday August 31, 2001.]

4

The Burgers Equation and the 1-D Scalar Case

The Burgers equation,

$$u_t + u\,u_x = 0 \text{ in } \mathbb{R} \times (0, \infty), \text{ with } u(\cdot, 0) = u_0 \text{ in } \mathbb{R}, \tag{4.1}$$

first appeared in the work of BATEMAN in 1915,[1] but it was forgotten. BURGERS reintroduced it in the 1940s with reference to turbulence, and Eberhard HOPF immediately pointed out that it was not the size of the velocity in the fluid that creates turbulence, as BURGERS had seemed to think, but the fluctuation in velocity in the fluid, because of Galilean invariance, and that is indeed a basic fact about turbulence that everyone agrees upon.[2] Eberhard HOPF first solved the equation in 1950, by considering for $\varepsilon > 0$ the equation

$$u_t + u\,u_x - \varepsilon\,u_{xx} = 0 \text{ in } \mathbb{R} \times (0, \infty), \text{ with } u(\cdot, 0) = u_0 \text{ in } \mathbb{R}, \tag{4.2}$$

now known as the *Burgers–Hopf equation*, and then letting ε tend to 0.

Although the added term corresponds to a viscosity effect, one calls this approach the method of artificial viscosity, because one often adds regularizing terms for purely mathematical reasons, in which case the qualifier artificial is a way to point out that one does not claim that the model used has a sound physical interpretation. It is sometimes useful to invent nonphysical models in order to overcome a technical difficulty that one has encountered in the study of a physical model, i.e. one which at some moment is supposed to give a good description of a part of physical reality; what one should avoid is to let such nonphysical models pass in describing a real situation, and when problems take many years to be solved it is useful to remind younger people about the reasons which had led to the introduction of the various models.

---

[1] Harry BATEMAN, English-born mathematician, 1882–1946. He had worked in Liverpool, in Manchester, England, in Bryn Mawr, PA, at Johns Hopkins University, Baltimore, MD, and at Caltech (California Institute of Technology), Pasadena, CA, named Throop College at the time he arrived.

[2] One may wonder then about the reasons why some people have recently used the term "Burgers turbulence"!

Eberhard HOPF found a way to transform (4.2) into a linear *heat equation*,[3] by what is now called the Hopf–Cole transformation,[4] because Julian COLE discovered it independently a little later.[5] This change of unknown comes naturally if one introduces the function U defined by

$$U(x,t) = \int_{-\infty}^{x} u(y,t)\,dy \text{ for } x \in \mathbb{R}, t \geq 0, \tag{4.3}$$

so that if the solution u is smooth and integrable, one obtains

$$U_x = u; \ U_t + \frac{U_x^2}{2} - \varepsilon\,U_{xx} = 0, \tag{4.4}$$

so that U is a potential associated to the conservation law given by the Burgers equation. Then, using the observation that if V satisfies $V_t - \varepsilon\,V_{xx} = 0$ then $f(V)$ satisfies $\big(f(V)\big)_t - \varepsilon\big(f(V)\big)_{xx} + \varepsilon\,f''(V)(V_x)^2 = 0$, one sees that the equation for U is satisfied by $U = f(V)$ if one has $2\varepsilon\,f'' = (f')^2$, which one integrates immediately into $\frac{2\varepsilon}{f'(V)} = C - V$, so that choosing $C = 0$ the formula for $u = U_x = f'(V)V_x$ becomes the Hopf–Cole transformation

$$u = \frac{-2\varepsilon\,V_x}{V}. \tag{4.5}$$

Another reason to use U is that $u_t + u\,u_x = 0$ transforms into a *Hamilton–Jacobi equation* $U_t + \frac{U_x^2}{2} = 0$, and one can then solve the equation by classical techniques of calculus of variations, like those of CARATHÉODORY,[6] who had introduced the method of dynamic programming, long before Richard BELLMAN made it popular.[7]

---

[3] Fourier's law, that the flux of heat is proportional (and opposite) to the gradient of temperature has been postulated, as well as Fick's law for diffusion of mass, and although the parabolic equations that they lead to are quite popular, one must notice the nonphysical effect that heat may travel arbitrarily fast, and one should consider that they are approximations corresponding to having let the velocity of light c tend to ∞, and this will be studied in more detail later.

[4] Julian David COLE, American mathematician, 1925–1999. He had worked at Caltech (California Institute of Technology), Pasadena, CA, at UCLA (University of California at Los Angeles), Los Angeles, CA, and at RPI (Rensselaer Polytechnic Institute), Troy, NY.

[5] Ten years after the work of BATEMAN, who had studied the case $\varepsilon \to 0$, FORSYTH had already introduced the "Hopf–Cole" transformation.

[6] Constantin CARATHÉODORY, German mathematician (of Greek origin), 1873–1950. He had worked in Göttingen, in Bonn, and in Hanover, Germany, in Breslau (then in Germany, now Wrocław, Poland), and in Berlin, Germany. After World War I, he worked in Athens, Greece, in Smyrna (then in Greece, now Izmir, Turkey), and in München (Munich), Germany.

[7] Richard Ernest BELLMAN, American mathematician, 1920–1984. He had worked at USC (University of Southern California), Los Angeles, CA.

This was the first approach which permitted one to see which discontinuous solutions were approached when ε tends to 0, but the Hopf–Cole transformation does not extend to scalar equations $u_t + \left(f(u)\right)_x = 0$ for a general f, which were studied for mathematical reasons,[8] but appeared to be a good training ground for understanding about admissibility conditions for shocks. The same approach of adding viscosity was also used by Olga OLEINIK, using more traditional compactness arguments for proving existence of a solution, and she proved uniqueness for solutions satisfying a one-sided inequality

$$u_x \leq \frac{1}{t}, \tag{4.6}$$

or more general inequalities of the type $u_x \leq E(t)$, which also hold for $u_t + \left(f(u)\right)_x = 0$ for some strictly convex f; for example, $f'' \geq \alpha > 0$ implies $u_x \leq \frac{1}{\alpha t}$. The maximum principle is used for proving this inequality, as well as others, and if u solves $u_t + u\,u_x - \varepsilon\,u_{xx} = 0$, then $v = u_x$ satisfies the equation $v_t + u\,v_x + v^2 - \varepsilon\,v_{xx} = 0$, so if $v(\cdot,0) \leq a_0$ one has $v(\cdot,t) \leq a(t)$, where $a' + a^2 = 0$ and $a(0) = a_0$, i.e. $a(t) = \frac{a_0}{1+t\,a_0}$ which is $\leq \frac{1}{t}$ if $a_0 > 0$ and ≤ 0 if $a_0 \leq 0$. This bound cannot be improved, because for an initial datum which is bounded and Lipschitz continuous, the solution of $u_t + u\,u_x = 0$ obtained by the method of characteristic curves is also bounded and Lipschitz continuous for an interval of time, and along a characteristic curve (which is the straight lines $x = y + t\,u_0(y)$ for $y \in \mathbb{R}$), the function v satisfies the differential equation $v'+v^2 = 0$, i.e. $v\big(x(t),t\big) = \frac{v_0}{1+t\,v_0}$. Of course, the existence part requires bounds, and if $u_0 \in L^\infty(\mathbb{R})$, then u stays bounded in $L^\infty(\mathbb{R})$, while if $u_0 \in BV(\mathbb{R})$, then u_x stays bounded in $L^1(\mathbb{R})$.

Another method is to use a numerical approximation, by *finite differences*, using the Lax–Friedrichs scheme. In finite-difference schemes one uses a mesh size Δx in space and a mesh size Δt in time, and one discovers that one cannot take Δt too large; it is standard to denote by U_i^n the approximation of $u(i\,\Delta x, n\,\Delta t)$, and the Lax–Friedrichs scheme for the equation $u_t + \left(f(u)\right)_x = 0$ is the explicit scheme

$$\frac{1}{\Delta t}\left(U_i^{n+1} - \frac{U_{i-1}^n + U_{i+1}^n}{2}\right) + \frac{1}{2\Delta x}\left(f(U_{i+1}^n) - f(U_{i-1}^n)\right) = 0, \tag{4.7}$$

which must be supplemented by giving the initial data U_i^0 for all i, for example,

$$U_i^0 = \frac{1}{\Delta x}\int_{i\Delta x - \frac{\Delta x}{2}}^{i\Delta x + \frac{\Delta x}{2}} u_0\,dx. \tag{4.8}$$

The condition that Δt should satisfy a bound in terms of Δx is called a Courant–Friedrichs–Lewy condition,[9] abbreviated as *CFL condition*. This

---

[8] Despite the popularity of this model, the Galilean invariance only occurs for $f(v) = \frac{v^2}{2} + a\,v + b$, so that one should wonder if other fs correspond to any physical situation.

[9] Hans LEWY, German-born mathematician, 1904–1988. He received the Wolf Prize in 1984, for initiating many, now classic and essential, developments in partial

elementary condition arises for numerical approximations of hyperbolic equations, where there is a finite speed of propagation, and expresses the necessary fact that the numerical domain of dependence must contain the exact domain of dependence if one wants the scheme to converge. For example, if one looks at a linear equation $u_t + a\, u_x = 0$, where the local speed of propagation is a, then if one has $\alpha \leq a(x,t) \leq \beta$, the solution $u(x,t)$ is equal to $u_0(y)$ where y is the base point of the characteristic curve going through (x,t) (assuming smoothness of a so that the characteristic curves are defined in a unique way), and one has $x - \beta t \leq y \leq x - \alpha t$; on the other hand a numerical scheme like the Lax–Friedrichs scheme has a speed of propagation $\frac{\Delta x}{\Delta t}$ as the value of U_i^{n+1} depends upon U_{i-1}^n and U_{i+1}^n, so that depends upon the values of U_j^0 for $i - n \leq j \leq i + n$ (but not all of them, and only the values with $i + j + n$ even are involved); the CFL condition in this case is $\frac{\Delta x}{\Delta t} \geq \max\{|\alpha|, |\beta|\}$, and if it is not true there is a constant a such that the sequence of approximations with Δx and Δt converging to 0 while keeping a fixed ratio (which is what one usually does for hyperbolic equations) does not converge to the solution. In the nonlinear problem $u_t + \big(f(u)\big)_x = 0$, one is lucky that a maximum principle holds and that if $M_- \leq u_0 \leq M_+$ then the desired solution satisfies $M_- \leq u(x,t) \leq M_+$ for almost every $x \in \mathbb{R}, t > 0$; as the problem is formally $u_t + f'(u)u_x = 0$, one takes $a = f'(u)$, so that one needs to look at the bounds of f' on the interval $[M_-, M_+]$, and this leads to the CFL condition

$$\left(\max_{v \in [M_-, M_+]} |f'(v)| \right) \Delta t \leq \Delta x, \tag{4.9}$$

and then one writes the (explicit) scheme as

$$U_i^{n+1} = \left(\frac{1}{2}U_{i-1}^n - \frac{\Delta t}{2\Delta x} f(U_{i-1}^n) \right) + \left(\frac{1}{2}U_{i+1}^n + \frac{\Delta t}{2\Delta x} f(U_{i+1}^n) \right) = G(U_{i-1}^n, U_{i+1}^n), \tag{4.10}$$

and one notices that $G(v,w)$ is order preserving in v and w (i.e. $G_v \geq 0$ and $G_w \geq 0$) if $M_- \leq v, w \leq M_+$, as the corresponding derivatives involve quantities like $\frac{1}{2} \pm \frac{f'(z)\Delta t}{2\Delta x}$, which is ≥ 0 for $z \in [M_-, M_+]$. A consequence is

$$\text{if } M_- \leq U_i^n \leq M_+ \text{ for all } i, \text{ then } M_- \leq U_i^{n+1} \leq M_+ \text{ for all } i; \tag{4.11}$$

indeed, due to the order-preserving property of G, the value of $U_i^{n+1} = G(U_{i-1}^n, U_{i+1}^n)$ is a minimum when U_{i-1}^n and U_{i+1}^n are replaced by M_-, which gives a value M_- for U_i^{n+1}, and a maximum when U_{i-1}^n and U_{i+1}^n are replaced by M_+, which gives a value M_+ for U_i^{n+1}. One has seen that the order relation on $\mathbb{R}$ plays a crucial role, and this will be seen again in obtaining a bound in $BV(\mathbb{R})$; unfortunately, nothing of that sort is known for the case of general systems.

---

differential equations, jointly with Kunihiko KODAIRA. He had worked in Göttingen, Germany, at Brown University, Providence, RI, and at UCB (University of California at Berkeley), Berkeley, CA.

It has been noticed that the semi-group defined by the equation $u_t + (f(u))_x = 0$ is a contraction semi-group in $L^1(\mathbb{R})$, by Barbara KEYFITZ,[10] and the work of KRUZHKOV is related; when I heard of the work of Philippe BENILAN on nonlinear contraction semi-groups in L^1,[11] I noticed that all the examples could be treated by techniques of order preserving, and when I arrived in Madison, WI, in the fall of 1974, I showed my argument to Michael CRANDALL,[12] and he proved the other part of our Lemma 4.1; later, with Andrew MAJDA he noticed applications to numerical schemes.[13]

Lemma 4.1. *(Crandall–Tartar)*[14] *Let Ω, Ω' be endowed with nonnegative Radon measures $d\mu, d\mu'$,[15] and let X be a subset of $L^1(\Omega; d\mu)$ stable by inf (or by sup); let S be a mapping from X into $L^1(\Omega'; d\mu')$ satisfying $\int_{\Omega'} S(v) \, d\mu' = \int_\Omega v \, d\mu$ for all $v \in X$; then the following two properties are equivalent:*

i) S is order preserving, i.e. $v, w \in X$ and $v \leq w$ almost everywhere (for $d\mu$) implies $S(v) \leq S(w)$ almost everywhere (for $d\mu'$),

ii) S is a contraction in L^1, i.e. $\int_{\Omega'} |S(v) - S(w)| \, d\mu' \leq \int_\Omega |v - w| \, d\mu$ for all $v, w \in X$.

Proof: For $v, w \in X$, let $z = \inf\{v, w\}$, so that $z \leq v$ and $z \leq w$, and (i) implies $S(z) \leq S(v)$ and $S(z) \leq S(w)$, from which one deduces $\int_{\Omega'} |S(v) - S(w)| \, d\mu' \leq \int_{\Omega'} |S(v) - S(z)| \, d\mu' + \int_{\Omega'} |S(z) - S(w)| \, d\mu'$, but $\int_{\Omega'} |S(v) - S(z)| \, d\mu' = \int_{\Omega'} \big(S(v) - S(z)\big) \, d\mu' = \int_\Omega (v - z) \, d\mu$ and $\int_{\Omega'} |S(z) - S(w)| \, d\mu' = \int_{\Omega'} \big(S(w) - S(z)\big) \, d\mu' = \int_\Omega (w - z) \, d\mu$ and $\int_\Omega (v - z) \, d\mu + \int_\Omega (w - z) \, d\mu = \int_\Omega |v - w| \, d\mu$.

---

[10] Barbara Lee KEYFITZ, Canadian-born mathematician, born in 1944. She worked at Columbia University, New York, NY, in Princeton, NJ, at Arizona State University, Tempe, AZ, in Houston, TX, and at the Fields Institute for Research in Mathematical Sciences, Toronto, Ontario.

[11] Philippe M. A. BENILAN, French mathematician, 1940–2001. He had worked in Besançon, France.

[12] Michael Grain CRANDALL, American mathematician, born in 1940. He worked at Stanford University, Stanford, CA, at UCLA (University of California at Los Angeles), Los Angeles, CA, at University of Wisconsin, Madison, WI, and he works now at UCSB (University of California at Santa Barbara), Santa Barbara, CA.

[13] Andrew Joseph MAJDA, American mathematician, born in 1949. He worked at UCB (University of California at Berkeley), Berkeley, CA, at Princeton University, Princeton, NJ, and at NYU (New York University), New York, NY.

[14] Luc Charles TARTAR, French-born mathematician, born in 1946. I worked at Université Paris IX Dauphine, Paris, France, at Université Paris-Sud, Orsay, at CEA (Commissariat à l'Énergie Atomique), Limeil, France, and at CMU (Carnegie Mellon University), Pittsburgh, PA.

[15] Johann RADON, Czech-born mathematician, 1887–1956. He had worked in Hamburg, in Greifswald and in Erlangen, Germany, in Breslau (then in Germany, now Wrocław, Poland) before World War II, and after 1947 in Vienna, Austria.

Assume ii) and let $v, w \in X$ with $v \leq w$, then $\int_{\Omega'} |S(w) - S(v)| \, d\mu' \leq \int_{\Omega} |w - v| \, d\mu = \int_{\Omega}(w - v) \, d\mu = \int_{\Omega'} (S(w) - S(v)) \, d\mu'$, so that one must have $S(w) \geq S(v)$ almost everywhere (for $d\mu'$). $\square$

The Lax–Friedrichs scheme is conservative, i.e. one has $\sum_i U_i^{n+1} = \sum_i U_i^n$ for every $n \geq 0$, so one uses Lemma 4.1 with counting measures and the fact that it is order preserving gives

$$\sum_i |U_i^{n+1} - V_i^{n+1}| \leq \sum_i |U_i^n - V_i^n| \text{ so that it is } \leq \sum_i |U_i^0 - V_i^0|. \quad (4.12)$$

Using the invariance by translation, for example taking $V_i^0 = U_{i+1}^0$ for every i, one has $V_i^n = U_{i+1}^n$ for every i and every $n \geq 0$, from which one deduces

$$\sum_i |U_{i+1}^{n+1} - U_i^{n+1}| \leq \sum_i |U_{i+1}^n - U_i^n| \text{ so that it is } \leq \sum_i |U_{i+1}^0 - U_i^0|, \quad (4.13)$$

giving a $BV(\mathbb{R})$ bound if the bounded variation of the initial approximation stays bounded; one deduces easily from the scheme a bound for an approximation of u_t.

[Taught on Wednesday September 5, 2001 (Monday September 3 was Labor Day).]

Notes on names cited in footnotes for Chapter 4, HOPKINS,[16] THROOP,[17] FICK,[18] RENSSELAER,[19] FORSYTH,[20] KODAIRA.[21]

---

[16] Johns HOPKINS, American financier and philanthropist, 1795–1873. Johns Hopkins University, Baltimore, MD, is named after him.

[17] Amos Gager THROOP, American businessman and politician, 1811–1894.

[18] Adolph Eugen FICK, German physiologist/physicist, 1829–1901. He had worked in Zürich, Switzerland, and in Würzburg, Germany.

[19] Kilean VAN RENSSELAER, Dutch merchant, c. 1580–1644. The Rensselaer Polytechnic Institute (RPI), Troy, NY, is named after him.

[20] Andrew Russell FORSYTH, Scottish mathematician, 1858–1942. He worked in Cambridge, England, UK, holding the Sadleirian chair of Pure Mathematics (1895–1910), and in London, England, UK.

[21] Kunihiko KODAIRA, Japanese mathematician, 1915–1997. He received the Wolf Prize in 1984, for his outstanding contributions to the study of complex manifolds and algebraic varieties, jointly with Hans LEWY. He had worked in Tokyo, Japan, at IAS (Institute for Advanced Study), Princeton, NJ, at Harvard University, Cambridge, MA, at Johns Hopkins University, Baltimore, MD, at Stanford University, Stanford, CA, and again in Tokyo.

5

The 1-D Scalar Case: the E-Conditions of Lax and of Oleinik

Although the equation $u_t + u\,u_x = 0$ was first introduced by BATEMAN, I shall follow the classical use and call it the Burgers equation, written $u_t + \left(\frac{u^2}{2}\right)_x = 0$ for the correct class of discontinuous solutions, and an important property is that it is invariant by Galilean transformations; it means that if one moves at constant velocity a, one replaces x by $x - a\,t$ and one replaces the velocity u by $u - a$, and one defines a function v related to u by

$$u(x,t) = a + v(x - a\,t, t), \qquad (5.1)$$

then, as is easily verified, the equation for v is also $v_t + v\,v_x = 0$.

For the sake of understanding in a better way which discontinuities should be accepted (and that should tell us more about how irreversibility occurs), it is useful to consider the more general equations $u_t + \left(f(u)\right)_x = 0$ for other functions f; the Galilean invariance only holds if f satisfies $f'(a+v) = a + f'(v)$ for all $a, v \in \mathbb{R}$, i.e. if $f(v) = \frac{v^2}{2} + \alpha v + \beta$ for all $v \in \mathbb{R}$, where $\alpha = f'(0)$ and $\beta = f(0)$.

If u is smooth and satisfies the equation $u_t + \left(f(u)\right)_x = 0$, i.e. $u_t + f'(u)u_x = 0$, then $w = f'(u)$ satisfies $w_t + w\,w_x = 0$, so one may think that the knowledge of the Burgers equation is sufficient for solving the more general equation, but that property only holds for smooth solutions, except if f is a polynomial of degree ≤ 2. Indeed, if u is a discontinuity jumping from a to $b \neq a$, the Rankine–Hugoniot condition says that u satisfies $u_t + \left(f(u)\right)_x = 0$ if and only if the discontinuity travels at velocity $s_1 = \frac{f(b)-f(a)}{b-a}$, while $f'(u)$ satisfies $\left(f'(u)\right)_t + \left(\frac{(f'(u))^2}{2}\right)_x = 0$ if and only if the discontinuity travels at velocity $s_2 = \frac{f'(a)+f'(b)}{2}$, and one has $s_1 = s_2$ for all a, b if and only if f' is affine.

The method of characteristic curves shows that if u_0 is a bounded Lipschitz continuous function, and f is assumed to have a locally bounded second derivative, then a solution of $u_t + \left(f(u)\right)_x = 0$ with $u(\cdot, 0) = u_0$ exists for an interval of time. Indeed, assuming that u is smooth, one defines the characteristic curves by $\frac{dx}{dt} = f'\big(u(x(t),t)\big)$ and $x(0) = y$, and as the equation implies that u is constant along such a curve, it is then a straight line, and that means that on the line $x = y + t\,f'(u_0(y))$, one has $u(x,t) = u_0(y)$; if one has $M_- \leq u_0 \leq M_+$, $|(u_0)_x| \leq A$ and $|f''(v)| \leq B$ for $v \in [M_-, M_+]$, then the solution is bounded (with $M_- \leq u \leq M_+$) and locally Lipschitz continuous for $x \in \mathbb{R}$ and $t \in [-T, +T]$ if $A\,B\,T < 1$ (with a bound $\frac{A}{1-ABt}$), because only one characteristic curve goes through each point (x,t) with $|t| \leq T$, and the mapping $(x,t) \mapsto y$ is Lipschitz continuous.

If u_0 is constant, the solution is constant, but one can construct an infinity of weak solutions with the same initial datum in the case where f is not an affine function. Indeed, because f is not affine, one can find a and b with $a < u_0 < b$ such that the chord joining $\big(a, f(a)\big)$ and $\big(b, f(b)\big)$ does not contain $\big(u_0, f(u_0)\big)$; in the case where the chord goes above $\big(u_0, f(u_0)\big)$, one has

$$s_1 = \frac{f(u_0) - f(a)}{u_0 - a} < s_2 = \frac{f(b) - f(a)}{b - a} < s_3 = \frac{f(b) - f(u_0)}{b - u_0}; \qquad (5.2)$$

choosing a point x_0 arbitrary, one defines u by

$$u(x,t) = \begin{cases} u_0 & \text{for } x < x_0 + s_1 t \\ a & \text{for } x_0 + s_1 t < x < x_0 + s_2 t \\ b & \text{for } x_0 + s_2 t < x < x_0 + s_3 t \\ u_0 & \text{for } x_0 + s_3 t < x \end{cases} \qquad (5.3)$$

and one checks easily that the Rankine–Hugoniot conditions are satisfied for each of the three discontinuities; in the case where the chord goes below $\big(u_0, f(u_0)\big)$, one has a similar construction.

These nonconstant weak solutions are rejected as nonphysical, the constant solution being considered the physical one, even though the model may not come from a reasonable modelling of physical reality.

The constant u_0 is solution of the regularized equation $u_t + u\,u_x - \varepsilon\,u_{xx} = 0$, and the choice $U_i^n = u_0$ for all $i \in \mathbb{Z}$, and $n \geq 0$ is a solution of the Lax–Friedrichs scheme, so that the limiting process selects the constant solution. It will be shown that for other initial data, the limit satisfies supplementary conditions (which are automatically satisfied for smooth solutions), namely that the discontinuities observed in the limiting weak solution must satisfy the *Oleinik E-condition*,[1] expressed below in the case of piecewise constant solutions.

---

[1] Although usually called an entropy condition, it is not related to thermodynamical entropy, and I follow the name E-condition used by Constantine DAFERMOS, referring to his book [4] for many more references and generalizations. My purpose here is not to describe everything about quasi-linear hyperbolic systems of conservation laws, but to observe that some discontinuities occur and that others do not, and to understand why it is so; this is important for the topic of kinetic theory, in relation with the question of how irreversibility occurs.

Definition 5.1. *A discontinuous function*

$$u(x,t) = \begin{cases} u_- & \text{for } x < x_0 + st \\ u_+ & \text{for } x > x_0 + st \end{cases} \tag{5.4}$$

which is a weak solution of $u_t + (f(u))_x = 0$, i.e. satisfies the Rankine-Hugoniot condition $f(u_+) - f(u_-) = s(u_+ - u_-)$, is said to satisfy the Oleinik E-condition *if*

> *either $u_- < u_+$ and the chord joining $(u_-, f(u_-))$ and $(u_+, f(u_+))$*
> *   is below the graph of f*
> *or $u_- > u_+$ and the chord joining $(u_-, f(u_-))$ and $(u_+, f(u_+))$* $\qquad$ (5.5)
> *   is above the graph of f.*

The weak solutions constructed before have three discontinuities, from u_0 to a, from a to b and from b to u_0, and the discontinuity from a to b fails to satisfy the Oleinik E-condition.

Before Olga OLEINIK had introduced her condition, which appeared to be the desired characterization for the general equations $u_t + (f(u))_x = 0$, Peter LAX had introduced a simpler *Lax E-condition*, which makes sense for systems.

Definition 5.2. *A discontinuous function*

$$U(x,t) = \begin{cases} U_- & \text{for } x < x_0 + st \\ U_+ & \text{for } x > x_0 + st \end{cases} \tag{5.6}$$

which is a weak solution of $U_t + (F(U))_x = 0$, i.e. satisfies the Rankine-Hugoniot condition $F(U_+) - F(U_-) = s(U_+ - U_-)$, is said to satisfy the Lax E-condition *if*

$$\text{one has } \lambda_j(U_-) \geq s \geq \lambda_j(U_+) \text{ for some } j \in \{1,\ldots,p\}. \tag{5.7}$$

For the scalar case, the intuition that such a condition must hold can be guessed easily from what happens for the Burgers equation.

If one looks for solutions of the Burgers equation depending upon $\frac{x}{t}$, one finds that $\frac{x}{t}$ is a special solution, and using invariance by translation in x or in t, one finds that for every $x_0, t_0 \in \mathbb{R}$, a particular solution is $\frac{x-x_0}{t-t_0}$, valid for $t \neq t_0$. If the initial datum is $u_0(x) = \alpha x + \beta$, which corresponds to $t_0 = -\frac{1}{\alpha}$ and $x_0 = -\frac{\beta}{\alpha}$, then the solution is $u(x,t) = \frac{\alpha x + \beta}{1 + \alpha t}$. If one restricts attention to increasing times, one sees that if $\alpha \geq 0$ the solution exists for all $t \geq 0$, and it corresponds to the solution computed using the method of characteristic curves, of course, while if $\alpha < 0$, the solution blows up for a critical time $T_c = t_0 = -\frac{1}{\alpha}$, and in that case all the characteristic lines pass through the point $\left(-\frac{\beta}{\alpha}, -\frac{1}{\alpha}\right)$.

Let $a < b$, and for $k > 0$, let us consider the initial datum u_0 defined by

$$u_0(x) = \begin{cases} a \text{ for } x \leq 0 \\ (1 - k\,x)a + k\,x\,b \text{ for } 0 \leq x \leq \frac{1}{k} \\ b \text{ for } x \geq \frac{1}{k}, \end{cases} \tag{5.8}$$

so that u_0 is bounded and Lipschitz continuous, and the solution u, computed by the method of characteristic curves, is given by

$$u(x,t) = \begin{cases} a \text{ for } x \leq a\,t \\ \frac{(1-k\,x)a+k\,x\,b}{1+k(b-a)t} \text{ for } a\,t \leq x \leq \frac{1}{k} + b\,t \\ b \text{ for } x \geq \frac{1}{k} + b\,t, \end{cases} \tag{5.9}$$

so that u is bounded and Lipschitz continuous for all $t \geq 0$. When one lets k tend to infinity, the sequence of initial data converges to u_0^* given by

$$u_0^*(x) = \begin{cases} a \text{ for } x \leq 0 \\ b \text{ for } x \geq 0, \end{cases} \tag{5.10}$$

and the sequence of solutions converges to u^* given by

$$u^*(x,t) = \begin{cases} a \text{ for } x \leq a\,t \\ \frac{x}{t} \text{ for } a\,t \leq x \leq b\,t \\ b \text{ for } x \geq b\,t, \end{cases} \tag{5.11}$$

so that one is led to decide that, for reasons of continuity,[2] one prefers the smooth solution u^*, i.e. a discontinuity corresponding to $u_- = a < b = u_+$ is unstable and creates a rarefaction wave as in the formula for u^*, and one rejects the solution with a discontinuity travelling at velocity $s = \frac{a+b}{2}$, which would be $\tilde{u}$ given by

$$\tilde{u}(x,t) = \begin{cases} a \text{ for } x \leq s\,t \\ b \text{ for } x \geq s\,t. \end{cases} \tag{5.12}$$

If one applies the method of characteristic curves to the discontinuous initial datum u_0^*, one observes that the characteristic lines with $y < 0$ cover the part of the (x,t) plane with $x < a\,t$ and the characteristic lines with $y > 0$ cover the part of the (x,t) plane with $x > b\,t$, leaving a gap, namely the sector $a\,t < x < b\,t$.

Conversely, in the case where $a > b$, and where one accepts the discontinuous solution $\tilde{u}$, the characteristic lines coming from $y < 0$ interact with those coming from $y > 0$, and the discontinuity is obtained as a compromise.

---

[2] One postulates that there is a natural topology for deciding if two initial data are near and that nearby initial data create nearby solutions, i.e. the mappings $S(t)$ defined by $S(t)(u_0) = u(\cdot, t)$ are continuous for $t \geq 0$, and define a continuous semi-group so that $S(t)(u_0) \to u_0$ as t tends to 0, and $S(s + t) = S(s)S(t)$ for $s, t \geq 0$. For the scalar case, the strong topology of $L^1(\mathbb{R})$ is such a topology, and it is believed to work also in the case of systems, at least in one space dimension, but as L^1 is not a good functional space for dealing with partial differential equations which do not satisfy the maximum principle, it seems that other spaces are needed, and I have proposed some interpolation spaces.

One is led then to consider that a discontinuity occurs because the left side of the discontinuity has information travelling faster than s, while the right side has information travelling slower than s, and that is what the Lax E-condition is about.

Another way to look at the problem is to consider in the (x, u) plane the graph of u_0^* to which one adds a vertical part from $(0, a)$ to $(0, b)$ at the discontinuity (so that one has a continuous curve), and to consider that each point (x, u) of the curve travels at velocity u for a small time Δt. If $a < b$ one then finds that the vertical part transforms into the graph of a Lipschitz continuous function, and that leads to accepting such a rarefaction wave (the method of characteristic curves with $y = 0$ and $u_0(y)$ moving from a to b generates characteristic lines which fill the gap that was observed before). Conversely, if $a > b$, one obtains a curve which is not the graph of a function, exactly as one observes for breaking waves on a sloping beach, and I had learnt about the influence of the variation of depth in my continuum mechanics course at École Polytechnique with Jean MANDEL,[3] although he had only treated linear effects, with the purpose of showing that momentum could be transported without any real transport of mass.[4] However, because the model does not allow for the breaking of waves, a compromise must be found, where the matter which has gone too fast at the top of the wave is used to help the matter which has gone too slow at the bottom of the wave, and the rule must be compatible with the fact that the integral of u should be conserved.

The crucial effect which explains why the Oleinik E-condition improves the Lax E-condition is that a discontinuity from a to b may well break up into smaller discontinuities separated by smooth parts.

Constantine DAFERMOS has found a simple way to analyse such a question by looking at functions f which are continuous and piecewise affine, and this

---

[3] Jean MANDEL, French mathematician, 1907–1982. I had him as a teacher in 1966–1967, for the course of continuum mechanics at École Polytechnique, Paris, France. He had worked in Saint-Étienne and in Paris, France.

[4] In the open sea, one observes some sinusoidal waves, with a profile depending only upon one direction, and these waves seem to move at a constant velocity, but if one linearizes the equation of hydrodynamics in an ocean of fixed depth H (around a zero velocity field), one finds that disturbances of the surface decompose as waves travelling at a velocity $V(H)$ (the same in every direction); it is the top of one of these unidimensional waves which travels at this velocity, and one can follow it easily with the eye, and this velocity is a phase velocity which does not correspond to any transport of mass (and a floating cork moves a little when the wave goes by, but does not drift), so that it transports linear momentum (felt at the end, on the beach where the waves break). If there is a sharp decrease in depth near a beach, for example due to the presence of a submerged coral reef, then the waves from the open sea arrive too fast compared to the local velocity favoured by the waves and one observes the breaking of waves, the delight of surfers, whose art is precisely about using the momentum transported by these waves.

simplifies the analysis of the smooth parts in the solutions (which, according to the analysis shown before for systems, come either from the portions where f is convex and u increases, or the portions where f is concave and u decreases). For this class of functions, if u_0 is piecewise constant, then the solution is piecewise constant at each later time, and a discontinuity from u_- on the left to u_+ on the right is accepted if and only if one cannot find v such that the discontinuity from u_- to v travels at a velocity s_1 strictly smaller than the velocity s_2 of the discontinuity from v to u_+ (in which case one would prefer to decompose the discontinuity into a succession of discontinuities which all satisfy the condition and move faster and faster); notice that the Lax E-condition is recovered as the limiting cases, where v is taken near u_- or near u_+. This manner of selecting discontinuities is precisely the Oleinik E-condition for this particular class of functions; for a general function f, one approaches f uniformly (on bounded intervals) by a sequence f_n of continuous piecewise affine functions, and one looks at the limit of the sequence u_n of solutions, noticing that although each u_n is piecewise constant, its limit may not be and that a smooth transition in the limiting solution is approached by parts containing plenty of discontinuities of small amplitude.

The analysis using continuous curves which are not necessarily graphs and then deducing a compromise to transform the result into a graph also leads one to discover why the Oleinik E-condition is natural, and it was followed by Yann BRENIER,[5] and his arguments look to me as if one accepted breaking of the waves but one was letting the gravity g tend to ∞.

The analysis of the regularized equation $u_t + \big(f(u)\big)_x - \varepsilon\, u_{xx} = 0$, and that of the Lax–Friedrichs finite-difference scheme, have relied too heavily upon the maximum principle, which is of little use for general systems of conservation laws. The order relation on $\mathbb{R}$ is also used in the method described above, where one first creates a curve which is not a graph and one then drops something which has gone too fast at the top, so one uses a direction on $\mathbb{R}$; it suggests that for systems one might need a vector field in $\mathbb{R}^p$ along which to push information, but the matter is made more difficult by the fact that for a system not all discontinuities satisfy a Rankine–Hugoniot condition, and that one needs more than one direction anyway for transporting information, as each of the directions of the eigenvectors r_j, $j = 1, \ldots, p$, have this role when the solution is smooth.

[Taught on Friday September 7, 2001.]

---

[5] Yann BRENIER, French mathematician, born in 1957. He has worked at UCLA (University of California at Los Angeles), Los Angeles, CA, at INRIA (Institut National de Recherche en Informatique et Automatique), Rocquencourt, at Université Paris VI (Pierre et Marie Curie), Paris, and at CNRS (Centre National de la Recherche Scientifique) at Université de Nice-Sophia-Antipolis, Nice, France.

6

Hopf's Formulation of the E-Condition of Oleinik

In the late 1960s, Eberhard HOPF found a nice analytical way for expressing the Oleinik E-condition (for a scalar equation), and his condition makes sense without assuming that the solution is smooth enough to have limits on each side of a discontinuity; Peter LAX then did the analysis for systems, and he quotes Eberhard HOPF but also KRUZHKOV as having the idea independently. Eberhard HOPF observed that if φ and ψ are related by $\psi' = f'\varphi'$ (and following Peter LAX one calls φ an entropy, and ψ an entropy flux, or (φ, ψ) an entropy/entropy flux pair), then any solution of $u_t + \left(f(u)\right)_x = 0$ which is piecewise smooth and satisfies the Oleinik E-condition for each of its discontinuities automatically satisfies the condition

$$\left(\varphi(u)\right)_t + \left(\psi(u)\right)_x \leq 0 \text{ in the sense of distributions/Radon measures,} \\ \text{for all convex entropies } \varphi, \tag{6.1}$$

and for a discontinuity from u_- to u_+, travelling at velocity $s = \frac{f(u_+)-f(u_-)}{u_+-u_-}$, this condition is equivalent to

$$\psi(u_+) - \psi(u_-) \leq s\left(\varphi(u_+) - \varphi(u_-)\right), \text{ for all convex entropies } \varphi. \tag{6.2}$$

Conversely, the above condition implies the Oleinik E-condition, and also the Rankine–Hugoniot condition for s, by choosing for φ all affine functions. This last equivalence is seen by noticing that every convex function can be approached (uniformly on bounded sets) by convex piecewise affine functions, and that shows that the conditions for all convex φ can be replaced by an equivalent condition where one uses the functions $\varphi(u) = \pm u$, corresponding to $\psi(u) = \pm f(u)$ and the family of entropy/entropy flux pairs indexed by $k \in \mathbb{R}$

$$\varphi_k(u) = \begin{cases} 0 & \text{for } u \leq k \\ u - k & \text{for } u \geq k \end{cases}; \quad \psi_k(u) = \begin{cases} 0 & \text{for } u \leq k \\ f(u) - f(k) & \text{for } u \geq k \end{cases}. \tag{6.3}$$

If $k \leq \min\{u_-, u_+\}$ or if $k \geq \max\{u_-, u_+\}$, the condition using the pair (φ_k, ψ_k) is trivially satisfied, and for $\min\{u_-, u_+\} < k < \max\{u_-, u_+\}$ the condition tells whether the point $(k, f(k))$ is above or below the chord joining $(u_-, f(u_-))$ and $(u_+, f(u_+))$, and using all these ks gives the Oleinik E-condition.

If u^ε solves $u_t^\varepsilon + (f(u^\varepsilon))_x - \varepsilon u_{xx}^\varepsilon = 0$ with a fixed initial datum u_0, then from the bounds obtained one may extract a subsequence u^η which converges almost everywhere to a limit u^0, and one wants to show that u^0 satisfies $(\varphi(u^0))_t + (\psi(u^0))_x \leq 0$ for all φ convex. Multiplying the equation for u^η by $\varphi'(u^\eta)$, one deduces that $(\varphi(u^\eta))_t + (\psi(u^\eta))_x - \eta(\varphi(u^\eta))_{xx} + \eta\varphi''(u^\eta)(u_x^\eta)^2 = 0$, and the sequence of nonnegative functions $\mu^\eta = \eta\varphi''(u^\eta)(u_x^\eta)^2$ converges in the sense of distributions (or in the sense of Radon measures) to a nonnegative Radon measure μ^0, and because u^η is bounded and converges almost everywhere to u^0, one has $g(u^\eta) \to g(u^0)$ in L_{loc}^p strong for every $p \in [1, \infty)$ and for every continuous function g, and passing to the limit in the equation one deduces that $(\varphi(u^0))_t + (\psi(u^0))_x + \mu^0 = 0$.

Peter LAX noticed that the same argument holds for a system of conservation laws

$$U_t + (F(U))_x = 0, \tag{6.4}$$

which one regularizes by an artificial viscosity term[1]

$$U_t^\varepsilon + (F(U^\varepsilon))_x - \varepsilon D U_{xx}^\varepsilon = 0, \tag{6.5}$$

if one selects $D = \alpha I$ with $\alpha > 0$. If φ is a convex entropy (and not all functions are entropies for systems), then one has

$$(\varphi(U^\varepsilon))_t + (\psi(U^\varepsilon))_x - \varepsilon\alpha(\varphi(U^\varepsilon))_{xx} + \varepsilon\alpha\varphi''(U^\varepsilon)[U_x^\varepsilon, U_x^\varepsilon] = 0, \tag{6.6}$$

where $\varphi''(u)[v, w]$ is the symmetric bilinear form defined by the Hessian matrix φ'' at u;[2] a particular difficulty is that one does not know enough a priori bounds to be able to extract a subsequence converging almost everywhere, and the mathematical result is then that if a subsequence exists which stays bounded in L^∞ and converges almost everywhere, then the limit U^0 satisfies

$$(\varphi(U^0))_t + (\psi(U^0))_x \leq 0 \text{ for all convex entropies } \varphi, \tag{6.7}$$

and one should limit the growth at infinity of the entropy functions used if one only has a bounded sequence in some L^p for $p < \infty$.

---

[1] For the system of gas dynamics, it means that one adds a term $-\varepsilon\, \varrho_{xx}$ in the equation of conservation of mass (corresponding to a diffusion of mass, as postulated by FICK), and a term $-\varepsilon(\varrho u)_{xx}$ in the equation of balance of momentum (corresponding to a viscosity effect quite different from that postulated by NAVIER), and a term $-\varepsilon\left(\frac{\varrho u^2}{2} + \varrho e\right)_{xx}$ in the equation of balance of energy (corresponding to a diffusion of heat of a much stranger form than that postulated by FOURIER).

[2] Ludwig Otto HESSE, German mathematician, 1811–1874. He had worked in Königsberg (then in Germany, now Kaliningrad, Russia), in Heidelberg, and in München (Munich), Germany.

The approximate solutions constructed by the Lax–Friedrichs scheme also approach weak solutions satisfying the Oleinik E-condition (under the formulation of Eberhard HOPF). Indeed, let $a = \min\{U^n_{i-1}, U^n_{i+1}\}$ and $b = \max\{U^n_{i-1}, U^n_{i+1}\}$ (and $a < b$, otherwise $U^{n+1}_i = a = b$ and there is nothing to prove); by the CFL condition one has imposed $\frac{\Delta t}{\Delta x}|f'(v)| \leq 1$ for $a \leq v \leq b$, and one wants to show that

$$\frac{U - \frac{a+b}{2}}{\Delta t} + \frac{f(b) - f(a)}{2\Delta x} = 0 \text{ implies } \frac{\varphi(U) - \frac{\varphi(a)+\varphi(b)}{2}}{\Delta t} + \frac{\psi(b) - \psi(a)}{2\Delta x} \leq 0 \tag{6.8}$$
for all convex entropies φ.

This is trivially satisfied if φ is affine, and it is equivalent then to show the implication for the special functions φ_k which were used before; because $a \leq U \leq b$, the condition is trivially satisfied if $k \leq a$ or if $k \geq b$; the case $a < k < b$ splits into two subcases, according to the position of U with respect to k. In the case $a \leq U \leq k < b$, one has $\varphi(a) = \varphi(U) = \psi(a) = 0$ and $\varphi(b) = b - k, \psi(b) = f(b) - f(k)$, and the inequality becomes $-\frac{b-k}{2\Delta t} + \frac{f(b)-f(k)}{2\Delta x} \leq 0$, which follows from the mean value theorem $f(b) - f(k) = (b-k)f'(v)$ for some $v \in (k, b)$; similarly for the case $a < k \leq U \leq b$, one has $\varphi(a) = \psi(a) = 0$ and $\varphi(U) = U - k, \varphi(b) = b - k, \psi(b) = f(b) - f(k)$, and the inequality becomes $\frac{U - \frac{k+b}{2}}{\Delta t} + \frac{f(b)-f(k)}{2\Delta x} \leq 0$, which after using the definition of U means $\frac{a-k}{2\Delta t} + \frac{f(a)-f(k)}{2\Delta x} \leq 0$, which follows from $f(a) - f(k) = (a - k)f'(w)$ for some $w \in (a, k)$.

Another approach for choosing or rejecting a discontinuity is to look for a viscous *shock profile*, i.e. a curve which describes intermediate values between U_- and U_+ (which are assumed to satisfy a Rankine–Hugoniot condition $F(U_+) - F(U_-) = s(U_+ - U_-)$), for a regularized equation of the form

$$U_t + \big(F(U)\big)_x - \varepsilon\big(D(U)U_x\big)_x = 0, \tag{6.9}$$

where the (artificial) viscosity matrix $D(U)$ is nonnegative, but there are other conditions to impose, in particular such a matrix should not destabilize constant states, as was noticed by Andrew MAJDA and Robert PEGO,[3] and one looks for a solution of the form

$$U(x, t) = V\left(\frac{x - st}{\varepsilon}\right) \text{ with } V(-\infty) = U_- \text{ and } V(+\infty) = U_+. \tag{6.10}$$

Notice that this is different from the Cauchy problem with an initial datum independent of ε. I have heard that this type of question had been initialized by GEL'FAND,[4] but Constantine DAFERMOS mentions that for gas dynamics

---

[3] Robert Leo PEGO, American mathematician. He worked at University of Michigan, Ann Arbor, MI, at University of Maryland, College Park, MD, and at CMU (Carnegie Mellon University), Pittsburgh, PA.

[4] Izrail Moiseevic GEL'FAND, Russian-born mathematician, born in 1913. He received the Wolf Prize in 1978, for his work in functional analysis, group represen-

such an idea had already been used by RANKINE and by Lord Rayleigh. The equation for V is then

$$-s V' + (F(V))' - (D(V)V')' = 0, \tag{6.11}$$

which one integrates immediately as

$$-s V + F(V) - D(V)V' = C, \tag{6.12}$$

where one must have

$$\begin{array}{l} C = F(U_-) - s U_- \text{ if } U \to U_- \text{ at } -\infty \\ C = F(U_+) - s U_+ \text{ if } U \to U_+ \text{ at } -\infty, \end{array} \tag{6.13}$$

so that there would be no solution if the Rankine–Hugoniot condition was not satisfied. If $D(U)$ is invertible, then one has an ordinary differential equation

$$V' = (D(V))^{-1}(F(V) - s V - C), \tag{6.14}$$

which has both U_- and U_+ as critical points, and one is looking for a connecting orbit. The existence of such a connecting orbit requires that U_- have an unstable manifold, so that at least one eigenvalue of $D(U_-)(\nabla F(U_-) - s I)$ has a nonnegative real part, and that U_+ have a stable manifold, so that at least one eigenvalue of $D(U_+)(\nabla F(U_+) - s I)$ has a nonpositive real part, and that is related to the Lax condition; it is, however, a difficult global question to decide if by leaving U_- along the unstable manifold, one is able to reach the stable manifold of U_+, and this question has been studied extensively by Charles CONLEY and Joel SMOLLER.[5] For a scalar equation, the criterion selects discontinuities such that between U_- and U_+ there is no other critical point of the differential equation, and this corresponds to the Oleinik E-condition.[6]

The search to determine which discontinuities are acceptable is certainly not over for the case of systems. One reason to be confident to have found the right condition for the scalar case is also that there is a uniqueness theorem (whose general form is probably due to KRUZHKOV), for solutions satisfying the Oleinik E-condition, expressed in the analytic form of Eberhard HOPF.

---

tation, and for his seminal contributions to many areas of mathematics and its applications, jointly with Carl L. SIEGEL. He worked in Moscow, Russia, and at Rutgers University, Piscataway, NJ.

[5] Charles Cameron CONLEY, American mathematician, 1933–1984. He had worked at University of Wisconsin, Madison, WI, where I met him during the year 1974–1975 that I spent there, and then at University of Minnesota, Minneapolis, MN.

[6] Some discontinuities satisfying the Oleinik E-condition may actually be obtained by putting together elementary discontinuities satisfying the condition with the same velocity s, and the viscous shock profile only selects elementary discontinuities satisfying the Oleinik E-condition.

In the late 1970s, I had initialized the study of oscillations (later called microstructures) in the nonlinear partial differential equations of continuum mechanics, and while the previous analysis for quasi-linear hyperbolic systems had concentrated on shocks, my analysis was to study oscillating sequences of solutions. When the model is supposed to represent physical reality, my approach is more suitable for following the physics behind the phenomena and for discovering if the laws have been averaged correctly and if more efficient effective equations should be derived, and this is related to the questions of homogenization which I have developed partly with François MURAT, generalizing an earlier approach of Sergio SPAGNOLO. I want to emphasize that our approaches use no periodicity assumptions, of course, and one should pay attention to the fact that, among those who only use periodic modulation ideas, many forget to refer to a general theory of homogenization, and they avoid mentioning Sergio SPAGNOLO for G-convergence, or François MURAT and myself for H-convergence, or even Évariste SANCHEZ-PALENCIA, who was the first to guess correct asymptotic expansions, or Ivo BABUŠKA, who was the first to use the method for engineering applications, and who first used the term homogenization in the mathematical literature,[7] to which I gave a more general meaning in my Peccot lectures,[8] in the beginning of 1977. It should not come as a surprise that most of those engaged in wide misattribution of ideas are also keen in advocating fake continuum mechanics or physics.

The question of wave breaking has been mentioned, and if an equation like the Burgers equation is supposed to describe the vertical displacement of the surface of the water, one knows that some situations create the breaking of waves, and one expects then to observe bubbles of air trapped for a while in the water and trying to move upward, and droplets of water in the air trying to move downward. A layer representing a mixture of air and water might be a good description for what is going on, but homogenization tells us that one cannot expect a too simple law for the evolution of such mixtures, because effective properties do not depend only upon proportions, and *H-measures* might be useful for computing corrections [18], and other mathematical tools may have to be developed for a complete understanding of that question. It is useful to observe then that the laws of thermodynamics have some limitations, not only because they have been derived from the observation of equilibria, but also because the effective properties of mixtures do not depend only upon proportions. One should then reject part of the rules of thermodynamics and explain why one does not follow them, like I do, but it remains an important mathematical question to settle here, which is to derive a new and better thermodynamics, by understanding more about the evolution of mixtures. Of course, this should be done without using probabilities, which are always used when one lacks information on the processes which must be understood.

---

[7] As pointed out to me by Michael VOGELIUS, the term homogenization was used previously by nuclear engineers.

[8] Claude Antoine PECCOT, French child prodigy, 1856–1876.

Although the introduction of probabilities is suitable for engineers, who must control situations for which one does not even know which equations to use, it is certainly not suitable for scientists, despite the bad example of those politically inclined physicists who coined the dogma that there are probabilities in the laws of nature.

Although the complete problem is too difficult at the moment, one can guess that if one lets the gravity g tend to ∞, the mixtures will separate quickly, with water on the bottom and air on the top, and the solution might look precisely like the discontinuous solutions of the Burgers equation which have been selected.

[Taught on Monday September 10, 2001.]

Notes on names cited in footnotes for Chapter 6, SIEGEL,[9] RUTGERS,[10] VOGELIUS,[11] and for the preceding footnotes, GEORGE II.[12]

---

[9] Carl Ludwig SIEGEL, German mathematician, 1896–1981. He received the Wolf Prize in 1978, for his contributions to the theory of numbers, theory of several complex variables, and celestial mechanics, jointly with Izrail GEL'FAND. He had worked at Georg-August University, Göttingen, Germany.

[10] Henry RUTGERS, American colonel, 1745–1830. Rutgers University, Piscataway, NJ, is named after him.

[11] Michael VOGELIUS, Danish-born mathematician. He worked at University of Maryland, College Park, MD, and at Rutgers University, Piscataway, NJ.

[12] Georg Augustus, 1683–1760. Duke of Brunswick-Lüneburg (Hanover), he became King of Great Britain and Ireland in 1727, under the name GEORGE II.

7

The Burgers Equation: Special Solutions

Let us now use our knowledge of which discontinuities to accept for the Burgers equation, i.e. $U_- \geq U_+$, to compute a few explicit solutions; in that case, the uniqueness result of Olga OLEINIK applies, based on the estimate

$$u_x \leq \frac{1}{t}, \tag{7.1}$$

which is inherited from the same inequality for the solution of the equation regularized by adding $-\varepsilon\, u_{xx}$.

Let u_0 be of the form

$$u_0(x) = \begin{cases} a \text{ for } x < x_1 \\ b \text{ for } x_1 < x < x_2 \\ a \text{ for } x > x_2, \end{cases} \tag{7.2}$$

where $a < b$. By using invariance by translation and Galilean invariance, one may assume that $a = 0$ and $x_1 = 0$, i.e. one puts $u(x,t) = a + v(x - x_1 - a\,t, t)$ and v satisfies the Burgers equation; the initial datum for v has a discontinuity at 0, where u jumps from 0 to $c = b - a > 0$, and this discontinuity transforms into a rarefaction wave $u = \frac{x}{t}$ and a discontinuity at $L = x_2 - x_1$ where u jumps from c to 0, and this discontinuity travels unchanged at velocity $\frac{c}{2}$, and this description is valid as long as the rarefaction wave has not caught up with the slower shock in front of it, i.e. for $t \leq \frac{2L}{c}$, so that one has

$$v(x,t) = \begin{cases} 0 \text{ for } x \leq 0 \\ \frac{x}{t} \text{ for } 0 \leq x \leq ct < 2L \\ c \text{ for } ct \leq x < L + \frac{ct}{2} \\ 0 \text{ for } L + \frac{ct}{2} < x \end{cases}, \text{ for } t < \frac{2L}{c}, \tag{7.3}$$

At $t = \frac{2L}{c}$, the solution has a triangular shape, $v = 0$ for $x \leq 0$, $v = \frac{x}{t}$ for $0 \leq x < 2L$, and $v = 0$ for $x > 2L$; afterwards the solution has a similar structure

$$v(x,t) = \begin{cases} 0 \text{ for } x \leq 0 \\ \frac{x}{t} \text{ for } 0 \leq x < z(t) \text{ , for } t \geq \frac{2L}{c}, \\ 0 \text{ for } x > z(t) \end{cases} \tag{7.4}$$

and besides $z\left(\frac{2L}{c}\right) = 2L$, the value of z is obtained by using the Rankine–Hugoniot condition

$$\frac{dz}{dt} = \frac{z}{2t}, \tag{7.5}$$

because the shock jumps from $\frac{z(t)}{t}$ to 0; integrating the differential equation gives $z(t) = k\sqrt{t}$, and as $k\sqrt{\frac{2L}{c}} = 2L$ one has $k = \sqrt{2Lc}$, so that

$$z(t) = \sqrt{2\sigma t}, \text{ with } \sigma = Lc, \text{ for } t \geq \frac{2L}{c}, \tag{7.6}$$

which can also be obtained by observing that the integral of u in x must be constant, and that gives $\frac{z^2}{2t} = Lc$, but one has to check the Rankine–Hugoniot condition then, in order to be sure that one has a solution.

One can observe on these particular solutions of the Burgers equation how shocks create irreversible effects. Indeed, if one takes different data of the preceding type at time 0, where for $j = 1, \ldots, m$, the function v_{j0} is 0 for $x < 0$, c_j for $0 < x < L_j$ and 0 for $x > L_j$, and if all the products $L_j c_j$ are equal to σ, then for t large enough, and more precisely for $t > \max_j \frac{2L_j}{c_j}$, then all the solutions coincide. Actually, letting L tend to 0, with $c = \frac{\sigma}{L}$, gives a sequence of initial data converging to $\sigma \delta_0$, so the formula for z can be understood as the solution with initial datum $\sigma \delta_0$.

There are other initial data which create the same triangular shaped solution at a later time; for example if u_0 is given by

$$u_0(x) = \begin{cases} 0 \text{ for } x < 0 \\ \alpha(L - x) \text{ for } 0 < x \leq L \\ 0 \text{ for } x \geq L, \end{cases} \tag{7.7}$$

with $\alpha > 0$, the solution is

$$u(x,t) = \begin{cases} 0 \text{ for } x < 0 \\ \frac{x}{t} \text{ for } 0 < x \leq \alpha Lt \\ \frac{\alpha(L-x)}{1-\alpha t} \text{ for } \alpha Lt \leq x \leq L \\ 0 \text{ for } x \geq L \end{cases}, \text{ for } 0 < t < \frac{1}{\alpha}, \tag{7.8}$$

giving for $t = \frac{1}{\alpha}$ a triangular profile corresponding to $\sigma = \int_{\mathbb{R}} u_0 \, dx = \frac{\alpha L^2}{2}$. Actually, if u_0 is 0 outside $(0, L)$ and is a piecewise constant nonincreasing nonnegative function on $(0, L)$, then after a finite time the solution has taken the triangular profile.

However, if $b > a > 0$ and

$$u_0(x) = \begin{cases} 0 \text{ for } x < 0 \\ b \text{ for } 0 < x < L \\ a \text{ for } x > L, \end{cases} \tag{7.9}$$

then after time $T_c = \frac{2L}{b-a}$ the solution takes the following form:

$$u(x,t) = \begin{cases} 0 \text{ for } x < 0 \\ \frac{x}{t} \text{ for } 0 < x < t\, y(t) \text{, } \quad \text{for } t > T_c = \frac{2L}{b-a}, \\ a \text{ for } x > t\, y(t) \end{cases} \tag{7.10}$$

where

$$y(t) > a \text{ and } \frac{d(t\, y)}{dt} = \frac{y+a}{2} \text{ for } t \geq T_c, \text{ and } y(T_c) = b, \tag{7.11}$$

giving after integration

$$y(t) = a + \frac{\alpha}{\sqrt{t}}, \text{ with } \alpha = \sqrt{T_c}(b-a) = \sqrt{2L(b-a)}. \tag{7.12}$$

The Burgers equation is invariant by some changes of scale, and if one makes the change

$$u(x,t) = a\, v(b\, x, a\, b\, t), \text{ or } u(x,t) = \frac{L}{T} v\left(\frac{x}{L}, \frac{t}{T}\right), \tag{7.13}$$

with $a, b \neq 0$, or $L, T \neq 0$, then u satisfies the Burgers equation if and only if v satisfies the Burgers equation; of course, this corresponds to changing the unit of length and the unit of time, and the unit for measuring velocities is automatically determined.

The solution of the Riemann problem corresponds to looking for a solution invariant by the change $(x,t) \mapsto (\lambda x, \lambda t)$, i.e. which uses the subgroup of transformations given by $a = 1$, so one looks for solutions of the form

$$u(x,t) = f\left(\frac{x}{t}\right), \tag{7.14}$$

and this gives the particular solution $\frac{x}{t}$, or the shocks with $u = U_-$ for $x < st$ and $u = U_+$ for $x > st$, with $U_- \geq U_+$ and $s = \frac{U_-+U_+}{2}$.

If one considers functions with compact support, whose integral must stay constant, one must consider the subgroup of transformations $b = a$, and a solution invariant by the transformations $a\, u(a\, x, a^2 t)$ must be of the form

$$\frac{1}{\sqrt{t}} f\left(\frac{x}{\sqrt{t}}\right), \tag{7.15}$$

which at first sight contains the already known solution $\frac{x}{t}$, apart from the fact that f must be integrable; a simple analysis gives the family that Peter LAX called N-waves,

$$u(x,t) = \begin{cases} 0 \text{ for } x < -\sqrt{2m_-t} \\ \frac{x}{t} \text{ for } -\sqrt{2m_-t} < x < \sqrt{2m_+t} \\ 0 \text{ for } x > \sqrt{2m_+t}, \end{cases} \qquad (7.16)$$

where $m_-, m_+ \geq 0$; as Peter LAX has noticed, there are invariants that can be defined for integrable solutions satisfying the Burgers equation, and two of them are the quantities m_- and m_+ corresponding to a breaking of the conserved quantity $\int_{\mathbb{R}} u\,dx = m_+ - m_-$ into two conserved quantities, the part moving to $-\infty$ and the part moving to $+\infty$.

One should notice that an L^∞ bound of the solution in $O\left(\frac{1}{\sqrt{t}}\right)$ can be derived from the L^1 norm of the initial datum u_0 (and the bound is sharp by looking at the preceding triangular-shaped solutions). One uses the fact that $u_x \leq \frac{1}{t}$, so if $u(x_0,t) = a > 0$ for example, then one must have $u(x,t) \geq \frac{x-x_0+at}{t}$ for $x_0 - at < x < x_0$, so that the integral of $|u|$ is greater than the corresponding surface $\frac{a^2t}{2}$, and must be $\leq \int_{\mathbb{R}} |u_0|\,dx$, giving the estimate

$$|u(x,t)| \leq \sqrt{\frac{2\int_{\mathbb{R}} |u_0|\,dx}{t}} \text{ almost everywhere for } x \in \mathbb{R}, t > 0. \qquad (7.17)$$

Peter LAX had also noticed that if the initial datum is periodic, then the solution is periodic of course, and it converges to the average over one period in $O\left(\frac{1}{t}\right)$, and this follows easily again from the estimate $u_x \leq \frac{1}{t}$. There are actually self-similar solutions which decay in $\frac{1}{t}$, and they correspond to the subgroup of transformations $b = 1$, for which one must look for solutions of the form

$$u(x,t) = \frac{1}{t} f(x), \qquad (7.18)$$

and one finds again the case $\frac{x}{t}$, but if one is interested in a global decay in $\frac{1}{t}$, one is led by a simple analysis to the following family of solutions, where one can choose an arbitrary family of disjoint intervals $I_j = (x_j - L_j, x_j + L_j)$ (with $L_j > 0$), finite or countable but with an upper bound on the L_j in the infinite case,

$$u(x,t) = \begin{cases} \frac{x-x_j+L_j}{t} \text{ for } x_j - L_j < x < x_j \\ \frac{x-x_j-L_j}{t} \text{ for } x_j < x < x_j < x_j + L_j \\ 0 \text{ outside the union of the intervals } I_j. \end{cases} \qquad (7.19)$$

In the summer of 1986, I had been asked by Roger CHÉRET to write a theoretical chapter for a book about shocks in solids,[1] [2], and I first wrote a set of notes (in French) for an introduction to quasi-linear hyperbolic systems

---

[1] Roger CHÉRET, French physicist. He worked at CEA (Commissariat à l'Énergie Atomique).

of conservation laws [17], from which he was going to select some material. I had been led to do other computations, and one of them was to wonder about what happens to a perturbation of a rarefaction wave (or compression wave in the case $\kappa < 0$), i.e. I considered the case

$$u_0(x) = \kappa\, x + v_0(x), \tag{7.20}$$

and I found that if one uses the transformation

$$u(x,t) = \frac{\kappa\, x}{1+\kappa t} + \frac{1}{1+\kappa t}v\left(\frac{x}{1+\kappa t}, \frac{t}{1+\kappa t}\right), \tag{7.21}$$

then v satisfies the Burgers equation with initial data v_0; if v_0 is bounded, one then has a correction $O(\frac{1}{t})$ if $\kappa > 0$, and one should notice that if v_0 has compact support the solution v decays in $O(\frac{1}{\sqrt{t}})$, but that does not make the correction $O(\frac{1}{t\sqrt{t}})$, because v is evaluated at time $\frac{t}{1+\kappa t}$ which tends to $\frac{1}{\kappa}$ as t tends to ∞.

It seemed to me at the time that the knowledge of explicit solutions should be used for improving numerical methods, and I was worried that the numerical methods are not Galilean invariant, for example, but I also thought about the idea of adding explicit solutions in the treatment of finite element approximations, like for domains with corners, and Louis BRUN had once mentioned to me that this idea had already been used by LAGRANGE.[2] After Jean OVADIA had told me that,[3] for numerical reasons, it was useful to consider the generalized Riemann problem

$$u_0(x) = \begin{cases} u_- + \kappa_- x & \text{for } x < 0 \\ u_+ + \kappa_+ x & \text{for } x > 0, \end{cases} \tag{7.22}$$

I computed the corresponding explicit solution, which in the case $u_- \le u_+$ is

$$u(x,t) \begin{cases} \frac{u_- + \kappa_- x}{1+\kappa_- t} & \text{for } x \le u_- t \\ \frac{x}{t} & \text{for } u_- t \le x \le u_+ t, \\ \frac{u_+ + \kappa_+ x}{1+\kappa_+ t} & \text{for } x \ge u_+ t \end{cases} \quad \text{as long as } 1 + \kappa_- t > 0, 1 + \kappa_+ t > 0, \tag{7.23}$$

and in the case $u_- > u_+$ is

$$u(x,t) = \begin{cases} \frac{u_- + \kappa_- x}{1+\kappa_- t} & \text{for } x < g(t) \\ \frac{u_+ + \kappa_+ x}{1+\kappa_+ t} & \text{for } x > g(t) \end{cases}, \quad \text{as long as } 1 + \kappa_- t > 0, 1 + \kappa_+ t > 0, \tag{7.24}$$

where g satisfies a differential equation, corresponding to the Rankine–Hugoniot condition

---

[2] Louis BRUN, French mathematician. He worked at CEA (Commissariat à l'Énergie Atomique).

[3] Jean OVADIA, French mathematician. He worked at CEA (Commissariat à l'Énergie Atomique).

$$\frac{dg}{dt} = \frac{1}{2}\left(\frac{u_- + \kappa_- g(t)}{1 + \kappa_- t} + \frac{u_+ + \kappa_+ x}{1 + \kappa_+ t}\right), \text{ and } g(0) = 0, \tag{7.25}$$

whose solution is

$$g(t) = t\,\frac{u_- \sqrt{1 + \kappa_- t} + u_+ \sqrt{1 + \kappa_+ t}}{\sqrt{1 + \kappa_- t} + \sqrt{1 + \kappa_+ t}}, \quad \text{as long as } 1 + \kappa_- t > 0, 1 + \kappa_+ t > 0, \tag{7.26}$$

showing that the amplitude of the shock at time t is

$$u(g(t)_-, t) - u(g(t)_+, t) = \frac{u_+ - u_-}{\sqrt{(1 + \kappa_- t)(1 + \kappa_+ t)}}, \tag{7.27}$$
$$\text{as long as } 1 + \kappa_- t > 0, 1 + \kappa_+ t > 0.$$

[Taught on Wednesday September 12, 2001.]

8

The Burgers Equation: Small Perturbations; the Heat Equation

The Burgers equation is a good example for pointing out that one should be careful about throwing away some terms because one thinks that they are small.

Let us assume that for some $\varepsilon > 0$, an initial datum satisfies

$$a - \varepsilon \leq u_0 \leq a + \varepsilon \text{ in } \mathbb{R}, \tag{8.1}$$

so that the correct solution of the Burgers equation

$$u_t + \left(\frac{u^2}{2}\right)_x = 0 \text{ in } \mathbb{R} \times (0, \infty); \ u(\cdot, 0) = u_0 \text{ in } \mathbb{R} \tag{8.2}$$

also satisfies

$$a - \varepsilon \leq u \leq a + \varepsilon \text{ in } \mathbb{R} \times (0, \infty). \tag{8.3}$$

With the usual tacit assumption that ε is a small quantity, it may seem reasonable to use the linearization

$$u = a + v, \text{ and } u^2 \approx a^2 + 2a\,v, \tag{8.4}$$

arguing that one keeps the term $2a\,v$ which is $O(\varepsilon)$, but one rejects the term v^2, which is $O(\varepsilon^2)$.

Mathematicians tend to be bothered by hasty simplifications which physicists and engineers do without guilt because they "know" that some terms are small; of course, in doing that they implicitly assume that the mathematical model is a good approximation of reality so that they believe that what one has observed in reality is a property that the model has, but mathematicians insist that one should prove that the mathematical model possesses some property and that one should not postulate it. After all, it could happen that the equation is not a good model of reality, and the only way to prove that it is not a good model is precisely to show that it does not possess a particular property which is observed. However, one should remember that every model has some limitations, and that a bad model may still be useful if one uses it under the conditions for which it is known to be good, and this knowledge may come from having performed a precise mathematical analysis.

Here one has a precise bound for the dropped term v^2, which one deems small because one norm of it is small, and one may think that v is near w given by

$$w_t + a\,w_x = 0; \ w(\cdot,0) = v_0, \text{ i.e. } w(x,t) = v_0(x - a\,t). \tag{8.5}$$

However, if one uses a Galilean transformation

$$u(x,t) = a + U(x - a\,t, t), \tag{8.6}$$

i.e. one moves with velocity a, then U solves the Burgers equation with initial data v_0, while the chosen approximation w is a constant equal to v_0, so that it is actually a quite bad approximation. One may think that the approximation is good for a small time, but looking more precisely into the matter shows that it is very dependent upon the regularity of v_0, and one then learns that before discarding a term one should first try to understand what norm one should use for measuring how small this term is. For example, if $v_0 = \varepsilon$, then $U = \varepsilon$, but if

$$v_0 = \varepsilon\,\varphi_0\Big(\frac{x}{\varepsilon^m}\Big), \tag{8.7}$$

then

$$U(x,t) = \varepsilon\,\varphi\Big(\frac{x}{\varepsilon^m}, \frac{t}{\varepsilon^{m-1}}\Big), \tag{8.8}$$

where φ is the solution of the Burgers equation with initial datum φ_0, so if φ_0 is periodic with average 0, φ decays in $O\big(\frac{1}{t}\big)$, giving the estimate $O\big(\frac{\varepsilon^m}{t}\big)$ for U, and one should observe that for $m \geq 1$ the function v_0 is small but not its derivative (high derivatives mean quick formation of discontinuities, which may imply rapid decay due to irreversible effects).

In 1987, I had suggested that noise in $\frac{1}{f}$, i.e. inversely proportional to frequency could be related to the tendency of the Burgers equation to create solutions with a triangular shape, and I had mentioned it first to James GLIMM and then in a letter sent for refereeing a note on the subject, to Paul GERMAIN and Alfred JOST,[1,2] because they correspond to Fourier transforms decaying in $\frac{1}{\xi}$. If a line of transmission is not exactly linear, but one makes the assumption that it is linear because the deviation from linearity seems small, one may have neglected small quadratic terms which have an effect which is not completely negligible, and in particular create small triangular-shaped

---

[1] Paul GERMAIN, French mathematician, born in 1920. He worked at Université Paris VI (Pierre et Marie Curie), Paris, and at ONERA (Office National d'Études et de Recherches Aéronautiques), Châtillon, France.

[2] Alfred JOST, French biologist, 1916–1991. He had worked in Paris, France, holding a chair (physiologie du développement, 1947–1987) at Collège de France, Paris.

structures in the solution, which it could be interesting to filter, but for which Fourier analysis does not seem the right thing to do.

Of course, there is a tendency to call noise whatever it is that one does not understand,[3] and the reason why one does not understand it is because one applies linear tools in situations where there are nonlinear effects, which one has too quickly postulated to be negligible, so that one finds oneself working on oversimplified equations that cannot explain what is observed. It is certainly not by an invocation of probabilistic methods that one can correct previous mistakes, and the dogma that there are probabilities in the laws of nature may be seen as a silly invention of people who wanted to hide the fact that they had not understood what they were doing.

The N-waves appearing in the Burgers equation have the curious effect of having propagated of a length of order $\sqrt{t}$ after time t, instead of a length of order t as one expects for linear waves, and I think that many observed effects of this type have been wrongly attributed to linear diffusion effects, like for the heat equation, postulated by FOURIER, or the effects of viscosity, considered by NAVIER.

In connection with the heat equation, there is an important probabilistic game which I have to describe for criticizing it, called "Brownian" motion,[4] which is not really related to what BROWN had observed, but it is related to my subject of kinetic theory. BROWN had observed pollen under a microscope, and he had noticed some erratic motions, which were wrongly interpreted as jumps in position, while they actually were the results of jumps in velocity due to collisions with much smaller particles. With the interpretation of random walks, which correspond to nonphysical jumps in position, the first mathematical analysis was the work of BACHELIER in 1900,[5] although the objection to jumps in position does not apply to his work, as he was interested in questions of finance, and the effects of buyers and sellers on the stock market are reasonably well simulated by a random walk model, but I heard from David HEATH that there seems to be a slight asymmetry,[6] and the stock prices seem to fall a little faster than they have risen. Of course, it is important to have a large number of customers, so that some kind of effective behaviour can be

---

[3] Real noise is related to acoustic effects, which seem to come from nonperiodic phenomena, which the human ear does not seem to process in the same way than music (although the sounds of nature like running water or wind are not classified as music, but are not described as noise either). The creation of sound in hydrodynamic phenomena is often neglected, but it is one way for the energy dissipated to be transported away.

[4] After having called it (mathematical) Brownian motion, I prefer to call it now "Brownian" motion, and the presence of quotes serves as a reminder that the name is not well chosen.

[5] Louis BACHELIER, French mathematician, 1870–1946. He had worked in Besançon, in Dijon, in Rennes, France, and in Besançon again.

[6] David HEATH, American mathematician. He worked at Cornell University, Ithaca, NY, and at CMU (Carnegie Mellon University), Pittsburgh, PA.

observed, and to improve the model one may need a better understanding of the possible behaviour of buyers and sellers. However, when EINSTEIN worked on the subject in 1905, I believe that he did not point out the nonphysical character of jumps in position, which require an infinite speed, while jumps in velocity are more realistic, although they require an infinite acceleration, so that instantaneous jumps in velocity are an idealization, as I shall discuss later in the study of collisions.

Since the work of POINCARÉ on relativity, before that of EINSTEIN, it is understood that no information can travel faster than the velocity of light c. Some physicists dispute that idea, mostly because they do not understand what errors have been made in inventing the rules of *quantum mechanics*. A first error is to rely on the *Schrödinger equation*, which has the same defect as the Fourier heat equation invented more than a century earlier, that it is a postulated equation which corresponds to physical models where one has (unknowingly) let the velocity of light c tend to $+\infty$, and that it is a logical mistake to use the real value of c in these models; a second error is that only waves exist at a microscopic level and that it is because physicists were sticking to 19th century ideas, like using particles for describing what happens in a gas, which were wrongly imposed on 20th century physics, that some of the silly games of quantum mechanics have been invented.[7]

According to the basic point of view on physics just described, there are no jumps in position in nature, but the "Brownian" motion is just one way to derive the heat equation

$$u_t - \kappa \, \Delta u = 0, \tag{8.9}$$

whose defects I shall emphasize. It was WIENER who developed the mathematical theory which most mathematicians call Brownian motion,[8] and which I call "Brownian" motion to recall that it is not related to what BROWN had observed,[9] and which probabilists like, a little too much in my opinion, but what BROWN had observed were jumps in velocity, which are related to the

---

[7] Mathematically, the problem is not to start from some Hamiltonian and derive a partial differential equation like the Schrödinger equation, in too many variables so that some other dogma has to be used for getting a reasonable equation with $\mathbf{x} \in \mathbb{R}^3$ or $(\mathbf{x}, \mathbf{v}) \in \mathbb{R}^3 \times \mathbb{R}^3$, but to start from semi-linear hyperbolic systems, like the Dirac equation (from which the Schrödinger equation can be derived by letting the velocity of light c tend to $+\infty$), and to show that in the limit of infinite frequencies waves are reasonably described by simpler models, for which one may use the interpretation of "idealized particles", elementary or not.

[8] Norbert G. WIENER, American mathematician, 1894–1964. He had worked at MIT (Massachusetts Institute of Technology), Cambridge, MA.

[9] It may have been called Brownian motion by EINSTEIN, who was obviously not such a good physicist that he could mistake jumps in velocity for jumps in position.

Fokker–Planck equation,[10] where the unknown is a function of position x, velocity v, and time t,

$$f_t - v.f_x - \kappa \, \Delta_v f = 0, \tag{8.10}$$

but there is an unfortunate tendency now, which may have started among probabilists, to call the Fokker–Planck equation any diffusion equation with a drift, even if there is only one variable x and no velocity variable v.[11] The Fokker–Planck equation can be created by a different probabilistic game, named after ORNSTEIN and UHLENBECK.[12,13]

Forgetting now about the question of physical relevance of the model, the heat equation can be solved by convolution (in x) of the initial datum with an elementary solution E, which one computes easily, either by using the Fourier transform, or by looking for a radial function of the form

$$E(x,t) = \begin{cases} \frac{1}{t^{N/2}} f\left(\frac{r}{\sqrt{\kappa t}}\right) \text{ for } t > 0 \\ 0 \text{ for } t < 0 \end{cases}, \tag{8.11}$$

and this form is guessed from the fact that the heat equation is invariant by rotation and by the scalings $a \, u(b \, x, b^2 \, t)$, and that to have the integral of u independent of t one must choose $a = b^N$; of course, κ has the dimension $length^2 \, time^{-1}$, and although mathematically the argument of scaling introduces the quantity $\frac{x}{\sqrt{t}}$, it is better to use $\frac{x}{\sqrt{\kappa t}}$, which is a dimensionless quantity; writing that E satisfies the heat equation for $t > 0$ gives a differential equation in f, easily integrated, and a constant of integration is imposed by the fact that one wants to have $\int_{\mathbb{R}^N} E(x,t) \, dx = 1$ for $t > 0$ (so that $E(\cdot, t)$ converges to δ_0 as t tends to 0), and this gives

$$E(x,t) = \begin{cases} \frac{1}{(4\pi \, \kappa \, t)^{N/2}} e^{-\frac{|x|^2}{4\kappa \, t}} \text{ for } t > 0 \\ 0 \text{ for } t < 0 \end{cases}. \tag{8.12}$$

---

[10] Adriaan Daniël FOKKER, Indonesian-born Dutch physicist and composer, 1887–1972. He had worked in Leiden, The Netherlands. He wrote music under the pseudonym Arie DE KLEIN.

[11] I wonder if this is intentional sabotage, because those who use this terminology forget to mention the difference between a diffusion in space and a diffusion in velocity, and their wrong use of a name from kinetic theory only induces students to talk about things that they do not know, as they are told nothing about kinetic theory.

[12] Leonard Samuel ORNSTEIN, Dutch physicist, 1880–1941. He had worked in Utrecht, The Netherlands.

[13] George Eugene UHLENBECK, Indonesian-born Dutch physicist, 1900–1988. He received the Wolf Prize (in Physics) in 1979, for his discovery, jointly with the late S. A. GOUDSMIT, of the electron spin, jointly with Giuseppe OCCHIALINI. He had worked at University of Michigan, Ann Arbor, MI, in Utrecht, The Netherlands, at Columbia University, New York, NY, at MIT (Massachusetts Institute of Technology), Cambridge, MA, at Princeton University, Princeton, NJ, and at the Rockefeller Institute, New York, NY.

At each positive time, the shape of E is that of a Gaussian, which is isotropic (i.e. invariant by rotations), but I shall show later a family of solutions of the heat equation which are anisotropic Gaussians, i.e. exponentials of quadratic functions whose principal part is not necessarily proportional to $|x|^2$.

It is important to notice that Gaussian functions arise naturally in a few mathematical problems, which are not all related, and one should not deduce that if a Gaussian occurs there must be a diffusion behind; actually, stationary solutions of the Boltzmann equation are Gaussian functions in velocity (this corresponds to independent work by MAXWELL and by BOLTZMANN), and the physical rule behind the Boltzmann equation is not diffusion in velocity, as for the Fokker–Planck equation, and although the linear Fokker–Planck equation does not have constant coefficient,[14] so that its Green function is not translation invariant,[15] it also involves Gaussians.

The heat equation has a smoothing effect, and this can be proven by observing that E has all its derivatives in x integrable (with the L^1 norms being powers of $\frac{1}{\sqrt{t}}$). Although the Fokker–Planck equation has only a diffusion in velocity, there is a combined effect of the transport part and the diffusion part for smoothing (less rapidly) in all directions. I first heard about this equation in the late 1960s at the Lions–Schwartz seminar, in a talk describing a work of Lars HÖRMANDER on a class of hypoelliptic operators (i.e. whose solutions must be smooth where the data is smooth), and although his work was very general, his article refers to a previous work of KOLMOGOROV on the smoothness of solutions of the Fokker–Planck equation.[16]

Another simple approach for solving the heat equation is to use a numerical scheme, and for simplification I consider the one-dimensional heat equation

$$u_t - \kappa\, u_{xx} = 0, \tag{8.13}$$

and the explicit difference scheme

---

[14] Physicists also use a nonlinear Fokker–Planck equation, either by deriving it from the Boltzmann equation, which is quite illogical, because the Fokker–Planck equation is about near collisions which are not well described by the postulated Boltzmann equation, or by postulating a nonconstant diffusion in velocity for the purpose of having Gaussians in velocity become solutions, which is dogma at its worst: dogmas are always introduced by people who fear rational thinking and want to force others to obey the rules that they believe in, the silly ones as well as the interesting ones.

[15] George GREEN, English mathematician, 1793–1841. He had been a miller and had never held any academic position.

[16] Andrey Nikolaevich KOLMOGOROV, Russian mathematician, 1903–1987. He received the Wolf Prize in 1980, for deep and original discoveries in Fourier analysis, probability theory, ergodic theory and dynamical systems, jointly with Henri CARTAN. He had worked in Moscow, Russia.

$$\frac{U_i^{n+1} - U_i^n}{\Delta t} - \kappa \frac{U_{i-1}^n - 2U_i^n + U_{i+1}^n}{(\Delta x)^2} = 0 \text{ for } i \in \mathbb{Z}, n \geq 0, \tag{8.14}$$

and the choice of the discretization is explained by using the Taylor expansion,[17] for a smooth function.[18] Writing the scheme as

$$U_i^{n+1} = \alpha\, U_{i-1}^n + (1 - 2\alpha)U_i^n + \alpha\, U_{i+1}^n, \text{ with } \alpha = \frac{\kappa \Delta t}{(\Delta x)^2}, \tag{8.15}$$

one sees that the condition

$$\frac{\kappa \Delta t}{(\Delta x)^2} \leq \frac{1}{2} \tag{8.16}$$

is a sufficient condition to obtain a nonnegative approximation for a nonnegative datum, i.e. that $U_i^0 \geq 0$ for all i implies $U_i^n \geq 0$ for all $i \in \mathbb{Z}, n \geq 0$, but also gives stability in $L^1(\mathbb{R})$ and $L^\infty(\mathbb{R})$, and therefore in $L^p(\mathbb{R})$ by interpolation for every $p \in [1, \infty]$; this condition will be seen to be necessary to obtain stability in $L^2(\mathbb{R})$, and the proof uses Fourier series.

There is a probabilistic interpretation of the preceding scheme, related to a random walk where one jumps from i to $i - 1$ or to $i + 1$ with probability α, and remaining at i with probability $(1 - 2\alpha)$, all these jumps being done independently, and U_i^n can be expressed as the expectation of U if one starts at $n = 0$ with the values U_j^0 for all j; I consider the idea of "Brownian" motion related to this remark.

It should be mentioned that FEYNMAN has developed probabilistic ideas for the Schrödinger equation, mixed with ideas of diagrams, which no mathematician understands.[19]

A first observation is that one can use difference schemes similar to (8.14) for the Schrödinger equation, and that there is a similar stability condition, but the analysis does not rely on positivity or probabilities, and it is based on L^2 estimates proven by using a Fourier series; actually, very few schemes used in numerical analysis can be interpreted by probabilistic arguments, and I see as an important limitation that probabilists only work on partial differential equations showing positivity properties.

A second observation comes from my work in the early 1980s about nonlocal effects induced by homogenization, which I had started because I thought

---

[17] Brook TAYLOR, English mathematician, 1685–1731. He had worked in London, England.

[18] If φ is smooth, then $\varphi(x \pm \Delta x) = \varphi(x) \pm \Delta x\, \varphi'(x) + \frac{(\Delta x)^2}{2} \varphi''(x) + O((\Delta x)^3)$, which shows that $\frac{\varphi(x-\Delta x) - 2\varphi(x) + \varphi(x+\Delta x)}{(\Delta x)^2} = \varphi''(x) + O(\Delta x)$.

[19] I have heard propositions to have a complex time in the construction of "Brownian" motion, which I find silly. I have asked James GLIMM what to read for learning about diagrams, and I interpreted his answer, that there was no article for that, as meaning that no one has definitions, and that it is a state of mind, hard to understand by mathematicians, who are not used to playing games before learning about their rules.

that the physicists' rules of spontaneous absorption or emission of "particles" that have been invented for explaining spectroscopy experiments are just a way of describing a nonlocal effect in an effective equation, whose form is still unknown. In the late 1980s, when I tried to attack a nonlinear model, I stumbled on a question of how to handle too many terms in an expansion, and how to explain the convergence of such an expansion, and I had the feeling that FEYNMAN's method of diagrams might have been his answer for describing a nonlocal effect and handling a similar bookkeeping problem, so that his use of probabilities had nothing to do with the way they are used in "Brownian" motion.

My approach is not to try to understand the precise rules of the games that physicists play, because they are able to play games without learning first about all the rules, so that they actually play lots of variants, and if a variant seems to give a good result they may incorporate some of its rules into their own game. Some mathematicians try to put some order into the list of all these rules that various physicists use, but that corresponds to being interested in physicists' problems and not necessarily in physics questions.

I advocate trying to understand what the physics problems are, behind the games that physicists play, and then developing the necessary mathematical tools for solving the problem that one has selected, and checking meanwhile if one still has reasons to believe that it will be useful for explaining a piece of physics. One may also discover something for one purpose and once it is partly done one may observe that it is useful for explaining something else, and for example the theory of distributions of Laurent SCHWARTZ came out of the question of defining $\sum_{n \in \mathbb{Z}} c_n e^{inx}$ for coefficients c_n with polynomial growth, and once it was understood it explained some of the formal computations of DIRAC.

There is a difference between research and development, and one should not be dogmatic about the way to do research, because one looks in part for new ideas, which no one may have thought of, but certainly one should be able to explain why one spends some time working on a problem.

[Taught on Friday September 14, 2001.]

Notes on names cited in footnotes for Chapter 8, GOUDSMIT,[20] OCCHIALINI,[21] ROCKEFELLER,[22]

---

[20] Samuel Abraham GOUDSMIT, Dutch physicist, 1902–1978. He had worked at University of Michigan, Ann Arbor, MI.

[21] Giuseppe OCCHIALINI, Italian physicist, 1907–1993. He received the Wolf Prize (in Physics) in 1979, for his contributions to the discoveries of electron pair production and of the charged pion, jointly with George Eugene UHLENBECK. He had worked in Genova (Genoa) and in Milano (Milan), Italy.

[22] John Davison ROCKEFELLER Sr., American industrialist and philanthropist, 1839–1937. The Rockefeller Institute, New York, NY, is named after him.

Henri CARTAN,[23] and for the preceding footnotes, É. CARTAN.[24]

---

[23] Henri Paul CARTAN, French mathematician, born in 1904. He received the Wolf Prize in 1980, for pioneering work in algebraic topology, complex variables, and homological algebra, and inspired leadership of a generation of mathematicians, jointly with Andrey N. KOLMOGOROV. He worked in Paris and at Université Paris-Sud, Orsay, France, retiring in 1975 just before I was hired there. Theorems attributed to CARTAN are often the work of his father E. CARTAN.

[24] Élie Joseph CARTAN, French mathematician, 1869–1951. He had worked in Paris, France.

9

Fourier Transform; the Asymptotic Behaviour for the Heat Equation

An important tool for studying partial differential equations with constant coefficients (like the heat equation $u_t - \kappa \Delta u = f$ in the whole space $x \in \mathbb{R}^N$, with usually $N = 1, 2, 3$ in applications, and with an initial condition) is the Fourier transform. Initially, using Laurent SCHWARTZ notation,[1] one defines the Fourier transform $\mathcal{F}$, as well as $\overline{\mathcal{F}}$, for functions $f \in L^1(\mathbb{R}^N)$ by

$$\mathcal{F}f(\xi) = \int_{\mathbb{R}^N} f(x)e^{-2i\pi(x.\xi)}\,dx \text{ and } \overline{\mathcal{F}}f(\xi) = \int_{\mathbb{R}^N} f(x)e^{+2i\pi(x.\xi)}\,dx, \quad (9.1)$$

so that $\overline{\mathcal{F}}\overline{f} = \overline{\mathcal{F}f}$ for every $f \in L^1(\mathbb{R}^N)$, and one notices that

$$|\mathcal{F}f(\xi)| \leq \int_{\mathbb{R}^N} |f(x)|\,dx \text{ for all } \xi \in \mathbb{R}^N, \text{ i.e. } ||\mathcal{F}f||_\infty \leq ||f||_1, \quad (9.2)$$

and the same properties hold for $\overline{\mathcal{F}}$ (which is actually the inverse of $\mathcal{F}$, an advantage of Laurent SCHWARTZ notation). A simple application of the Lebesgue dominated convergence theorem shows that $\mathcal{F}f$ is continuous; once one has shown that $\mathcal{F}f$ tends to 0 at infinity when $f \in C_c^\infty(\mathbb{R}^N)$ (or $\mathcal{S}(\mathbb{R}^N)$) one deduces by an argument of density that

$$f \in L^1(\mathbb{R}^N) \text{ implies } \mathcal{F}f \in C_0(\mathbb{R}^N), \quad (9.3)$$

where $C_0(\mathbb{R}^N)$ is the space of continuous functions tending to 0 at infinity, which is a Banach space,[2] when equipped with the sup norm; the same property holds for $\overline{\mathcal{F}}$.

---

[1] Specialists of harmonic analysis do not put the coefficient 2π in the definition, and a factor involving π appears in their inverse Fourier transform; one should then be careful in comparing formulas from various books, but at the end one finds the same solutions, of course. I do prefer the symmetry which appears in Laurent SCHWARTZ notation, but I am a little biased, because I was a student of Laurent SCHWARTZ at École Polytechnique in 1965–1966, and I have always used his notation.

[2] Stefan BANACH, Polish mathematician, 1892–1945. He had worked in Lwów (then in Poland, now Lvov, Ukraine). There is now a Stefan Banach International

If f is of class C^1 with f and $\frac{\partial f}{\partial x_j}$ belonging to $L^1(\mathbb{R}^N)$, an integration by parts shows that

$$\mathcal{F}\frac{\partial f}{\partial x_j}(\xi) = 2i\pi\xi_j\mathcal{F}f(\xi) \text{ for all } \xi \in \mathbb{R}^N, \tag{9.4}$$

while if f is of class C^1 with f and $|x|f$ belonging to $L^1(\mathbb{R}^N)$, Lebesgue dominated convergence shows that $\mathcal{F}f$ is of class C^1 and

$$\mathcal{F}(-2i\pi x_j f)(\xi) = \frac{\partial(\mathcal{F}f)}{\partial\xi_j}(\xi) \text{ for } j = 1,\dots,N \text{ and all } \xi \in \mathbb{R}^N. \tag{9.5}$$

Reiteration of these two properties led Laurent SCHWARTZ to introduce the Schwartz space $\mathcal{S}(\mathbb{R}^N)$ of C^∞ functions f such that for every derivative D^α of any order and every polynomial P of any degree one has $PD^\alpha f \in L^1(\mathbb{R}^N)$ (or equivalently all these products are asked to be bounded); then the Fourier transform maps continuously $\mathcal{S}(\mathbb{R}^N)$ into itself (as well as its inverse $\overline{\mathcal{F}}$).[3] Laurent SCHWARTZ extended then Fourier transform to $\mathcal{S}'(\mathbb{R}^N)$ (whose elements are called tempered distributions), the dual of $\mathcal{S}(\mathbb{R}^N)$, by noticing the formula

$$\int_{\mathbb{R}^N}(\mathcal{F}f)g\,d\xi = \int_{\mathbb{R}^N\times\mathbb{R}^N}f(x)g(\xi)e^{-2i\pi(x.\xi)}\,dx\,d\xi = \int_{\mathbb{R}^N}(\mathcal{F}g)f\,dx \atop \text{for all } f,g \in L^1(\mathbb{R}^N), \tag{9.6}$$

which is proven by Fubini theorem,[4] and defined the extension by

$$\langle\mathcal{F}T,\varphi\rangle = \langle T,\mathcal{F}\varphi\rangle \text{ for all } T \in \mathcal{S}'(\mathbb{R}^N) \text{ and all } \varphi \in \mathcal{S}(\mathbb{R}^N). \tag{9.7}$$

As derivations and multiplication by polynomials map $\mathcal{S}(\mathbb{R}^N)$ into itself, one deduces that

$$\mathcal{F}\frac{\partial T}{\partial x_j} = 2i\pi\xi_j\mathcal{F}T \text{ and } \mathcal{F}(-2i\pi x_j T) = \frac{\partial(\mathcal{F}T)}{\partial\xi_j} \atop \text{for all } T \in \mathcal{S}'(\mathbb{R}^N) \text{ and } j = 1,\dots,N. \tag{9.8}$$

As $1 \in \mathcal{S}'(\mathbb{R}^N)$ and $\frac{\partial 1}{\partial x_j} = 0$ for $j = 1,\dots,N$, one deduces that $\xi_j\mathcal{F}1 = 0$ for $j = 1,\dots,N$, and this implies that $\mathcal{F}1 = A\delta_0$, and one finds that $A = 1$ by using the Gaussian function $G(x) = e^{-\pi|x|^2}$, whose Fourier transform is

---

Mathematical Centre in Warsaw, Poland. The term Banach space was introduced by FRÉCHET.

[3] The natural topology for the Schwartz space $\mathcal{S}(\mathbb{R}^N)$, defined by the family of norms $||x^\beta D^\alpha f||_{L^1(\mathbb{R}^N)}$ for all multi-indices α, β, does not make it a Banach space, but only a Fréchet space, i.e. a locally convex space which is a complete metric space (I do not know who introduced the term Fréchet space). The Fourier transform $\mathcal{F}$ is continuous, together with its inverse $\overline{\mathcal{F}}$.

[4] Guido FUBINI, Italian-born mathematician, 1879–1943. He had worked in Catania, in Genova (Genoa), and in Torino (Turin), Italy, and then in New York, NY.

$\mathcal{F}G(\xi) = e^{-\pi |\xi|^2}$; this last result follows from the fact that $\frac{\partial G}{\partial x_j} = -2\pi\, x_j\, G$ for $j = 1, \ldots, N$, which implies that $\frac{\partial(\mathcal{F}G)}{\partial \xi_j} = -2\pi\, \xi_j\, \mathcal{F}G$ for $j = 1, \ldots, N$, so that $\mathcal{F}G(\xi) = B\, e^{-\pi |\xi|^2}$, and the fact that $B = 1$ follows by taking $\xi = 0$ and using $\int_{\mathbb{R}} e^{-\pi\, x^2}\, dx = 1.^5$ The fact that the inverse of $\mathcal{F}$ is $\overline{\mathcal{F}}$, either from $\mathcal{S}(\mathbb{R}^N)$ into itself or from $\mathcal{S}'(\mathbb{R}^N)$ into itself, is equivalent to the Plancherel formula[6]

$$\int_{\mathbb{R}^N} \mathcal{F}f(\xi)\overline{\mathcal{F}g(\xi)}\, d\xi = \int_{\mathbb{R}^N} f(x)\overline{g(x)}\, dx \text{ for all } f, g \in \mathcal{S}(\mathbb{R}^N), \qquad (9.9)$$

which permits us to extend $\mathcal{F}$ to $L^2(\mathbb{R}^N)$ into an isometry with inverse $\overline{\mathcal{F}}$. Indeed, using $h(\xi) = \overline{\mathcal{F}g(\xi)}$, the Plancherel formula means $\mathcal{F}\overline{\mathcal{F}}g = \overline{g}$ for every g, i.e. $\mathcal{F}\overline{\mathcal{F}} = I$ on $\mathcal{S}(\mathbb{R}^N)$ (which is the same as $\overline{\mathcal{F}}\mathcal{F} = I$ by complex conjugation); one has $\overline{\mathcal{F}}\mathcal{F}u(x) = \int_{\mathbb{R}^N} \mathcal{F}u(\xi)e^{+2i\pi (x.\xi)}\, d\xi = \int_{\mathbb{R}^N} e^{+2i\pi (x.\xi)}\left(\int_{\mathbb{R}^N} u(y)e^{-2i\pi (\xi.y)}\, dy\right)d\xi$, but the hypotheses of the Fubini theorem are not satisfied for exchanging the order of integrations,[7] and one notices instead that

$$\overline{\mathcal{F}}\mathcal{F}u(x) = \lim_{n\to\infty} \int_{\mathbb{R}^N} G\!\left(\frac{\xi}{n}\right)\mathcal{F}u(\xi)e^{+2i\pi (x.\xi)}\, d\xi$$
$$= \lim_{n\to\infty} \int_{\mathbb{R}^N} \int_{\mathbb{R}^N} G\!\left(\frac{\xi}{n}\right)u(y)e^{2i\pi (x-y.\xi)}\, dy\, d\xi \qquad (9.10)$$
$$= \lim_{n\to\infty} \int_{\mathbb{R}^N} u(y)n^N G\big(n(y-x)\big)\, dy = u(x),$$

where one has used a simple scaling property, that if $f \in L^1(\mathbb{R}^N)$ and $\lambda \neq 0$ and $g(x) = f(\lambda\, x)$, then $\mathcal{F}g(\xi) = \frac{1}{\lambda^N}\mathcal{F}\!\left(\frac{\xi}{\lambda}\right)$ for all $\xi \in \mathbb{R}^N$, a particular case of a linear change of variable in the definition of the Fourier transform, whose general form shows that for $f \in L^1(\mathbb{R}^N)$ and $A \in \mathcal{L}(\mathbb{R}^N; \mathbb{R}^N)$ invertible,

$$\text{if } g(x) = f(A\, x) \text{ then } \mathcal{F}g(\xi) = \frac{1}{|det(A)|}\mathcal{F}\big((A^{-1})^T\xi\big) \text{ for } \xi \in \mathbb{R}^N; \quad (9.11)$$

one has also used that G is its own Fourier transform, and that the intermediate result is the convolution of u by a smoothing sequence.

One important property of the Fourier transform is that it transforms convolution into multiplication[8]

---

[5] If $I = \int_{\mathbb{R}} e^{-\pi\, x^2}\, dx$, then $I^2 = \int_{\mathbb{R}\times\mathbb{R}} e^{-\pi(x^2+y^2)}\, dx\, dy$, and using polar coordinates (and $dx\, dy = r\, dr\, d\theta$) it is $\int_0^\infty \int_0^{2\pi} e^{-\pi\, r^2} r\, dr\, d\theta = \int_0^\infty 2\pi\, r\, e^{-\pi\, r^2}\, dr = 1$.

[6] Michel PLANCHEREL, Swiss mathematician, 1885–1967. He had worked in Genève (Geneva), in Fribourg, and at ETH (Eidgenössische Technische Hochschule), Zürich, Switzerland.

[7] It would make the formal quantity $\int_{\mathbb{R}^N} e^{-2i\pi(y-x.\xi)}\, d\xi$ appear, which physicists write as $\delta(y - x)$, and this corresponds to what happens if one wants to compute $\mathcal{F}1$ as if it was defined by an integral; however, the result is consistent with the fact that $\mathcal{F}1 = \delta_0$, which had to be proven otherwise, and for what concerns linear computations one can often (but not always!) transform a formal consideration into a proof by using a suitable regularization.

[8] As noticed by Laurent SCHWARTZ, derivations are convolutions with distributions having their support at 0, so that the property of $\mathcal{F}$ transforming derivations

for $f, g \in L^1(\mathbb{R}^N)$ and $h(x) = f \star g(x) = \int_{\mathbb{R}^N} f(y)g(x-y)\,dy$ (9.12)
one has $\mathcal{F}h(\xi) = \mathcal{F}f(\xi)\,\mathcal{F}g(\xi)$ for all $\xi \in \mathbb{R}^N$,

which is an easy consequence of the Fubini theorem, writing

$$
\begin{aligned}
\mathcal{F}h(\xi) &= \int_{\mathbb{R}^N} \left(\int_{\mathbb{R}^N} f(x-y)g(y)\,dy \right) e^{-2i\pi(x.\xi)}\,dx \\
&= \int_{\mathbb{R}^N} \left[g(y)e^{-2i\pi(y.\xi)} \left(\int_{\mathbb{R}^N} f(x-y)e^{-2i\pi(x-y.\xi)}\,dx \right) \right]\,dy \\
&= \int_{\mathbb{R}^N} g(y)e^{-2i\pi(y.\xi)}\mathcal{F}f(\xi)\,dy = \mathcal{F}f(\xi)\mathcal{F}g(\xi).
\end{aligned}
$$
(9.13)

The formula $\mathcal{F}(f \star g) = (\mathcal{F}f)(\mathcal{F}g)$ shows that $\mathcal{F}L^1(\mathbb{R}^N)$ is a multiplicative algebra (continuously embedded into $C_0(\mathbb{R}^N)$);[9] the formula extends to $f, g \in L^2(\mathbb{R}^N)$, and using the fact that $\mathcal{F}$ is a surjective isometry from $L^2(\mathbb{R}^N)$ onto itself, one deduces that $L^2(\mathbb{R}^N) \star L^2(\mathbb{R}^N) = \mathcal{F}L^1(\mathbb{R}^N)$.

The same formula holds for $\overline{\mathcal{F}}$, and using $u = \overline{\mathcal{F}}f$ and $v = \overline{\mathcal{F}}g$, one deduces that $\overline{\mathcal{F}}(f \star g) = \overline{\mathcal{F}}f\,\overline{\mathcal{F}}g = u\,v$, so that $\mathcal{F}(uv) = \mathcal{F}\overline{\mathcal{F}}(f \star g) = f \star g = \mathcal{F}u \star \mathcal{F}v$, but one should be careful about the hypotheses for u and v for such a formula $\mathcal{F}(uv) = \mathcal{F}u \star \mathcal{F}v$ to hold; the preceding proof works for $u, v \in \mathcal{F}L^1(\mathbb{R}^N)$, and it is easily proven also for $u, v \in L^2(\mathbb{R}^N)$.[10] As products and convolution products are not defined for all distributions, it may happen that one side of the equality is defined while there is no clear way of defining the other side directly; for example, the Heaviside function H is defined by $H(x) = 0$ for $x < 0$ and $H(x) = 1$ for $x > 0$, so that $H \in L^\infty(\mathbb{R}) \subset \mathcal{S}'(\mathbb{R})$, and as the derivative of H is δ_0, one finds that $2i\pi\xi\,\mathcal{F}H(\xi) = 1$ so that $\mathcal{F}H(\xi) = \frac{1}{2i\pi}pv.\left(\frac{1}{\xi}\right) + C\,\delta_0$, and the value of C is found if one notices that $H - \frac{1}{2}$ is real and odd,[11] so that its Fourier transform is purely imaginary and odd, giving $C = \frac{1}{2}$; although the product of H by itself is well defined, the definitions of convolutions for distributions do not allow both distributions to be $pv.\frac{1}{\xi}$.

Another approach for studying partial differential equations with constant coefficients like the heat equation (or generalizations like $u_t - \sum_{i,j=1}^{N} D_{i,j}u_{x_ix_j}$

---

into multiplications is the same property as $\mathcal{F}$ transforming convolutions into products.

[9] The word algebra comes from an Arabic word in the treatise *Hisab al-jabr w'al-muqabala* by AL KHWARIZMI, whose name has also been used for coining the word algorithm.

[10] The formula is also true for $u, v \in \mathcal{F}L^1(\mathbb{R}^N) + L^2(\mathbb{R}^N)$, in which case the two sides of the formula belong to $\mathcal{F}L^1(\mathbb{R}^N) + L^2(\mathbb{R}^N) + L^1(\mathbb{R}^N)$.

[11] For a smooth function f one defines $\check{f}$ by $\check{f}(x) = f(-x)$, and for a distribution T one defines $\check{T}$ by $\langle \check{T}, \varphi \rangle = \langle T, \check{\varphi} \rangle$ for every $\varphi \in C_c^\infty(\mathbb{R}^N)$, so that T even means $\check{T} = T$ and T odd means $\check{T} = -T$; then δ_0 is even and $pv.\left(\frac{1}{x}\right)$ is odd; pv. stands for *principal value*, and $pv.\left(\frac{1}{x}\right)$ is the only odd distribution T in $\mathbb{R}$ satisfying $x\,T = 1$, and it is defined more precisely by $\left\langle pv.\left(\frac{1}{x}\right), \varphi \right\rangle = \lim_{n \to \infty}\left(\int_{-\infty}^{-1/n} \frac{\varphi(x)}{x}\,dx + \int_{1/n}^{+\infty} \frac{\varphi(x)}{x}\,dx \right)$ for every $\varphi \in C_c^\infty(\mathbb{R})$, and this type of definition goes back to CAUCHY, and was generalized by HADAMARD before Laurent SCHWARTZ put it in the framework of distributions.

$= 0$, where the matrix D with entries $D_{i,j}$ is symmetric and positive definite, which corresponds to diffusion in an anisotropic medium), is the method of elementary solutions; first I shall deduce it by using the Fourier transform, in the simple case of the heat equation in an isotropic medium $u_t - \kappa \Delta u = 0$, in order to simplify the computations. One applies a partial Fourier transform, in x alone; this means that one is looking for a solution u which is continuous in t,[12] with values in $\mathcal{S}'(\mathbb{R}^N)$ for example,[13] and one finds that

$$(\mathcal{F}u)_t + 4\kappa\,\pi^2|\xi|^2\mathcal{F}u = 0, , \text{ with initial datum } \mathcal{F}u\,|_{t=0} = \mathcal{F}u_0, \qquad (9.14)$$

giving

$$\mathcal{F}u(\xi,t) = \mathcal{F}u_0(\xi)e^{-4\kappa\,\pi^2|\xi|^2 t} \text{ for } t \geq 0, \qquad (9.15)$$

and this means that

$$u(\cdot,t) = u_0 \star E(\cdot,t), \qquad (9.16)$$

where the elementary solution E is the inverse Fourier transform of the Gaussian $e^{-4\kappa\,\pi^2|\xi|^2 t}$,[14] as it is $G(\lambda\,\xi)$ with $\lambda = \sqrt{4\kappa\,\pi\,t}$, one finds

$$E(x,t) = \frac{1}{(4\kappa\,\pi\,t)^{N/2}}e^{-|x|^2/(4\kappa\,t)}. \qquad (9.17)$$

The elementary solution permits us to write the solution of $u_t - \kappa\Delta u = f$ with $u\,|_{t=0} = u_0$, by using the invariance of the equation by translations (in x and in t) and the linearity of the equation, and one obtains

$$u(x,t) = \int_{\mathbb{R}^N} E(x-y,t)u_0(y)\,dy + \int_0^t \left(\int_{\mathbb{R}^N} E(x-y,t-s)f(y,s)\,dy\right)ds. \qquad (9.18)$$

Laurent SCHWARTZ explained such formulas in the framework of his theory of distributions by noticing that, once one extends functions by 0 for $t < 0$, i.e.

$$\tilde{u}(x,t) = \begin{cases} 0 \text{ for } t < 0 \\ u(x,t) \text{ for } t > 0, \end{cases} \qquad (9.19)$$

one has

$$\tilde{u}_t - \kappa\Delta\tilde{u} = \tilde{f} + u_0 \otimes \delta_0 \qquad (9.20)$$

---

[12] Some kind of continuity is required in order to give a meaning to the value at a particular time, so that imposing an inital datum makes sense.

[13] One could also look for a solution in $\mathcal{S}'(\mathbb{R}^N \times \mathbb{R})$. One should notice that the functions $e^{\alpha\,t + \sum_j \beta_j x_j}$ are solutions of the equation if $\alpha = \kappa\sum_j \beta_j^2$, but except if $\alpha = 0$ and $\beta_j = 0$ for all j, these functions are not tempered distributions, and one must be careful about talking of their Fourier transform.

[14] One should say an elementary solution, but if one asks that it be 0 for $t < 0$ and that it belongs to $\mathcal{S}'(\mathbb{R}^N \times \mathbb{R})$, then one finds only one.

and that the elementary solution is satisfying[15]

$$\widetilde{E}_t - \kappa \, \Delta \, \widetilde{E} = \delta_{(0,0)} = \delta_0 \otimes \delta_0, \qquad (9.21)$$

and the formula for u is just the same as

$$\widetilde{u} = \widetilde{E} \star_{(x,t)} (\widetilde{f} + u_0 \otimes \delta_0), \qquad (9.22)$$

where $\star_{(x,t)}$ serves to emphasize that the convolution is in both variables x and t. Another way to write such formulas is the semi-group approach, where one defines $S(t)$ acting on functions in x for $t > 0$ by

$$S(t)v = E(\cdot, t) \star_x v, \qquad (9.23)$$

and the formula becomes

$$u(\cdot, t) = S(t)u_0 + \int_0^t S(t - s)f(\cdot, s) \, ds. \qquad (9.24)$$

There is another way to discover the formula for the elementary solution E, which is to observe that the equation is invariant by the group of transformations $b \, u(a \, P \, x, a^2 t)$, where $a, b \in \mathbb{R}$ and $P \in SO(N)$ is a rotation, and that by formally integrating in x one wants the integral $\int_{\mathbb{R}^N} u(x, t) \, dx$ to be independent of t, and the integral is conserved by the transformation if $b = a^N$; a self-similar solution is then such that[16]

$$u(x,t) = a^N u(a \, P \, x, a^2 t) \text{ for all } a, P \text{ means } u(x,t) = \frac{1}{t^{N/2}} f\Big(\frac{|x|}{\sqrt{\kappa \, t}}\Big). \quad (9.25)$$

Using the notation $\sigma = \frac{x}{\sqrt{\kappa \, t}}$, the equation is $-\frac{N}{2}f - \frac{\sigma}{2}f' - f'' - \frac{N-1}{\sigma}f' = 0$, and it is natural to start with the case $N = 1$, which can be integrated immediately into $\frac{\sigma}{2}f - f' = A$, giving $f = B \, e^{-\sigma^2/4}$ for $A = 0$, the other solutions being singular at the origin; then one observes that the coefficient of N is 0 for that particular solution, so it is a solution for every N.

There is a larger class of explicit solutions of the heat equation, which I call anisotropic Gaussians, for which the action of the group of translations

---

[15] In a tensor product $\mu \otimes \nu$, μ is a Radon measure or a distribution in x while ν is a Radon measure or a distribution in t; of course, one may write $\delta_0(x)$ and $\delta_0(t)$ in order to see more clearly what variables are used, but it is better to avoid the physicists' notation of Dirac masses as if they were functions. The uncertainty about notation is analogous to that created by denoting by 0 the null vector in every vector space, but engineers and physicists often like to write $\underline{u}$ for a vector and $\underline{\underline{u}}$ for a tensor, etc., while mathematicians only mention a problem of notation when a formula could be naturally interpreted in different ways.

[16] The reason why one prefers $\frac{x}{\sqrt{\kappa \, t}}$ to $\frac{x}{\sqrt{t}}$ is that it is a quantity without dimension, because the only parameter κ appearing in the equation has a dimension $length^2 \, time^{-1}$.

and rotations is nontrivial, and I shall describe them in a moment, but the reason why I was led to study this class in the mid 1980s was a remark found in a book by ZEL'DOVICH & RAIZER,[17,18] concerning a quasi-linear version of the equation (used for high-temperature phenomena). Considering the case $f = 0$, one has a first $L^\infty(\mathbb{R}^N)$ bound

$$|u(x,t)| \leq \frac{C_0}{t^{N/2}} \text{ for } x \in \mathbb{R}^N, t > 0, \tag{9.26}$$

and C_0 can be taken proportional to $\|u_0\|_{L^1(\mathbb{R}^N)}$, but ZEL'DOVICH & RAIZER argued that for large time all the information contained in the initial datum has diffused far away, and from afar the support of u_0 looks like a point, so the solution looks like the elementary solution, and this idea leads to the better approximation for large t,

$$|u(x,t) - M\,E(x,t)| \leq \frac{C_1}{t^{(N+1)/2}} \text{ for } x \in \mathbb{R}^N, t > 0, \text{ where } M = \int_{\mathbb{R}^N} u_0\,dx; \tag{9.27}$$

then they argued that there is no reason to put the information at an arbitrary point like 0, and because in the case where u_0 is thought of as a density of mass the only natural point is the centre of mass, one is led in the case $M \neq 0$ to a better approximation for large t,

$$|u(x,t) - M\,E(x - x_*, t)| \leq \frac{C_2}{t^{(N+2)/2}} \text{ for } x \in \mathbb{R}^N, t > 0, \text{ where } M\,x_{*k} = \int_{\mathbb{R}^N} x_k u_0(x)\,dx \text{ for } k = 1, \dots, N; \tag{9.28}$$

then they argued that when t is large there is no reason to select the particular time $t = 0$, and that by comparing u to $M\,E(x - x_*, t - t_*)$ with a suitable t_* one may improve the asymptotic estimate; however, they were doing their computation in one space variable (for a nonlinear version $u_t - (u^m)_{xx} = 0$), but I was working in N space variables, and I found that their trick of choosing t_* does not always work, because of a question of anisotropy. In order to do this last step correctly, and to prove in a rigourous way the preceding statements, I found the following explanation.

Lemma 9.1. *If u_0 and v_0 are such that $\int_{\mathbb{R}^N}(1 + |x|^{k+1})\,|u_0|\,dx < \infty$, $\int_{\mathbb{R}^N}(1 + |x|^{k+1})\,|v_0|\,dx < \infty$ and*

$$\int_{\mathbb{R}^N}(u_0 - v_0)P\,dx = 0 \text{ for all polynomials } P \text{ of degree } \leq k, \tag{9.29}$$

then, denoting by u and v the solutions with initial data u_0 and v_0, one has

$$|u(x,t) - v(x,t)| \leq \frac{C_k}{t^{(N+k+1)/2}} \text{ for } x \in \mathbb{R}^N, t > 0. \tag{9.30}$$

---

[17] Yakov Borisovich ZEL'DOVICH, Russian physicist, 1914–1987. He had worked at Lomonosov State University, Moscow, Russia.
[18] Yuri Petrovich RAIZER, Russian physicist.

Proof: One has $|u(x,t) - v(x,t)| \leq \int_{\mathbb{R}^N} |\mathcal{F}(u-v)(\xi,t)| \, d\xi$, and $\mathcal{F}(u-v)(\xi,t) = e^{-4\kappa|\xi|^2 t} \mathcal{F}(u_0 - v_0)(\xi)$. The hypothesis implies that the function $w = \mathcal{F}(u_0 - v_0)$ has bounded derivarives of order $k+1$ and that all its derivatives of order $\leq k$ at 0 are 0, so that $|\mathcal{F}(u_0 - v_0)(\xi)| \leq C|\xi|^{k+1}$ for all $\xi \in \mathbb{R}^N$, and then one needs to compute $\int_{\mathbb{R}^N} |\xi|^{k+1} e^{-4\kappa|\xi|^2 t} \, d\xi$, for which the change of variable $\xi = \frac{\eta}{\sqrt{t}}$ gives the desired estimate. $\square$

Lemma 9.1 is also valid if v_0 is a Radon measure with finite total mass, and the choice $v_0 = M \delta_a$ with $M = \int_{\mathbb{R}^N} u_0 \, dx$ serves to have the same moments of order 0, whatever the point a is, but the moments of order 1 agree if and only if one chooses $a = x_*$ (in the case $M \neq 0$), and in order to have the second moments agree, one must look at the analogue of a matrix of inertia J given by

$$J_{j,k} = \int_{\mathbb{R}^N} (x - x_{*j})(x - x_{*k}) u_0(x) \, dx, \text{ for } j, k = 1, \dots, N, \tag{9.31}$$

which cannot agree with those of a Gaussian function $v_0(x) = M E(x - x_*, -t_*)$ (for a choice $t_* < 0$), unless J is proportional to I. For more general cases it is useful then to know explicit solutions for which the moments of order up to 2 are known, and I used for that a family of solutions of the form $v = e^{\text{quadratic}(x)}$. Besides $\int_{\mathbb{R}} e^{-\pi x^2} \, dx = 1$, which gives $\int_{\mathbb{R}} e^{-\alpha x^2} \, dx = \sqrt{\frac{\pi}{\alpha}}$ by rescaling, one also uses $\int_{\mathbb{R}} x^2 e^{-\alpha x^2} \, dx = \frac{\sqrt{\pi}}{2\alpha^{3/2}}$ by integration by parts; for A symmetric positive definite and B symmetric, one deduces the following formulas:

$$\begin{aligned} \int_{\mathbb{R}^N} e^{-(A x.x)} \, dx &= \frac{\pi^{N/2}}{\sqrt{\det(A)}} \\ \int_{\mathbb{R}^N} (B x.x) e^{-(A x.x)} \, dx &= \frac{\pi^{N/2}}{2\sqrt{\det(A)}} \, trace(A^{-1} B), \end{aligned} \tag{9.32}$$

these integrals being easily computed in an orthonormal basis of eigenvectors of A (using the invariance by rotations of the Lebesgue measure), the second integral appearing to be the first one multiplied by $\sum_{i=1}^{N} \frac{B_{i,i}}{2\alpha_i}$ where $\alpha_1, \dots, \alpha_N$ are the eigenvalues of A, which one must write in an intrinsic way as $\frac{1}{2} trace(A^{-1} B)$ in order for the formula to be valid in any orthonormal basis. The choice

$$v_0(x) = a \, e^{-\left(A(x - x_*).(x - x_*)\right)} \tag{9.33}$$

has the same moments of order up to 2 than u_0 if $a \frac{\pi^{N/2}}{\sqrt{\det(A)}} = M$ and $a \frac{\pi^{N/2}}{2\sqrt{\det(A)}} trace(A^{-1} B) = trace(J B)$ for every B, i.e. $a \frac{\pi^{N/2}}{2\sqrt{\det(A)}} A^{-1} = J$, giving $\frac{J}{M} = \frac{A^{-1}}{2}$ or $A = \frac{M}{2} J^{-1}$, and then $a = \frac{M}{\sqrt{\det(J)}} \left(\frac{M}{2\pi}\right)^{N/2}$.

It remains to compute in an explicit way the solution with initial datum v_0, and this can easily be done by Fourier transform, but by looking directly at solutions of the form $e^{\text{quadratic}(x)}$ with a quadratic form having coefficients in t, one can solve more general equations of the form[19]

$$u_t - \sum_{i,j=1}^{N} D_{i,j} u_{x_i x_j} + \sum_{k=1}^{N} B_k u_{x_k} + C u = 0, \tag{9.34}$$

with coefficients $D_{i,j} \in \mathcal{P}_0$, $B_k \in \mathcal{P}_1$ and $C \in \mathcal{P}_2$, where $\mathcal{P}_m$ denotes the space of polynomials in x of degree at most m, with coefficients depending upon t; this is due to the fact that if one uses the change of unknown function

$$u = e^{\varphi}, \tag{9.35}$$

then φ satisfies a nonlinear partial differential equation

$$\varphi_t = \sum_{i,j=1}^{N} D_{i,j}(\varphi_{x_i x_j} + \varphi_{x_i}\varphi_{x_j}) - \sum_{k=1}^{N} B_k \varphi_{x_k} - C, \tag{9.36}$$

and the right side shows a nonlinear operator that maps $\mathcal{P}_2$ into itself, and the partial differential equation becomes an ordinary differential equation when restricted to $\mathcal{P}_2$, which one may solve explicitly.[20] We write φ as

$$\varphi(x,t) = -(A(t)x.x) + 2(b(t).x) + c(t) \tag{9.37}$$

with $A(t)$ symmetric, and then if

$D(t)$ is symmetric
$B_k(x,t) = \sum_{j=1}^{N}(B_1)_{j,k}(t)x_j + (B_0)_k(t)$ for $k = 1,\ldots,N$ (9.38)
$C(x,t) = (C_2(t)x.x) + 2(C_1(t).x) + C_0(t)$ with $C_2(t)$ symmetric,

one has

$$\sum_{i,j} D_{i,j}\frac{\partial^2 \varphi}{\partial x_i \partial x_j} = 2\sum_{i,j} D_{i,j}A_{i,j} = 2trace(A\,D)$$
$$\sum_{i,j=1}^{N} D_{i,j}\frac{\partial \varphi}{\partial x_i}\frac{\partial \varphi}{\partial x_j} = \sum_{i,j=1}^{N} 4D_{i,j}(-A\,x+b)_i(-A\,x+b)_j$$
$$= 4(A\,D\,A\,x.x) - 8(A\,D\,b.x) + 4(D\,b.b) \tag{9.39}$$
$$\sum_{k=1}^{N} B_k\varphi_{x_k} = \sum_{k=1}^{N} 2\Big(\sum_{j=1}^{N}(B_1)_{j,k}(t)x_j + (B_0)_k(t)\Big)(-A\,x+b)_k$$
$$= -2(B_1 A\,x.x) + 2(B_1 x.b) - 2(A\,x.B_0) + 2(B_0.b)$$

---

[19] This form englobes the linear Fokker–Planck equation $f_t - v.f_x - \kappa\,\Delta_v f = 0$, where the "space" variable is $(\mathbf{x}, \mathbf{v})$, and in this case D is nonnegative but degenerate, and some coefficients B_k are linear in $\mathbf{v}$; one may also add terms in $(\mathbf{E}+\mathbf{v}\times\mathbf{B}).f_\mathbf{v}$ corresponding to the Lorentz force, with $\mathbf{E} \in \mathcal{P}_1$ and $\mathbf{B} \in \mathcal{P}_0$. The computations in this case permit one to write in an explicit way what the Green function is.

[20] One could also do these computations after using a partial Fourier transform in x, but I prefer to show the computations in the way I present them here, because the idea extends to some nonlinear equations, for which the Fourier transform is not a good tool; for example, it can also be used in linear cases with negative diffusion, where the solutions (local in time) may not belong to $\mathcal{S}'(\mathbb{R}^N)$.

and one deduces that A, b, c satisfy the differential system

$$\begin{aligned} \frac{dA}{dt} &= -4A\,D\,A + B_1 A + A\,B_1^T - C_2 \\ \frac{db}{dt} &= -4A\,D\,b - B_1^T b + A\,B_0 - C_1 \\ \frac{dc}{dt} &= -2trace(A\,D) + 4(D\,b.b) - 2(B_0.b) - C_0. \end{aligned} \quad (9.40)$$

The first equation gives A, the second gives b and then the third gives c; in the case where $B_k = 0$ for $k = 1, \ldots, N$ and $C = 0$, the equation for A becomes $\frac{dA}{dt} = -4A\,D\,A$, or

$$\frac{d(A^{-1})}{dt} = 4D, \text{ i.e. } A^{-1}(t) = A^{-1}(0) + 4\int_0^t D(s)\,ds, \quad (9.41)$$

showing that if $A(0)$ is positive definite and D nonnegative then $A(t)$ is positive definite, but in the case where A is nonnegative and degenerate, then $A(t)$ stays degenerate in the same subspace and positive definite on the orthogonal space; in this particular case, one also deduces

$$\frac{d(A^{-1}b)}{dt} = 0, \quad (9.42)$$

which is the stationarity of x_* (which is not always valid if there are lower-order terms in the equation). In the mid 1980s, Jean-Pierre GUIRAUD had pointed out that my computations were variants of what was usually done,[21] which must be to use the elementary solutions E and its derivatives of order 1 or 2, i.e. E multiplied by suitable polynomials.

[Taught on Monday September 17, 2001.]

Notes on names cited in footnotes for Chapter 9, FRÉCHET,[22] AL KHWARIZMI,[23] HADAMARD,[24] LOMONOSOV,[25] and for the preceding footnotes, AL MA'MUN.[26]

---

[21] Jean-Pierre GUIRAUD, French mathematician. He worked at Université Paris VI (Pierre et Marie Curie), Paris, and at ONERA (Office National d'Études et de Recherches Aéronautiques), Châtillon, France.

[22] Maurice René FRÉCHET, French mathematician, 1878–1973. He had worked in Poitiers, in Strasbourg and in Paris, France.

[23] Abu Ja'far Muhammad ibn Musa AL KHWARIZMI (or KHAWARIZMI), "Iraqi" mathematician, 780–850. It is not known where he was born, but he had worked in an academy (bayt al-hikmah = house of wisdom) that the Caliph AL MAMUN had set up in his capital Baghdad (now in Iraq), with the goal of translating Greek philosophical and scientific works into Arabic.

[24] Jacques Salomon HADAMARD, French mathematician, 1865–1963. He had worked in Bordeaux, in Paris, France, holding a chair (mécanique analytique et mécanique céleste, 1909–1937) at Collège de France, Paris.

[25] Mikhail Vasilievich LOMONOSOV, Russian scientist, 1711–1765. He had worked in Moscow, Russia. Lomonosov State University, Moscow, Russia, is named after him.

[26] Abu al-Abbas Abd Allah AL MA'MUN ibn Harun, 7th Caliph of the Abbasid dynasty, 786–833. He had ruled over the Muslim world from Baghdad, now in Iraq.

10

Radon Measures; the Law of Large Numbers

Gaussian functions occur in many situations, sometimes related to their appearance for the heat equation, and a classical example is related to probabilities and the accumulated errors in independent experiments which is the subject of this lecture, but there are other reasons, and their appearance for the Boltzmann equation is different.

The bases of probability theory were laid by the work of FERMAT,[1] PASCAL,[2] and D. BERNOULLI,[3] who were concerned with discrete events, and this corresponds to using Radon measures having a finite number of Dirac masses, while more general questions may involve Radon measures, or *Borel measures*.

Having studied partial differential equations, I prefer Radon measures, which are included in the theory of distributions of Laurent SCHWARTZ as distributions of order ≤ 0, and for an open set Ω, $\mathcal{M}(\Omega)$ is the dual of $C_c(\Omega)$, the space of continuous functions with compact support in Ω. Actually, Radon measures do not require the differential structure of $\mathbb{R}^N$, and they can be defined on a locally compact topological space which is σ-compact (i.e. a countable union of compact subsets). For a compact set K, $\mathcal{M}(K)$ is the dual of $C(K)$, which is a Banach space with the sup norm, and a probability on K is any nonnegative Radon measure (i.e. $\langle \mu, \varphi \rangle \geq 0$ whenever $\varphi \geq 0$ in K) with total mass 1 (i.e. $\langle \mu, 1 \rangle = 1$). In dealing with probabilities in the context of Radon measures in an open set Ω, one restricts attention to test functions in $C_0(\Omega)$, the space of bounded continuous functions tending to 0 at the boundary and at infinity, which is a Banach space with the sup norm, and its dual is the space $\mathcal{M}_b(\Omega)$ of Radon measures with finite total mass. An

---

[1] Pierre DE FERMAT, French mathematician, 1601–1665. He had worked (as a lawyer and government official) in Toulouse, France.

[2] Blaise PASCAL, French mathematician and philosopher, 1623–1662. The Université de Clermont-Ferrand II, Aubière, France, is named after him.

[3] Daniel BERNOULLI, Swiss mathematician, 1700–1782. He had worked in St Petersburg, Russia, and in Basel, Switzerland.

important difference between Radon measures and Borel measures is that one cannot use them on spaces of continuous paths (because infinite-dimensional Banach spaces are not locally compact), and although FEYNMAN has used such spaces of paths in his computations, I think that it is a mistake to consider that such notions are important in physics, and as "Brownian" motion can be avoided in dealing with the heat equation, I conjecture that using spaces of paths will be avoided, once one has understood what physicists are really after when they play such games. Until proven wrong, I shall continue to teach that it is Radon measures which are adapted to questions in continuum mechanics or physics.

If for example one throws a coin n times and one counts the number j of heads (and $n - j$ of tails), an event is a list of length n of heads or tails (which one may consider as being 0 or 1, or view such a list as a vertex of a cube $\{0, 1\}^n$), and there are 2^n events, all equally probable with probability 2^{-n} if the coin tossing is not biased. To compute the probability of having j heads one must count the number of subsets with j elements in a set of n elements, and there are $\binom{n}{j}$ of them, so the probability is $\pi(j) = 2^{-n}\binom{n}{j}$, for $j = 0, \ldots, n$; of course, one has $\sum_{j=0}^n \pi(j) = 1$, which also follows from the Newton binomial formula $(1 + x)^n = \sum_{j=0}^n \binom{n}{j} x^j$ by taking $x = 1$. The average value of j is $\sum_0^n j \, \pi(j)$, which is obviously $\frac{n}{2}$, as $\pi(j) = \pi(n - j)$ for every j, but a more analytic way is to derive $(1 + x)^n$ so that $n(1 + x)^{n-1} = \sum_{j=0}^n \binom{n}{j} j \, x^{j-1}$, and taking $x = 1$ gives $\sum_{j=0}^n j \binom{n}{j} = n \, 2^{n-1}$, or $\sum_{j=0}^n j \, 2^{-n} \binom{n}{j} = \frac{n}{2}$. Similarly, $n(n-1)(1+x)^{n-2} = \sum_{j=0}^n \binom{n}{j} j(j-1) \, x^{j-2}$ gives $\sum_{j=0}^n j^2 \binom{n}{j} = \sum_{j=0}^n j \binom{n}{j} + n(n-1)2^{n-2} = n \, 2^{n-1} + n(n-1)2^{n-2} = n(n+1)2^{n-2}$, so that if one computes the average of $\left(j - \frac{n}{2}\right)^2$, one finds $2^{-n}\left(n(n+1)2^{n-2} - n(n \, 2^{n-1}) + \frac{n^2}{4}2^n\right) = \frac{n}{4}$; one sees then that the important values of j are of the form $\frac{n}{2} + O(\sqrt{n})$, and this type of scaling is general when one repeats a process with the same probabilities at each stage, independent of what happened before, and this is related to the law of large numbers. Hidden behind these computations is the fact that one has used repeated convolutions of a Radon measure by itself, here the Radon measure $\frac{1}{2}\delta_0 + \frac{1}{2}\delta_1$, and it is useful to review convolutions and Fourier transform of Radon measures (when they are tempered distributions).

A Radon measure μ on an open set $\Omega \subset \mathbb{R}^N$ is a linear form $\varphi \mapsto \langle \mu, \varphi \rangle$ defined on $C_c(\Omega)$, the space of continuous functions with compact support, which satisfies the (continuity) condition

for every compact $K \subset \Omega$ there exists a constant C_K such that
$$|\langle \mu, \varphi \rangle| \leq C_K \max_x |\varphi(x)| \text{ for all } \varphi \in C_c(\Omega) \text{ having their support in } K.$$
(10.1)

A Radon measure is said to have finite total mass if one can take C_K independent of K, and the total mass is then the supremum of $|\langle \mu, \varphi \rangle|$ for all

$\varphi \in C_c(\Omega)$ of norm ≤ 1 (i.e. satisfying $|\varphi(x)| \leq 1$ for all $x \in \Omega$);[4] in this case one finds easily that the mapping $\varphi \mapsto \langle \mu, \varphi \rangle$ extends to $\varphi \in C_b(\Omega)$, the Banach space of bounded continuous functions, equipped with the sup norm. For μ a Radon measure with finite total mass in $\mathbb{R}^N$, one can define its Fourier transform $\mathcal{F}\mu$ by

$$\mathcal{F}\mu(\xi) = \langle \mu, e^{-2i\pi(\cdot.\xi)} \rangle, \tag{10.2}$$

and the Lebesgue dominated convergence theorem shows that

$$\mathcal{F}\mu \in C_b(\mathbb{R}^N). \tag{10.3}$$

For $\mu = \delta_a$, the Dirac mass at a, one has

$$\mathcal{F}\delta_a(\xi) = e^{-2i\pi(a.\xi)} \text{ for all } \xi \in \mathbb{R}^N, \tag{10.4}$$

showing that the Fourier transform may not tend to 0 at infinity.[5] One has $\mathcal{F}\delta_0 = 1$, but in order to give a meaning to the formula $\mathcal{F}1 = \delta_0$, one must use the extension by Laurent SCHWARTZ of the Fourier transform to the space of tempered distributions $\mathcal{S}'(\mathbb{R}^N)$.[6]

If μ_1 and μ_2 are two Radon measures with finite total mass, the convolution product $\mu_1 \star \mu_2$ is defined by $\langle \mu_1 \star \mu_2, \varphi \rangle = \langle \mu_1 \otimes \mu_2, \varphi(x + y) \rangle$, where the tensor product $\mu_1 \otimes \mu_2$ is defined by $\langle \mu_1 \otimes \mu_2, \varphi_1 \otimes \varphi_2 \rangle = \langle \mu_1, \varphi_1 \rangle \langle \mu_2, \varphi_2 \rangle$ for all $\varphi_1, \varphi_2 \in C_c(\mathbb{R}^N)$.[7] The convolution product cannot be defined for all

---

[4] Some people use the term "bounded measure" for saying that a measure has a finite total mass, and this comes from the point of view of measuring sets, so that for all Borel sets A one has $|\mu(A)| \leq K$, and in that case they call its norm the "total variation" of the measure, as it is $\sup_A \mu(A) - \inf_B \mu(B)$. The point of view of measuring sets is not adapted to questions of continuum mechanics or physics, and many who use this point of view have actually advocated questions of fake mechanics.

[5] There are cases of Radon measures with finite total mass which are not of the form $f \, dx$ for $f \in L^1(\mathbb{R}^N)$ but for which the Fourier transform nevertheless tends to 0 at infinity.

[6] This extension does not give a Fourier transform for all smooth functions; in particular if $f(x) = e^x$, one has $f' = f$, and if the Fourier transform could be defined with the usual formula $\mathcal{F}f'(\xi) = 2i\pi\xi \mathcal{F}f(\xi)$, one would have $(2i\pi\xi - 1)\mathcal{F}f = 0$, i.e. the support of $\mathcal{F}f$ would be included in the set where $2i\pi\xi = 1$, which is empty, so $\mathcal{F}f$ would be 0, and an extension of the Fourier transform which is not invertible is useless (of course, one may then look for extensions which do not give distributions).

[7] For continuous functions, $\Phi = \varphi_1 \otimes \varphi_2$ means $\Phi(x,y) = \varphi_1(x)\varphi_2(y)$ for all $(x,y) \in X \times Y$, where X and Y are open sets of finite-dimensional spaces for example; the definition for Radon measures is a natural extension. Every continuous function with compact support in $X \times Y$ can be approached uniformly by linear combinations of tensor products, for example by using the Weierstrass theorem (that on a compact set of $\mathbb{R}^N$ every continuous function can be approached uniformly by polynomials, which are linear combinations of tensor products) and

Radon measures, and it is an extension of the particular case $\delta_a \star \delta_b = \delta_{a+b}$, which shows that the group structure of $\mathbb{R}^N$ plays a crucial role (while it plays no role in the definition of the tensor product), and this definition extends immediately as a bilinear mapping by $(\sum_i \alpha_i \delta_{a_i}) \star (\sum_j \beta_j \delta_{b_j}) = \sum_{i,j} \alpha_i \beta_j \delta_{a_i+b_j}$ if one imposes $\sum_i |\alpha_i| < \infty$ and $\sum_j |\beta_j| < \infty$; then every Radon measure with finite total mass can be approached (in weak $\star$ topology) by combinations of Dirac masses, and the definition extends by continuity. For functions $f, g \in L^1(\mathbb{R}^N)$ the convolution formula is $(f \star g)(x) = \int_{\mathbb{R}^N} f(y) g(x-y) \, dy$, and it corresponds to the fact that the choice $\mu_1 = f \, dx$ and $\mu_2 = g \, dx$ gives $\mu_1 \star \mu_2 = (f \star g) dx$, and the property that the Lebesgue measure dx is invariant by translation is crucial (for other locally compact groups, one needs to use a Haar measure for the group,[8] and this general approach may have been done by WEIL).

From a probabilistic point of view, one wants a probability μ to act on some (measurable) sets, and one writes $\mu(A)$ for what is written as $\langle \mu, \chi_A \rangle$ when μ is a nonnegative Radon measure with total mass 1, where χ_A is the characteristic function of A; of course, one must use the Lebesgue extension that one can use test functions which are not necessarily continuous, and A must then be μ-measurable (i.e. a Borel set modulo a set of μ-measure 0).

Tensor products are then natural for dealing with independent probabilities, i.e. one has two probabilities μ_1 and μ_2 on sets X_1 and X_2 and one wants to define a probability μ on $X_1 \times X_2$ such that $\mu(A_1 \times A_2) = \mu_1(A_1)\mu_2(A_2)$ for all measurable subsets $A_1 \subset X_1$ and $A_2 \subset X_2$.

Convolution products appear natural when one deals with an Abelian (i.e. commutative) group G and $X_1 = X_2 = G$; if one measures a first value $z_1 \in G$ with a probability μ_1 and independently a second value $z_2 \in G$ with a probability μ_2 and one wants to compute the law corresponding to $z_1 + z_2$, then one finds that it is $\mu_1 \star \mu_2$.

If $G = \mathbb{R}^N$, and one measures n times the value z according to the same probability μ, each measurement being independent of the preceding ones, then one is dealing with $\mu \star \mu \star \ldots \star \mu$ (with n terms in the convolution product); if one averages the result obtained $\frac{z_1 + \ldots + z_n}{n}$, one must use a rescaling. It is then natural to wonder if $\frac{z_1 + \ldots + z_n}{n}$ is a good approximation of a suitably defined averaged value, and in what sense the sequence of rescaled laws converges.

---

truncation, and this shows that there is at most one Radon measure satisfying the imposed conditions. In order to compute $\langle \mu_1 \otimes \mu_2, \Phi \rangle$ for $\Phi \in C_c(X \times Y)$, one lets μ_2 act on the function $\Phi(x, \cdot)$, and this gives $\Psi_1(x)$ for a function $\Psi_1 \in C_c(X)$, and one then lets μ_1 act on Ψ_1, and by the uniqueness part the same result is obtained by letting μ_1 act on the function $\Phi(\cdot, y)$, giving $\Psi_2(y)$, and letting μ_2 act on Ψ_2.

[8] Alfréd HAAR, Hungarian mathematician, 1885–1933. He had worked in Göttingen, Germany, in Kolozsvár (then in Hungary, now Cluj-Napoca, Romania), in Budapest and in Szeged, Hungary.

If μ is a nonnegative Radon measure of total mass 1 on $\mathbb{R}^N$, then in order to define an average value, one asks that $|x|\,\mu$ has finite total mass, then one can define its average, or centre of mass by

$$average(\mu) = a \text{ means } a_j = \langle \mu, x_j \rangle \text{ for } j = 1, \ldots, N \text{ (if } \langle \mu, 1 \rangle = 1) \quad (10.5)$$

If $(1 + |x|)\mu_1$ and $(1 + |x|)\mu_2$ have finite total mass, then

$$x(\mu_1 \star \mu_2) = (x\,\mu_1) \star \mu_2 + \mu_1 \star (x\,\mu_2)$$
$$average(\mu_1 \star \mu_2) = average(\mu_1) + average(\mu_2) \text{ (if } \langle \mu_1, 1 \rangle = \langle \mu_2, 1 \rangle = 1).$$
$$(10.6)$$

If μ is a nonnegative Radon measure on $\mathbb{R}^N$ with $\langle \mu, 1 \rangle = 1$, and one defines $\mu^n = \mu \star \ldots \star \mu$ (with n terms in the convolution product), then assuming that $(1 + |x|)\mu$ has finite total mass and $a = \langle \mu, x \rangle$, the centre of mass of μ^n is at $n\,a$, and it is then natural to rescale μ^n and define ν^n by

$$\langle \nu^n, \varphi \rangle = \langle \mu^n, \varphi_n \rangle = \langle \mu \star \ldots \star \mu, \varphi_n \rangle \text{ with } \varphi_n = \varphi\left(\frac{\cdot}{n}\right), \varphi \in C_b(\mathbb{R}^N), \quad (10.7)$$

so that ν^n is a (bounded) sequence of nonnegative Radon measures, with total mass 1 and centre of mass at a. This gives us the framework for the first part of the law of large numbers.

Lemma 10.1. *Under the preceding hypotheses* ($\mu \geq 0$, $\langle \mu, 1 \rangle = 1$, $\langle \mu, |x| \rangle < \infty$, $\langle \mu, x \rangle = a$), *the sequence* ν^n *converges to* δ_a *in the weak $\star$ topology of X', with $X = C_b(\mathbb{R}^N)$, i.e.*

$$\lim_{n \to \infty} \langle \nu^n, \varphi \rangle = \varphi(a) \text{ for all } \varphi \in X = C_b(\mathbb{R}^N). \quad (10.8)$$

Proof: One must notice that X is not separable,[9] and that the dual X' contains elements which are not distributions (in agreement with the fact that $C_c^\infty(\mathbb{R}^N)$ is not dense in X),[10] so one starts by considering test functions $\varphi \in Y = C_0(\mathbb{R}^N)$; as $C_c^\infty(\mathbb{R}^N)$ is dense in Y, the dual Y' is a space of distributions, and it is actually the space of Radon measures with finite total mass. As Y is separable, on bounded sets of Y' the weak $\star$ topology is metrizable, and one can extract a subsequence ν^m which converges in Y' weak $\star$ to ν^∞;

---

[9] For example, in the case $N = 1$, one may consider all functions f such that $f(n) \in \{-1, +1\}$ for $n \in \mathbb{Z}$, the functions being extended to be affine continuous in each of the intervals $[m, m+1]$; the distance of two different functions of this family is equal to 2, and because the family is not countable, one cannot cover X by a countable number of balls of radius < 1 (as such a ball contains at most one function f from the family), so that X is not separable.

[10] The sequence δ_n is bounded in X', so that it belongs to a weakly $\star$ compact subset of X' by Alaoglu theorem, but none of the accumulation points of the sequence is a distribution, because for every $\varphi \in Y = C_0(\mathbb{R}^N)$ one has $\langle \delta_n, \varphi \rangle \to 0$, but 0 is not an accumulation point because $\langle \delta_n, 1 \rangle \to 1$.

once one will have shown that $\nu^\infty = \delta_a$, one will deduce that all the sequence ν^n converges to δ_a in Y' weak $\star$.

The identification of ν^∞ follows from the use of the Fourier transform, which is well defined because $Y' \subset \mathcal{S}'(\mathbb{R}^N)$. One has $\mathcal{F}\mu^n(\xi) = (\mathcal{F}\mu(\xi))^n$, and $\mathcal{F}\nu^n(\xi) = \mathcal{F}\mu^n\left(\frac{\xi}{n}\right) = \left(\mathcal{F}\mu\left(\frac{\xi}{n}\right)\right)^n$; the hypotheses imply that $|\mathcal{F}\mu(\xi)| \leq 1$ for all $\xi \in \mathbb{R}^N$, $\mathcal{F}\mu(0) = 1$, and $\mathcal{F}\mu$ is of class C^1, with $\frac{\partial(\mathcal{F}\mu)}{\partial\xi_j}(0) = -2i\pi a_j$ for $j = 1,\ldots,N$, so that $\mathcal{F}\mu(\xi) = 1 - 2i\pi(\xi.a) + o(|\xi|)$ for $|\xi|$ small, so that $\mathcal{F}\nu^n(\xi) \to e^{-2i\pi(\xi.a)}$ for every $\xi \in \mathbb{R}^N$, and because $|\mathcal{F}\nu^n(\xi)| \leq 1$, one deduces from the Lebesgue dominated convergence theorem that $\mathcal{F}\nu^n(\xi) \to e^{-2i\pi(\xi.a)} = \mathcal{F}\delta_a(\xi)$ in $L^p_{loc}(\mathbb{R}^N)$ strong for $1 \leq p < \infty$ and $L^\infty(\mathbb{R}^N)$ weak $\star$ and therefore in $\mathcal{S}'(\mathbb{R}^N)$ weak $\star$, so that $\nu^\infty = \delta_a$.

Let $\varphi_0 \in C_c(\mathbb{R}^N)$ be such that $0 \leq \varphi_0 \leq 1$ and $\varphi_0(x) = 1$ for $|x| \leq |a|$, then $\langle \nu^n, \varphi_0 \rangle \to \langle \delta_a, \varphi_0 \rangle = \varphi_0(a) = 1$, so that $\langle \nu^n, \varphi_0 \rangle = 1 - \varepsilon_n$ with $0 < \varepsilon_n$ and $\varepsilon_n \to 0$. Then for $\varphi \in C_b(\mathbb{R}^N)$ one has $\langle \nu^n, \varphi \rangle = \langle \nu^n, \varphi\varphi_0 \rangle + \langle \nu^n, \varphi(1 - \varphi_0) \rangle$ and $\langle \nu^n, \varphi\varphi_0 \rangle \to \langle \delta_a, \varphi\varphi_0 \rangle = \varphi(a)\varphi_0(a) = \varphi(a)$ because $\varphi\varphi_0 \in C_0(\mathbb{R}^N)$ and $\langle \nu^n, \varphi(1-\varphi_0) \rangle \leq ||\varphi||_{C_b(\mathbb{R}^N)}\langle \nu^n, (1-\varphi_0) \rangle = \varepsilon_n||\varphi||_{C_b(\mathbb{R}^N)} \to 0$ because $\nu^n \geq 0$ and $1 - \varphi_0 \geq 0$; this shows that for every $\varphi \in C_b(\mathbb{R}^N)$ one has $\langle \nu^n, \varphi \rangle \to \varphi(a)$. $\quad\square$

If μ is a nonnegative Radon measure of total mass 1 on $\mathbb{R}^N$, and $|x|^2\mu$ has finite total mass, then $|x|\mu$ has finite total mass by the Cauchy–Bunyakovsky–Schwarz inequality,[11,12] but apart from this first appearance, I shall call this inequality as it is known, the Cauchy–Schwarz inequality,[13] and one defines the matrix of inertia J of μ by

$$J(\mu)_{j,k} = \langle \mu, (x-a)_j(x-a)_k \rangle \text{ for } j,k = 1,\ldots,N \text{ (if } \langle \mu, 1 \rangle = 1 \text{ and } \langle \mu, x \rangle = a). \tag{10.9}$$

If $(1 + |x|)^2\mu_1$ and $(1 + |x|)^2\mu_2$ have finite total mass, then

$$\begin{aligned} x_jx_k(\mu_1 \star \mu_2) &= (x_jx_k\mu_1) \star \mu_2 + \mu_1 \star (x_jx_k\mu_2) \\ &+ (x_j\mu_1) \star (x_k\mu_2) + (x_k\mu_1) \star (x_j\mu_2) \\ J(\mu_1 \star \mu_2) &= J(\mu_1) + J(\mu_2) \text{ (if } \langle \mu_1, 1 \rangle = \langle \mu_2, 1 \rangle = 1). \end{aligned} \tag{10.10}$$

Then the matrix of inertia of μ^n is that of μ multiplied by n, and it is then natural to translate and rescale μ^n and define π^n by

$$\langle \pi^n, \psi \rangle = \langle \mu^n, \psi_n \rangle \text{ with } \psi_n(na + \sqrt{n}\,\sigma) = \psi(\sigma), \psi \in C_b(\mathbb{R}^N), \tag{10.11}$$

[11] Viktor Yakovlevich BUNYAKOVSKY, Ukrainian-born mathematician, 1804–1889. He had worked in St Petersburg, Russia.

[12] Hermann Amandus SCHWARZ, German mathematician, 1843–1921. He had worked at ETH (Eidgenössische Technische Hochschule), Zürich, Switzerland, and in Berlin, Germany.

[13] It should indeed be attributed to BUNYAKOVSKY, who had studied with CAUCHY in Paris (1825), and had proven the "Cauchy–Schwarz inequality" in 1859, 25 years before SCHWARZ.

so that π^n is a (bounded) sequence of nonnegative Radon measures, with total mass 1, centre of mass at 0 and matrix of inertia $J(\mu)$. This gives us the framework for the second part of the law of large numbers.

Lemma 10.2. *Under the preceding hypotheses (i.e. adding $\langle \mu, |x|^2 \rangle < \infty$, $J(\mu) = \langle \mu, (x-a) \otimes (x-a) \rangle$), the sequence π^n converges to π^∞ in the weak $\star$ topology of X', with $X = C_b(\mathbb{R}^N)$, where π^∞ is the (anisotropic) Gaussian defined by*

$$\mathcal{F}\pi^\infty = e^{-2\pi^2(J(\mu)\xi.\xi)} \ for \ \xi \in \mathbb{R}^N. \tag{10.12}$$

Proof: One starts by considering test functions $\psi \in Y = C_0(\mathbb{R}^N)$, one extracts a subsequence converging to π^∞ in Y' weak $\star$ by the Banach theorem,[14] and one identifies π^∞ by its Fourier transform. The relation between ψ and ψ_n consists in translating μ of $-a$ and μ^n of $-n\,a$, so one may assume that $a = 0$, and in that case the hypotheses imply that $\mathcal{F}\mu$ is of class C^2 with $\mathcal{F}\mu(0) = 1$, $\frac{\partial(\mathcal{F}\mu)}{\partial\xi_j}(0) = 0$ for $j = 1, \ldots, N$, $\frac{\partial^2(\mathcal{F}\mu)}{\partial\xi_j\partial\xi_k}(0) = 4\pi^2 J(\mu)_{j,k}$ for $j, k = 1, \ldots, N$, so that $\mathcal{F}\mu(\xi) = 1 - 2\pi^2(J(\mu)\xi.\xi) + o(|\xi|^2)$ for $|\xi|$ small. Then (because $a = 0$) one has $\mathcal{F}\pi^n(\xi) = \left(\mathcal{F}\mu\left(\frac{\xi}{\sqrt{n}}\right)\right)^n \to e^{-2\pi^2(J(\mu)\xi.\xi)}$, and $|\mathcal{F}\pi^n(\xi)| \leq 1$, giving $\mathcal{F}\pi^\infty(\xi) = e^{-2\pi^2(J(\mu)\xi.\xi)}$ for $\xi \in \mathbb{R}^N$. One concludes in the case $\psi \in C_b(\mathbb{R}^N)$ by a truncation argument, like in Lemma 10.1. $\qquad\square$

[Taught on Wednesday September 19, 2001.]

Notes on names cited in footnotes for Chapter 10, WEIERSTRASS,[15] ALAO-GLU.[16]

---

[14] Y being a separable Banach space, the weak $\star$ topology on a bounded set of Y' is separable.

[15] Karl Theodor Wilhelm WEIERSTRASS, German mathematician, 1815–1897. He had first taught in high schools in Münster and in Braunsberg, Germany, and then worked in Berlin, Germany.

[16] Leonidas ALAOGLU, Canadian-born mathematician.

11

A 1-D Model with Characteristic Speed $\frac{1}{\varepsilon}$

In Lemma 10.1 and Lemma 10.2, describing the law of large numbers, the hypothesis that μ is nonnegative is not really necessary, and the proof has mostly used $\langle \mu, 1 \rangle = 1$ and $|\mathcal{F}\mu(\xi)| \leq 1$ for all $\xi \in \mathbb{R}^N$; if $\mu \geq 0$, then μ^n is nonnegative and has total mass 1, and this was used for showing weak $\star$ convergence with test functions in $C_0(\mathbb{R}^N)$ and then in $C_b(\mathbb{R}^N)$, while if one replaces nonnegativity by $|\mathcal{F}\mu(\xi)| \leq 1$ for all $\xi \in \mathbb{R}^N$, one deduces from $\mathcal{F}\mu^n(\xi) = \big(\mathcal{F}\mu(\xi)\big)^n$ that $\mathcal{F}\mu^n$ is bounded in $L^\infty(\mathbb{R}^N)$, and the proof holds for test functions in $\mathcal{F}L^1(\mathbb{R}^N)$, which is included in $C_0(\mathbb{R}^N)$.

This remark is useful when approximating partial differential equations with constant coefficients by (one step) explicit finite-difference schemes of the form

$$U_i^{n+1} = \sum_j a_j U_{i+j}^n \text{ for } i \in \mathbb{Z}^N, n \geq 0, \tag{11.1}$$

where there is only a finite number of $j \in \mathbb{Z}^N$ in the sum and the coefficients a_j depend explicitly on Δx and Δt; usually one lets Δx and Δt tend to 0 in such a way that the coefficients a_j do not change,[1] and one must check the consistency and stability of the scheme.

Consistency means that the scheme is adapted to the equation that one wants to solve, and it is checked by using the Taylor expansion of a smooth solution, or by verifying that the scheme is exact for a precise family of polynomial solutions of the equations; for example, if one wants to solve $u_t - \kappa \Delta u = 0$, one wants the scheme to be exact for the function 1 and for the functions $x_k, k = 1, \dots, N$, and this condition is the same for an equation $u_t - \sum_{i,j} D_{i,j} \frac{\partial^2 u}{\partial x_i \partial x_j} = 0$, and then one wants the scheme to be exact for all the polynomial solutions $\sum_{i,j} c_{i,j} x_i x_j - (2\kappa \sum_i c_{i,i})t$, but this condition

---

[1] In a hyperbolic setting it means that $\frac{\Delta t}{\Delta x}$ is fixed, and one must impose a CFL condition, but in a parabolic setting it usually means that $\frac{\Delta t}{(\Delta x)^2}$ is constant, and small enough for a stability condition to hold.

depends upon which diffusion tensor appears in the equation that one wants to solve.

Stability consists in showing in advance that the numbers generated by the algorithm (11.1) give a bounded sequence of approximations in a suitable norm, and the ℓ^2 norm $\left(\sum_i |U_i^n|^2\right)^{1/2}$ plays a special role, in part because it gives a stability condition in $L^2(\mathbb{R}^N)$ after defining a suitable function by interpolation, but mainly because one has a necessary and sufficient condition of stability in that norm (if the coefficients a_j are kept fixed), which is that the function M defined by $M(\xi) = \sum_j a_j e^{ij\xi}$ satisfies $|M(\xi)| \leq 1$ for all $\xi \in \mathbb{R}^N$.[2] One then finds the same type of condition which appeared in the proof of the law of large numbers, with the difference that this approach works for many partial differential equations with constant coefficients, even if Gaussian functions play no role in their solution.

It has been seen that the Lax–Friedrichs scheme with the natural CFL condition can be interpreted in terms of a random walk, and this classical association of random walks and heat equation or other diffusion equations is often considered but it is rarely mentioned that this has not much to do with the physics of the phenomena which one tries to describe by diffusion models.

As I mentioned before, jumps in position are not physical, because they involve infinite velocities, while jumps in velocity are reasonably good approximations of what happens in collisions or in almost collisions, when the velocities involved are very small compared to the velocity of light c. It was considered natural by FOURIER to postulate that the heat flux is proportional and opposite to the gradient of the temperature, because he knew that heat flows from hot regions to cold regions, and he could hardly have argued that temperature is a statistical concept which has no meaning at a microscopic level, because such ideas only appeared at least fifty years after his work, with the introduction of ideas in kinetic theory of gases by MAXWELL and by BOLTZMANN; a particle does not have a temperature, but it has a velocity and it is the fact that not all particles have the same velocity which creates the need for mesoscopic/macroscopic quantities like internal energy, to which temperature is related; although everyone has a clear intuition of what is hot and what is cold, it does not mean that one understands much about what

---

[2] This is done by considering the U_j^n as the coefficients of a Fourier series $\sum_{j \in \mathbb{Z}^N} U_j^n e^{2i\pi(j.\xi)}$, defining a periodic function $f^n(\xi)$ (with period the unit cube); the algorithm has then the form $f^{n+1}(\xi) = M(2\pi\,\xi)f^n(\xi)$; the ℓ^2 norm is adopted because of the Parseval theorem which states that it coincides with the L^2 norm on a period (BESSEL being only credited for proving an inequality).

is really going on at a microscopic level,[3] and the Fourier law is only found natural because the class of equations considered is too restrictive.[4]

What happens with particles is not that they jump but that they change their velocity, because of interactions with their environment; for simplification, one uses models of collisions which are instantaneous (and whose result uses probabilities), and it is the fact that one has neglected the time of interaction which creates the impression that an infinite acceleration has occurred, resulting in a jump in velocity, and I shall come back to this question in more detail later.

The subject of this lecture is to show how a linear model in which there are large velocities and jumps in velocities attributed to scattering, approximates a diffusion in space, once one lets a characteristic velocity tend to infinity. It is precisely the source of illogical statements made by physicists, that they do not appreciate that $c = +\infty$ in some postulated model, and that these models cannot show that something travels faster than the real velocity of light. From a logical point of view, I am not sure why some physicists insist on using a postulated equation like the Schrödinger equation for pretending that something may travel faster than the velocity of light c, when it is already a feature of the Fourier heat equation, which was postulated a century before! From a mathematical point of view it seems better to observe that the Fourier heat equation or the Schrödinger equation appears when one lets c tend to infinity in more precise models, like the equation of radiative transfer, or the *Dirac equation*. Of course, physicists consider that the Dirac equation is for one relativistic electron, but a scenario in which an electron would stop before entering a laboratory to ask if the experience it will participate in is relativistic or not (so that it will choose between solving the Dirac equation or solving the Schrödinger equation) does not seem too serious! Of course, physicists may object that "quantum particles" are quite strange objects, which may not be bothered by the silliness of the games that one attributes them, but now that mathematicians have tools for understanding more about localized solutions of hyperbolic systems, it is quite obvious that there are no particles at all, only waves.

---

[3] If one puts one hand on a marble table, one finds it cold, and one does not have the same sensation with the other hand on a wooden table, although both tables are at equilibrium at the temperature of the room; a classical explanation is that marble is a good conductor (of electricity and heat) and that it takes the heat away from the hand, which is at a higher temperature than the room (so that I have assumed the room to be at a much lower temperature than 37 degrees Celsius), and therefore it is not a difference in temperature that the two hands have felt, but a difference in heat flux, created by a difference in conductivity.

[4] Even with a more general class like pseudo-differential operators, one should observe that one does not understand much yet about nonlinear effects of a micro-local nature, and one should consider most macroscopic models as approximations which must be improved later, once new and more adapted mathematical tools have been introduced.

I consider the following model, which I first learnt how to treat in a mathematical way in lectures by Jacques-Louis LIONS,

$$\frac{\partial u_\varepsilon}{\partial t} + \frac{1}{\varepsilon}\frac{\partial u_\varepsilon}{\partial x} + \frac{a}{\varepsilon^2}(u_\varepsilon - v_\varepsilon) = 0 \text{ in } \mathbb{R} \times (0,\infty), \; u_\varepsilon(\cdot,0) = \varphi \text{ in } \mathbb{R}$$
$$\frac{\partial v_\varepsilon}{\partial t} - \frac{1}{\varepsilon}\frac{\partial v_\varepsilon}{\partial x} - \frac{a}{\varepsilon^2}(u_\varepsilon - v_\varepsilon) = 0 \text{ in } \mathbb{R} \times (0,\infty), \; v_\varepsilon(\cdot,0) = \psi \text{ in } \mathbb{R}. \tag{11.2}$$

I learnt much later about what such models are supposed to represent (in a caricatural way), and the coefficient $\frac{1}{\varepsilon}$ serves as velocity of light c, and $u_\varepsilon(x,t), v_\varepsilon(x,t)$ represent (nonnegative) densities of "photons" moving along the x axis, in the positive or negative direction; the term in $u_\varepsilon - v_\varepsilon$ describes a *scattering effect*, which makes some "photons" change their direction by interacting at some high rate with the material environment. I must say that physicists' ideas concerning photons look strange to me, because the quantification $h\nu$ for the energy of a "photon of frequency ν" only makes sense if light interacts with matter, because the Planck constant h is a coupling parameter between light and matter; when there is no matter, photons follow the Maxwell–Heaviside equation which is linear and they are propagated without interacting, and this is why I mention a material environment, without which I cannot understand the origin of scattering. Physicists say that photons are bosons, i.e. they follow Bose–Einstein statistics,[5] so that they may appear spontaneously with a higher probability when there are already photons present, but I am still not sure about what this rule means; physicists use this argument for a computation attributed to EINSTEIN, to "explain" Planck's law for black body radiation (which has dependence on frequency and on temperature, and seems to fit well with experimental measurements, apart from the missing frequencies of absorption in a gas), and that may be the origin of that strange dogma.

In more general models of radiative transfer, one has an unknown function $f(x,t,\omega,\nu)$ which is a (nonnegative) density of photons at the location $x \in \mathbb{R}^3$ and time t, moving in the direction of the unit vector $\omega \in \mathbb{S}^2$ and having frequency ν; the equation contains a free transport term $\frac{\partial f}{\partial t} + c\sum_{j=1}^{3}\omega_j\frac{\partial f}{\partial x_j}$, terms of absorption/emission, and terms of scattering; assuming a linear scattering effect independent of the frequency as a simplification,[6] there would be a nonnegative kernel $K(\omega' \mapsto \omega)$ for switching from direction ω' to direction ω, and the equation would contain a term $(\int_{\mathbb{S}^2} K(\omega \mapsto \xi)\,d\xi)f(x,t,\omega,\nu) - \int_{\mathbb{S}^2} K(\omega' \mapsto \omega)f(x,t,\omega',\nu)\,d\omega'$.

For $\varepsilon > 0$ in our simplified model, there is a unique solution for initial data $\varphi,\psi \in L^1_{loc}(\mathbb{R})$ if the coefficient a is measurable and (essentially) bounded on compact sets, by using the finite speed of propagation property. If $\varphi,\psi \in L^p(\mathbb{R})$ for some $p \in [1,\infty]$, then if a is measurable and (essentially) bounded there is a unique solution $u_\varepsilon, v_\varepsilon$ and the norms $||u_\varepsilon(\cdot,t)||_{L^p(\mathbb{R})}$ and

---

[5] Satyendra Nath BOSE, Indian physicist, 1894–1974. He had worked in Dhaka (now capital of Bangladesh), and in Calcutta, India.

[6] Uniformity of the stationary solutions in direction requires that $\int_{\mathbb{S}^2} K(\omega \mapsto \xi)\,d\xi = \int_{\mathbb{S}^2} K(\xi \mapsto \omega)\,d\xi$ for all $\omega \in \mathbb{S}^2$.

$||v_\varepsilon(\cdot, t)||_{L^p(\mathbb{R})}$ grow at most in $e^{C t/\varepsilon^2}$. If one assumes that $a \geq 0$, and $\varphi, \psi \geq 0$, then one has $u_\varepsilon, v_\varepsilon \geq 0$ for $t > 0$, but without knowing the sign of φ, ψ one has $\int_\mathbb{R} [\Phi(u_\varepsilon(x, t)) + \Phi(v_\varepsilon(x, t))]\, dx \leq \int_\mathbb{R} [\Phi(\varphi(x)) + \Phi(\psi(x))]\, dx$ for every convex Φ.

Lemma 11.1. *If $0 < \alpha \leq a(x, t) \leq \beta < \infty$ a.e. $x \in \mathbb{R}, t > 0$, and $\varphi, \psi \in L^2(\mathbb{R})$, then u_ε and v_ε converge weakly to z in $L^2(\mathbb{R} \times (0, T))$ for every $T > 0$ as $\varepsilon \to 0$, where z is the solution of*

$$z_t - \left(\tfrac{1}{2a} z_x\right)_x = 0 \text{ in } \mathbb{R} \times (0, \infty)$$
$$z(\cdot, 0) = \tfrac{1}{2}(\varphi + \psi), \tag{11.3}$$

and $\frac{u_\varepsilon - v_\varepsilon}{\varepsilon}$ converges weakly to $-\frac{1}{a} z_x$ in $L^2(\mathbb{R} \times (0, \infty))$.

Proof: Multiplying the first equation by u_ε and the second equation by v_ε, and integrating from 0 to T one obtains

$$\tfrac{1}{2} \int_\mathbb{R} (|u_\varepsilon(x, T)|^2 + |v_\varepsilon(x, T)|^2)\, dx + \alpha \int_0^T \int_\mathbb{R} \left|\tfrac{u_\varepsilon - v_\varepsilon}{\varepsilon}\right|^2 dx\, dt$$
$$\leq \tfrac{1}{2} \int_\mathbb{R} (|\varphi(x)|^2 + |\psi(x)|^2)\, dx. \tag{11.4}$$

This shows that $q_\varepsilon = \frac{u_\varepsilon - v_\varepsilon}{\varepsilon}$ stays in a bounded set of $L^2(\mathbb{R} \times (0, \infty))$ and that for every $T < \infty$, u_ε and v_ε stay in a bounded set of $L^2(\mathbb{R} \times (0, T))$. One may then extract a subsequence $\eta \to 0$ such that u_η and v_η converge weakly to z in $L^2(\mathbb{R} \times (0, T))$ for every $T < \infty$, and q_η converges weakly to q_0 in $L^2(\mathbb{R} \times (0, \infty))$, and the reason why the weak limits of u_η and v_η coincide is that $u_\eta - v_\eta = \eta q_\eta$ converges strongly to 0. From the fact that the limit z will be identified, and that $q_0 = -\frac{1}{a} z_x$, one deduces that the whole sequence converges weakly.

Adding the equations gives $(u_\eta + v_\eta)_t + (q_\eta)_x = 0$, which shows at the limit that $2z_t + (q_0)_x = 0$ and $z|_{t=0} = \frac{1}{2}(\varphi + \psi)$. Subtracting the equations and multiplying by η gives $\eta(u_\eta - v_\eta)_t + (u_\eta + v_\eta)_x + 2a\, q_\eta = 0$, which shows at the limit that $2z_x + 2a\, q_0 = 0$, from which the equation for z follows. $\quad\square$

The result extends to $\varphi, \psi \in L^1(\mathbb{R}) + C_0(\mathbb{R})$, because a function in such a functional space can be decomposed into an element of $L^2(\mathbb{R})$, a small term in $L^1(\mathbb{R})$, and a small term in $L^\infty(\mathbb{R})$, and the fact that $a \geq 0$ is important for showing that small initial data in $L^1(\mathbb{R})$ or $L^\infty(\mathbb{R})$ give uniformly small solutions in these spaces.

[Taught on Friday September 21, 2001.]

Notes on names cited in footnotes for Chapter 11. PARSEVAL.[7]

---

[7] Marc-Antoine PARSEVAL DES CHÊNES, French mathematician, 1755–1836.

12

A 2-D Generalization; the Perron–Frobenius Theory

A simple generalization to a situation in $\mathbb{R}^2$ is to consider the following system, where indexing the solutions with ε has been omitted for simplification:

$$
\begin{aligned}
\frac{\partial u_1}{\partial t} + \frac{1}{\varepsilon}\frac{\partial u_1}{\partial x} + \frac{a}{\varepsilon^2}(+3u_1 - u_2 - u_3 - u_4) &= 0 \text{ in } \mathbb{R}^2 \times (0,\infty) \\
\frac{\partial u_2}{\partial t} - \frac{1}{\varepsilon}\frac{\partial u_2}{\partial x} + \frac{a}{\varepsilon^2}(-u_1 + 3u_2 - u_3 - u_4) &= 0 \text{ in } \mathbb{R}^2 \times (0,\infty) \\
\frac{\partial u_3}{\partial t} + \frac{1}{\varepsilon}\frac{\partial u_3}{\partial y} + \frac{a}{\varepsilon^2}(-u_1 - u_2 + 3u_3 - u_4) &= 0 \text{ in } \mathbb{R}^2 \times (0,\infty) \\
\frac{\partial u_4}{\partial t} - \frac{1}{\varepsilon}\frac{\partial u_4}{\partial y} + \frac{a}{\varepsilon^2}(-u_1 - u_2 - u_3 + 3u_4) &= 0 \text{ in } \mathbb{R}^2 \times (0,\infty) \\
u_j \mid_{t=0} &= v_j \text{ in } \mathbb{R}^2.
\end{aligned}
\tag{12.1}
$$

Of course, this is a caricature of a plane situation where only four directions are allowed for "photons" to move and where the scattering tends to equilibrate all four directions, but the limiting equation as ε tends to 0 does not inherit a biased behaviour towards the directions of the axes, and an isotropic diffusion term will appear.

If the initial data v_j belong to $L^2(\mathbb{R}^2)$ for all j, and $0 < \alpha \le a(x,y,t) \le \beta < \infty$, then multiplying the equation #j by u_j, summing in j and integrating from 0 to T gives the estimate

$$
\begin{aligned}
&\frac{1}{2}\int_{\mathbb{R}^2}\left(\sum_{j=1}^4 |u_j(x,y,T)|^2\right) dx\, dy \\
&\quad + \frac{\alpha}{\varepsilon^2}\int_0^T\int_{\mathbb{R}^2}\left[4\sum_{j=1}^4 |u_j|^2 - \left(\sum_{k=1}^4 u_k\right)^2\right] dx\, dy\, dt \\
&\quad \le \frac{1}{2}\int_{\mathbb{R}^2}\left(\sum_{j=1}^4 |v_j|^2\right) dx\, dy,
\end{aligned}
\tag{12.2}
$$

and if one observes that $4\sum_j |u_j|^2 - (\sum_k u_k)^2 = \frac{1}{2}\sum_{j,k}|u_j - u_k|^2$, one sees that u_j stays in a bounded set of $L^\infty(0,T;L^2(\mathbb{R}^2))$ for all j and all $T < \infty$, and $\frac{u_j - u_k}{\varepsilon}$ stays in a bounded set of $L^2(\mathbb{R}^2 \times (0,\infty))$ for all j,k. One then extracts a subsequence $u_{j,\eta}$ such that

$$
\begin{aligned}
u_{j,\eta} &\rightharpoonup z \text{ in } L^2(\mathbb{R}^2 \times (0,T)) \text{ weak, for all } j \text{ and all } T < \infty \\
\frac{u_{1,\eta} - u_{2,\eta}}{\eta} &\rightharpoonup q_1 \text{ and } \frac{u_{3,\eta} - u_{4,\eta}}{\eta} \rightharpoonup q_2 \text{ in } L^2(\mathbb{R}^2 \times (0,\infty)) \text{ weak,}
\end{aligned}
\tag{12.3}
$$

and the fact that the whole sequence converges weakly will come from the identification of z, q_1, q_2. Adding the four equations gives $\left(u_{1,\eta} + u_{2,\eta} + u_{3,\eta} + u_{4,\eta}\right)_t + \left(\frac{u_{1,\eta} - u_{2,\eta}}{\eta}\right)_x + \left(\frac{u_{3,\eta} - u_{4,\eta}}{\eta}\right)_y = 0$, and letting η tend to 0 one deduces $4z_t + (q_1)_x + (q_2)_y = 0$ and $z\,|_{t=0} = \frac{1}{4}(v_1 + v_2 + v_3 + v_4)$; subtracting the second equation from the first and multiplying by η gives $\eta(u_{1,\eta} - u_{2,\eta})_t + (u_{1,\eta} + u_{2,\eta})_x + 4a\,\frac{u_{1,\eta} - u_{2,\eta}}{\eta} = 0$ and letting η tend to 0 one deduces $2z_x + 4a\,q_1 = 0$; subtracting the fourth equation from the third and multiplying by η gives $\eta(u_{3,\eta} - u_{4,\eta})_t + (u_{3,\eta} + u_{4,\eta})_y + 4a\,\frac{u_{3,\eta} - u_{4,\eta}}{\eta} = 0$ and letting η tend to 0 one deduces $2z_y + 4a\,q_2 = 0$. This gives

$$z_t - \left(\frac{1}{8a}\,z_x\right)_x - \left(\frac{1}{8a}\,z_y\right)_y = 0, \text{ i.e. } z_t - div\left(\frac{1}{8a}\,grad(z)\right) = 0, \qquad (12.4)$$
$$z\,|_{t=0} = \frac{v_1 + v_2 + v_3 + v_4}{4};\ q_1 = -\frac{z_x}{2a};\ q_2 = -\frac{z_y}{2a}.$$

In order to generalize the preceding example to a more general situation, with $x \in \mathbb{R}^N$, with a finite number of large velocities and with general transition probabilities between the different families travelling at one of these velocities, it is useful to recall some results concerning discrete Markov processes,[1] which can be derived from results in linear algebra, due to PERRON,[2] and to FROBENIUS.

A discrete Markov process is a probabilistic setting in which there are only a finite number of states, numbered from 1 to m, and probabilities of transitions $P_{i,j}$ from state #j to state #i, from time n to time $n + 1$, and what happens at time n is independent of what happened before, and what the integer n is.[3] If at time n the probability of being in the state #j is x_j, for $j = 1, \ldots, m$, then at time $n + 1$ the probability of being in the state #i is $\sum_{j=1}^m P_{i,j} x_j$. If one denotes by P the matrix with entries $P_{i,j}$, and if z_i^n is the probability of being in state #i at time n, and z^n is the vector with components $z_i^n, i = 1, \ldots, m$, then one has $z^{n+1} = P z^n$ for every n, so that $z^n = P^n z^0$. One has $P_{i,j} \geq 0$ for $i, j = 1, \ldots, m$, and $\sum_{i=1}^m P_{i,j} = 1$ for $j = 1, \ldots, m$, expressing that one must necessarily be in one of the m

---

[1] Andrei Andreyevich MARKOV, Russian mathematician, 1856–1922. He had worked in St Petersburg, Russia.

[2] Oskar PERRON, German mathematician, 1880–1975. He had worked in Tübingen, in Heidelberg, and in München (Munich), Germany.

[3] This is in essence the same idea used in semi-group theory, that the state of a system at time t is the only information that one needs in order to predict the future evolution of the system: if $u(t)$ is the state of the system at time t, the state of the system at time $t + s$ is $S\big(s; u(t)\big)$, and $S(s; v)$ is the state at time s if one starts at time 0 with the system in state v; of course, one then has $S(0; v) = v$ for all v and $S\big(s; S(t; v)\big) = S(s + t; v)$ for all v and all $s, t \geq 0$, which in the linear case is written as $S(t; v) = S(t)v$ with $S(0) = I$ and $S(s + t) = S(s)S(t)$ for all $s, t \geq 0$ (the term semi-group comes from the fact that if this was true for all $s, t \in \mathbb{R}$ one would have a group of transformations, but the transformations are only defined for $t \geq 0$).

states; the last condition is written as $P^T \mathbf{1} = \mathbf{1}$, where $\mathbf{1}$ is the vector with all components equal to 1, which is the same as having $(z^{n+1}.\mathbf{1}) = (z^n.\mathbf{1})$ whatever z^n is.

Describing the asymptotic behaviour when n tends to ∞ requires then an understanding of the eigenvalues of P with maximum modulus; 1 is an eigenvalue of P^T and so of P, and by the Hadamard–Gershgorin theorem,[4,5] the eigenvalues belong to the union of the closed discs D_i, centred at $P_{i,i}$ and with radius $R_i = \sum_{k \neq i} |P_{k,i}|$, which is $1 - P_{i,i}$ because the entries of P are nonnegative, and all the eigenvalues then have a modulus ≤ 1. The Perron theorem gives more precise information if $P_{i,j} > 0$ for all $i, j = 1, \ldots, m$, that the only eigenvalue of modulus 1 is 1, that it is a simple eigenvalue, and that an eigenvector e has all its coefficients > 0; if one normalizes e by $(e, \mathbf{1}) = 1$, then as n tends to ∞ the sequence z^n converges to e (because z^0 has nonnegative coefficients, and $(z^0, \mathbf{1}) = 1$). The case where some entries $P_{i,j}$ are 0 requires an improvement due to FROBENIUS, and as it is better to describe this more general case, I need to recall some results from linear algebra.[6]

Definition 12.1. *A $m \times m$ matrix A, with entries in an arbitrary ring, is reducible if $\{1, \ldots, m\} = I \cup J$, with I, J, nonempty and disjoint and $A_{i,j} = 0$ for all $i \in I$ and all $j \in J$. A is irreducible if it is not reducible.*

In order to check if a matrix A is irreducible, one associates to it an oriented graph with m vertices numbered from 1 to m, by putting an oriented arc from vertex #i to vertex #j if and only if $A_{i,j} \neq 0$. It is easy to check that A is irreducible if and only if there exists a closed path following the oriented arcs and going at least once through each of the vertices.[7]

---

[4] HADAMARD remarked that if A is diagonally dominant, i.e. $|A_{i,i}| > \sum_{j \neq i} |A_{i,j}|$ for all i (or $|A_{i,i}| > \sum_{j \neq i} |A_{j,i}|$ for all i), then A is invertible, while GERSHGORIN expressed the same idea in a more geometrical way: if A is an $m \times m$ matrix with complex coefficients and λ is an eigenvalue of A, then there exists $i \in \{1, \ldots, m\}$ such that $|\lambda - A_{i,i}| \leq \sum_{j \neq i} |A_{i,j}|$; indeed, if x is a corresponding eigenvector and i is such that $|x_i| \geq |x_j|$ for all $j \neq i$, one has $\lambda x_i = (A x)_i = \sum_j A_{i,j} x_j$, so that $|\lambda - A_{i,i}| |x_i| = |\sum_{j \neq i} A_{i,j} x_j| \leq \sum_{j \neq i} |A_{i,j}| |x_j| \leq (\sum_{j \neq i} |A_{i,j}|) |x_i|$.

[5] Semyon Aranovich GERSHGORIN, Belarusian-born mathematician, 1901–1933. He had worked in Petrograd/Leningrad, Russia.

[6] I was not taught these results as a student, perhaps because algebraists are not so interested in them, probably because they use the order relation on $\mathbb{R}$. It was in my first year as an assistant professor at Université Paris IX Dauphine in 1971 that I learnt about them, because I had been asked to teach complements of linear algebra from a book by GANTMACHER (translated into French). I realized afterwards that probabilists do learn about these questions, probably mixed with ideas about Markov processes, but it is useful to see that they are results of linear algebra, which should be taught independently of any probabilistic framework.

[7] One defines an equivalence relation by saying that i is equivalent to j if and only if either $i = j$ or $i \neq j$ and there exists an oriented path going from vertex #i

Lemma 12.2. *Assume that an $m \times m$ matrix A has real nonnegative entries, then A is irreducible if and only if $\tilde{A} = I + A + \ldots + A^{m-1}$ has all its entries positive.*

Proof: One has $(A^2)_{i,k} = \sum_j A_{i,j} A_{j,k} \geq 0$, and $(A^2)_{i,k} > 0$ if and only if there exists j with $A_{i,j} > 0$ and $A_{j,k} > 0$, i.e. if and only if there is a oriented path of length 2 going from vertex #i to vertex #k, and more generally $(A^p)_{i,k} > 0$ if and only if there is a oriented path of length p going from vertex #i to vertex #k. If A is irreducible, then for $i \neq j$ there is a path going from vertex #i to vertex #j, and one may ensure that it has length $\leq m - 1$ by cutting off the loops so that it goes at most once through each of the vertices, so $\tilde{A}_{i,j} \neq 0$, and for $i = j$, one has $\tilde{A}_{i,i} \geq I_{i,i} = 1$. If A is reducible, then $A_{i,j} = 0$ for all $i \in I$ and all $j \in J$ with I and J disjoint implies that for every p one has $(A^p)_{i,j} = 0$ for all $i \in I$ and all $j \in J$.   □

For a vector x, the notation $x \geq 0$ will mean that $x_i \geq 0$ for all i, and $x > 0$ will mean that $x_i > 0$ for all i (so that, if $m > 1$, it is not the same thing as $x \geq 0$ and $x \neq 0$). The *spectral radius* $\rho(A)$ of a matrix with complex entries is $\max_j |\lambda_j|$, where the λ_j are the eigenvalues of A.

Proposition 12.3. *Let A be irreducible with nonnegative entries. Then $r = \rho(A)$ is a simple eigenvalue of A, for an eigenvector $e > 0$. If $Ax \geq \alpha x$ with $x \geq 0$ and $x \neq 0$, then $\alpha \leq r$; if $Ay \leq \beta y$ with $y \geq 0$ and $y \neq 0$, then $r \leq \beta$; if $Az = \lambda z$ with $z \geq 0$ and $z \neq 0$, then $\lambda = r$.*

Proof: Let $\Sigma = \{\xi \geq 0, (\xi, \mathbf{1}) = 1\}$, and $\tilde{\Sigma} = \{\frac{\eta}{(\eta.1)} \in \Sigma \mid \eta = \tilde{A}\xi$ with $\xi \in \Sigma\}$, which is well defined because $\xi \geq 0$ and $\xi \neq 0$ implies $\tilde{A}\xi > 0$. One defines φ on $\tilde{\Sigma}$ by $\varphi(\eta) = \min_j \frac{(A\eta)_j}{\eta_j}$, which is well defined and > 0 because $\eta \in \tilde{\Sigma}$ implies $\eta > 0$. The function φ is continuous on $\tilde{\Sigma}$ and $\tilde{\Sigma}$ is compact so φ attains its maximum at a point $e \in \tilde{\Sigma}$ with $\varphi(e) = r > 0$. By definition one has $Ae = re + f$ with $f \geq 0$, and one must have $f = 0$; indeed, one applies $\tilde{A}$ (which commutes with A), giving $A\tilde{A}e = r\tilde{A}e + \tilde{A}f$ and if one had $f \neq 0$ it would imply $\tilde{A}f > 0$ and therefore $\tilde{A}f \geq \varepsilon \tilde{A}e$ for some $\varepsilon > 0$, implying $\varphi(\eta) \geq r + \varepsilon$ for $\eta = \frac{\tilde{A}e}{(\tilde{A}e,1)} \in \tilde{\Sigma}$, contradicting the maximality of r.

If $Ax \geq \alpha x$ with $x \geq 0$ and $x \neq 0$, then one applies $\tilde{A}$ and $\eta = \frac{\tilde{A}x}{(\tilde{A}x.1)} \in \tilde{\Sigma}$ satisfies $\alpha \leq \varphi(\eta) \leq r$. If λ is an eigenvalue of A, with eigenvector u, then for every j one has $|\lambda| |u_j| = |\sum_k A_{j,k} u_k| \leq \sum_k A_{j,k} |u_k|$, so that the vector x defined by $x_j = |u_j|$ for all j satisfies $x \geq 0$, $x \neq 0$ and $Ax \geq |\lambda| x$, so that $|\lambda| \leq r$, showing that r must be the spectral radius $\rho(A)$.

to vertex #j and also an oriented path going from vertex #j to vertex #i. A is irreducible if and only if there is only one equivalence class, and when there is more than one equivalence class it gives a way to choose what I and J are for showing that A is reducible.

As A^T has nonnegative entries and is irreducible, there exists an eigenvector $e' > 0$ such that $A^T e' = r\, e'$ (as A^T has the same eigenvalues and the same spectral radius than A). If $A\, y \leq \beta\, y$ with $y \geq 0$ and $y \neq 0$, then taking the scalar product with e' gives $r\,(e'.y) = (A^T e'.y) = (A\, y, e') \leq \beta\,(e'.y)$, showing $r \leq \beta$.

The eigenspace for r is one-dimensional, because if $A\, f = r\, f$ and $f \neq 0$, then one may consider that f is a real vector (or one takes the real part or the imaginary part of f) and then one can choose $t \in \mathbb{R}$ such that $g = f + t\, e \geq 0$ with one component $g_i = 0$, but this implies that $\sum_j A_{i,j} g_j = r\, g_i = 0$, and therefore $A_{i,j} \neq 0$ implies $g_j = 0$; one deduces that $g_k = 0$ if one can join i to k by following an oriented path on the graph associated to A, and as A is irreducible one finds $g = 0$. The algebraic multiplicity of r is one, because if there was an associated Jordan block,[8] there would exist f ($\neq 0$) such that $A\, f = r\, f + e$, and for $t \in \mathbb{R}$ large enough, one would have $g = f + t\, e \geq 0$, $g \neq 0$ and $A\, g = r\, g + e$, and applying $\widetilde{A}$ would show that $\eta = \frac{\widetilde{A} g}{(\widetilde{A} g.1)} \in \widetilde{\Sigma}$ satisfies $\varphi(\eta) > r$, contradicting the maximality of r.   □

PERRON had proven the preceding result in the case where $A_{i,j} > 0$ for all i, j, and the fact that the irreducible character of A implies the same result is the work of FROBENIUS, but PERRON had also shown that the only eigenvalue of modulus r is r itself, and that is not always the case in the irreducible case, but it only happens when the nonzero entries show a special pattern.

Proposition 12.4. *Let A be irreducible with nonnegative entries, and such that A has an eigenvalue different from r which has modulus r. Then there exists an integer $p \geq 2$ and a partition of $\{1, \ldots, m\}$ into p nonempty subsets $I_1, \ldots, I_p$ such that all the nonzero entries $A_{i,j}$ satisfy $i \in I_k$ and $j \in I_{k+1}$ for some k (with $I_{p+1} = I_1$). In that case the spectrum of A is invariant by rotation of $2\pi/p$, and if μ is any eigenvalue of A and z is any pth root of unity $(e^{2ij\pi/p}$ for $j = 0, \ldots, p-1)$ then $z\, \mu$ is an eigenvalue of A with the same algebraic multiplicity than μ, so there are at least $p-1$ simple eigenvalues of modulus r which are distinct from r, and if m is not a multiple of p, 0 must be an eigenvalue of A with an algebraic multiplicity n such that $n = m$ (mod p).*

Proof: Let $\lambda = r\, e^{2i\pi\theta}$ with $0 < \theta < 2\pi$, and let u be a corresponding eigenvector normalized by $\sum_j |u_j| = 1$, and let $x \in \Sigma$ be defined by $x_j = |u_j|$ for all j. Then one has $r\, x_j = |\lambda\, u_j| = |\sum_k A_{j,k} u_k| \leq \sum_k A_{j,k} x_k$ for all j, i.e. $A\, x \geq r\, x$, and this implies $A\, x = r\, x$ and therefore $x = e$, because if it was not true then $\eta = \frac{\widetilde{A} x}{(\widetilde{A} x.1)} \in \widetilde{\Sigma}$ would satisfy $\varphi(\eta) > r$. In order to have equality in the triangle inequality that has been used, it is necessary that all the nonzero terms $A_{j,k} u_k$ have the same argument than $\lambda\, u_j$, and this means that the

---

[8] Marie Ennemond Camille JORDAN, French mathematician, 1833–1922. He had worked in Paris, France, holding a chair (mathématiques, 1883–1883) at Collège de France, Paris.

argument of u_j increases of θ each time one follows one oriented arc along the graph associated to A, and there must be a smaller integer $p > 1$ such that $p\theta$ is a multiple of 2π. Multiplying u by a complex number of modulus 1 so that $u_1 > 0$, I_j is then defined as the subset of indices k such that the argument of u_k is $(j-1)\theta$ modulo 2π, and A has the required structure for its nonzero entries. Such a structure implies that if z is a pth root of unity the characteristic polynomial $P(\lambda) = det(\lambda I - A)$ satisfies $P(z\lambda) = z^m P(\lambda)$ for all $\lambda \in \mathbb{C}$,[9] and that shows that the characteristic polynomial has the form $\lambda^n Q(\lambda^p)$ with a polynomial Q such that $Q(0) \neq 0$, and $n = m$ (mod p), and that μ and $z\mu$ always have the same algebraic multiplicity, and in particular all the eigenvalues of modulus r are simple.   □

Definition 12.5. *If A is irreducible with nonnegative entries, it is said to be* primitive *if the only eigenvalue of modulus $r = \rho(A)$ is r, and* imprimitive *with index $p \geq 2$ if there are eigenvalues of modulus r different from r and if p is the largest integer such that $r\,e^{2i\pi/p}$ is an eigenvalue.*

Lemma 12.6. *If A is irreducible with nonnegative entries, and q is the gcd (greatest common divisor) of the length of loops on the graph associated to A, then A is primitive if and only if $q = 1$, and if $q > 1$ then A is imprimitive of index q.*

Proof: One has seen that if A has an eigenvalue $r\,e^{2i\pi\theta}$ with $0 < \theta < 2\pi$, then there exists an integer $p \geq 2$ and A has a block structure which implies that all loops have a length that is a multiple of p, so that the gcd of the length of the loops is a multiple of p. If the gcd of the length of the loops is $q > 1$, one defines the subsets I_j, with $j = 1, \ldots, q$ by putting $i \in I_j$ if and only if there exists a path (along the graph associated to A) going from 1 to i and with length equal to $j - 1$ modulo q; the definition makes sense because if ℓ_1 and ℓ_2 are the lengths of two such paths and ℓ_3 is the length of a path going from j to 1 (which exists because A is irreducible), then one has a loop of length $\ell_1 + \ell_3$ and a loop of length $\ell_2 + \ell_3$, both of which are multiples of q and therefore $\ell_1 = \ell_2$ (mod q); this shows that A has a block structure which implies that its characteristic polynomial is of the form $\lambda^n Q(\lambda^q)$, so that $r\,e^{2i\pi/q}$ is an eigenvalue of A.   □

If one coefficient $A_{i,i} \neq 0$ then there is a loop of length 1 and A is primitive. If A is imprimitive of index 2, then if a_1, a_2 are the sizes of I_1, I_2, one has $a_1 + a_2 = m$ and there are at most $2a_1 a_2$ nonzero entries of A, and the

---

[9] One starts from $\lambda I - A$ and one multiplies the rows and columns by powers of z in the following way: row i is multiplied by z^k if $i \in I_k$ and column j is multiplied by z^{1-k} if $j \in I_k$, then one ends up with the diagonal entries being multiplied by z, and the entries with $i \in I_k$ and $j \in I_{k+1}$ being multiplied by 1 (and one needs $z^p = 1$ so that the entries with $i \in I_p$ and $j \in I_1$ are not changed), and it does not matter what the other entries are multiplied by as they are 0. One ends up with the matrix $z\lambda I - A$, and the determinant has been multiplied by z^m.

maximum possible for $2a_1a_2$ is $\frac{m^2}{2}$ if m is even (and both sizes are $\frac{m}{2}$) and $\frac{m^2-1}{2}$ if m is odd (and the sizes are $\frac{m-1}{2}$ and $\frac{m+1}{2}$); if A is imprimitive of index p (with $2 < p \leq m$) and $a_1, \ldots, a_p$ are the sizes of $I_1, \ldots, I_p$, then there are at most $a_1a_2 + a_2a_3 + \ldots + a_pa_1$ nonzero entries of A, and the maximum possible for real nonnegative a_j of sum m is when they are all equal to $\frac{m}{p}$ and the number of nonzero entries is then $\leq \frac{m^2}{p}$. Therefore one can conclude that A is necessarily primitive if the number of its nonzero entries is $> \frac{m^2}{2}$ (i.e. more than half of the entries are different from 0).[10]

The fact that A is primitive gives a simple description of the asymptotic behaviour of the sequences obtained by iterating A.

Lemma 12.7. *If A is irreducible with nonnegative entries and primitive, then for any $w^0 \geq 0$ and $w^0 \neq 0$ the sequence $w^n = A^n w^0$ satisfies $\frac{w^n}{r^n} \to ce$ with $c > 0$, and $c = \frac{(e'.w^0)}{(e'.e)}$, where e and e' are positive eigenvectors of A and of A^T for the eigenvalue $r = \rho(A) = \rho(A^T)$.*

Proof: One decomposes $\mathbb{R}^m$ into two subspaces which are invariant by A, the one-dimensional span of e and the subspace $X = \{x \mid (e'.x) = 0\}$; the restriction of A to X has a spectral radius $r' = \rho(A \mid_X) < r$, because by hypothesis all the eigenvalues of A different from r have a modulus $< r$. One decomposes $w^0 = ce + x^0$ with $x^0 \in X$, and one has $(e'.w^0) = c(e'.e)$, showing that $c > 0$; as $w^n = A^n w^0 = cr^n e + A^n x^0$ and $\limsup_{n\to\infty} ||A^n||^{1/n}_{\mathcal{L}(X;X)} = r' < r$ one deduces that $r^{-n}||A^n x^0|| \to 0$.   □

In the case where A is imprimitive with index $p \geq 2$, one must introduce the eigenvectors e_j for the eigenvalues $r\, e^{2ij\pi/p}$ for $j = 1, \ldots, p-1$ and $w^0 = ce + \sum_{j=1}^{p-1} c_j e_j + y$, where $y \in Y$, a subspace invariant by A where the eigenvalues have a modulus $< r$; then $w^n = A^n w^0 = r^n(ce + \sum_{j=1}^{p-1} c_j e^{2ijn\pi/p}e_j) + A^n y$, so that $\omega^n = r^{-n} w^n$ looks like $ce + \sum_{j=1}^{p-1} c_j e^{2ijn\pi/p}e_j$ and may have no limit if some coefficient c_j is not 0. If one averages on p successive iterates, one finds that $\omega^n + \ldots + \omega^{n+p-1} \to pce$ as $n \to \infty$, and if one does not know the value of p one finds that $\frac{1}{n}(\omega^1 + \ldots + \omega^n) \to ce$ as $n \to \infty$.

---

[10] Another characterization is that A is primitive if and only if A^k has all its entries positive for some integer $k \geq 1$. If $A^k > 0$, then A is irreducible and as A^k has only one eigenvalue of maximum modulus r^k, A cannot have more than one eigenvalue of modulus r and it is primitive. Conversely if A is primitive, there are two loops on the graph of lengths ℓ_1, ℓ_2 with gcd 1, and if L is the length of a loop going at least once through all the vertices, there are such loops with length $L + a_1\ell_1 + a_2\ell_2$ for all nonnegative integers a_1, a_2 and this covers all the integers $\geq N$ for some integer N; for each i, j there exists n with $1 \leq n \leq m - 1$ such that $A^n_{i,j} > 0$ because there is a path of length n from i to j and therefore there are paths of length $n + n'$ for all $n' \geq N$, so that one can find k such that for all $i, j = 1, \ldots, m$, there is a path of length k from i to j, and this gives $A^k > 0$.

[Taught on Monday October 1, 2001 (during the preceding week, I attended a conference in Salamanca, Spain).]

Notes on names cited in footnotes for Chapter 12, GANTMAKHER.[11]

---

[11] Feliks Ruvimovich GANTMAKHER, Ukrainian mathematician, 1908–1964.

13

A General Finite-Dimensional Model with Characteristic Speed $\frac{1}{\varepsilon}$

In the examples already studied of a linear hyperbolic system with velocities in $\frac{1}{\varepsilon}$ and scattering terms in $\frac{1}{\varepsilon^2}$, there were a few special circumstances that made the proof easy for showing that an isotropic diffusion appeared in the limit. We want to consider now a more general situation in $\mathbb{R}^N$ with an arbitrary number m of velocities and general probabilities of transition between the various families:

$$\frac{\partial u_i}{\partial t} + \frac{1}{\varepsilon}(C_i.grad_x(u_i)) + \frac{a}{\varepsilon^2}\sum_{j=1}^m M_{i,j}u_j = 0 \text{ in } \mathbb{R}^N \times (0,T);$$
$$u_i(\cdot,0) = v_i \text{ in } \mathbb{R}^N, i = 1,\ldots,m, \tag{13.1}$$

where the C_i are constant vectors and the $M_{i,k}$ are constant, but a may depend upon x and t (and I omit an index ε for the u_i). We first make the hypothesis that

$$a \in L^\infty(\mathbb{R}^N \times (0,T)), \tag{13.2}$$

so that for any $p \in [1,\infty]$ one can deduce existence and uniqueness theorems for data in $(L^p(\mathbb{R}^N))^m$, and adding the hypothesis

$$a \geq 0; \ M_{i,j} \leq 0 \text{ for all } i \neq j, \tag{13.3}$$

one deduces that nonnegative data create nonnegative solutions for $t \geq 0$. We assume that

$$\sum_{i=1}^m M_{i,j} = 0 \text{ for } j = 1,\ldots,m \tag{13.4}$$

so that one has conservation of mass for bounded data with compact support and this extends to give a uniform bound in L^1 for nonnegative integrable data. In order to obtain uniform bounds in L^∞ and in L^2 which are independent of $\varepsilon > 0$, one assumes that

$$M = (M_{i,j})_{i,j=1,\ldots,m} \text{ is irreducible}, \tag{13.5}$$

and one uses the Perron–Frobenius theory.

Lemma 13.1. *There exists a vector e with positive components such that $M\,e = 0$.*

Proof: One considers $A = s\,I - M$, with $s \geq \max_i M_{i,i}$ so that A is irreducible with nonnegative entries. By hypothesis one has $M^T\mathbf{1} = 0$, so that $A^T\mathbf{1} = s\,\mathbf{1}$, showing that $\rho(A^T) = s$, and therefore there exists $e > 0$ such that $A\,e = s\,e$, i.e. $M\,e = 0$. □

This helps obtain uniform L^∞ estimates (i.e. independent of ε), as one has

$$m_-e_i \leq v_i(x) \leq m_+e_i \text{ a.e. } x \in \mathbb{R}^N, \text{ for } i = 1,\dots,m \text{ implies} \atop m_-e_i \leq u_i(x,t) \leq m_+e_i \text{ a.e. } x \in \mathbb{R}^N, t \in (0,T), \text{ for } i = 1,\dots,m, \quad (13.6)$$

and theorems for describing more general forward invariant sets will be shown in the following lecture. Uniform L^2 estimates follow from the following result.

Lemma 13.2. *There exists $\gamma > 0$ such that*

$$\sum_{i,j=1}^{m} \frac{1}{e_i} M_{i,j}\xi_i\xi_j \geq \gamma \sum_{i=1}^{m}\left|\xi_i - \frac{(\xi.\mathbf{1})}{(e.\mathbf{1})}e_i\right|^2 \text{ for every } \xi \in \mathbb{R}^m. \quad (13.7)$$

Proof: One considers the matrix B with entries $B_{i,j} = s\,\delta_{i,j} - M_{i,j}\frac{\sqrt{e_j}}{\sqrt{e_i}}$ with $s \geq \max_i M_{i,i}$, so that B is irreducible with nonnegative entries. If one defines f by $f_i = \sqrt{e_i}$ then one has $B\,f = s\,f$ and $B^T f = s\,f$. The symmetric matrix $B_{sym} = \frac{B+B^T}{2}$ is irreducible with nonnegative entries and has an eigenvector f with positive components so s is the spectral radius of B_{sym} and because it is symmetric its eigenvalues are real, and therefore the eigenvalues different from s are $\leq s - \beta$ for some $\beta > 0$. This implies that for each vector η one has the inequality $\left((s\,I - B_{sym})\eta.\eta\right) \geq \beta|\eta'|^2$ where η' is the projection of η on the orthogonal of f, i.e. $\eta' = \eta - \frac{(\eta.f)}{|f|^2}f$; the left side of the inequality is $\left((s\,I - B)\eta.\eta\right) = \sum_{i,j=1}^{m} M_{i,j}\frac{\sqrt{e_j}}{\sqrt{e_i}}\eta_i\eta_j$, and if one chooses $\eta_i = \frac{\xi_i}{\sqrt{e_i}}$ then it is $\sum_{i,j=1}^{m}\frac{1}{e_i}M_{i,j}\xi_i\xi_j$; in order to evaluate the right side of the inequality, one notices that $|f|^2 = \sum_i e_i = (e.\mathbf{1})$ and therefore $\eta'_i = \eta_i - \frac{\sum_j \eta_j\sqrt{e_j}}{(e.\mathbf{1})}\sqrt{e_i} = \frac{\xi_i}{\sqrt{e_i}} - \frac{(\xi.\mathbf{1})}{(e.\mathbf{1})}\sqrt{e_i} = \frac{1}{\sqrt{e_i}}\left(\xi - \frac{(\xi.\mathbf{1})}{(e.\mathbf{1})}e\right)_i$, and the right side is $\beta\sum_i\frac{1}{e_i}\left(\xi - \frac{(\xi.\mathbf{1})}{(e.\mathbf{1})}e\right)_i^2$, which is $\geq \gamma\left|\xi - \frac{(\xi.\mathbf{1})}{(e.\mathbf{1})}e\right|^2$ if $\gamma = \min_i\frac{\beta}{e_i}$. □

We then deduce a uniform L^2 estimate by multiplying the ith equation by $\frac{u_i}{e_i}$ and summing in i, which gives

$$\frac{\partial}{\partial t}\left(\sum_{i=1}^{m}\frac{|u_i|^2}{2e_i}\right) + \sum_{j=1}^{N}\frac{\partial}{\partial x_j}\left(\sum_{i=1}^{m}\frac{(C_i)_j|u_i|^2}{2\varepsilon\,e_i}\right) + \frac{\gamma}{\varepsilon^2}\sum_{i=1}^{m}\left(u_i - \frac{(u.\mathbf{1})}{(e.\mathbf{1})}e_i\right)^2 \leq 0, \quad (13.8)$$

implying by integration

$$\int_{\mathbb{R}^N} \Big(\sum_{i=1}^m \frac{|u_i(x,t)|^2}{e_i}\Big)\, dx \le \int_{\mathbb{R}^N} \Big(\sum_{i=1}^m \frac{|v_i(x)|^2}{e_i}\Big)\, dx \text{ for } 0 \le t \le T, \qquad (13.9)$$

and

$$\int_0^T \int_{\mathbb{R}^N} \frac{1}{\varepsilon^2}\Big(u_i - \frac{(u.1)}{(e.1)}e_i\Big)^2 dx\, dt \le C \text{ (independent of } \varepsilon) \text{ for } i = 1,\dots,m.$$
$$(13.10)$$

Using the uniform bounds obtained one can extract a subsequence indexed by η such that u_i^η converges weakly to u_i^0 for $i = 1,\dots,m$, and $\frac{1}{\eta}\big(u_i^\eta - \frac{(u^\eta.1)}{(e.1)}e_i\big)$ converges weakly to q_i. Denoting $z = \frac{(u^0.1)}{(e.1)}$, one finds that $u_i^0 = z\, e_i$ for $i = 1,\dots,m$. In order to obtain the limiting equation, one sums all the equations and, because of the hypothesis of conservation of mass, one obtains

$$\frac{\partial}{\partial t}\Big(\sum_{i=1}^m u_i^\eta\Big) + \sum_{j=1}^N \frac{\partial}{\partial x_j}\Big(\sum_{i=1}^m \frac{(C_i)_j}{\eta} u_i^\eta\Big) = 0, \qquad (13.11)$$

and one is led to impose the condition

$$\sum_{i=1}^m (C_i)_j e_i = 0 \text{ for } j = 1,\dots,N, \text{ i.e. } \sum_{i=1}^m C_i\, e_i = 0, \qquad (13.12)$$

so that the equation can be written as

$$\frac{\partial}{\partial t}\Big(\sum_{i=1}^m u_i^\eta\Big) + \sum_{j=1}^N \frac{\partial}{\partial x_j}\Big(\sum_{i=1}^m \frac{(C_i)_j}{\eta}\Big(u_i^\eta - \frac{(u^\eta.1)}{(e.1)}e_i\Big)\Big) = 0, \qquad (13.13)$$

and gives at the limit $\eta \to 0$

$$(e.1)\frac{\partial z}{\partial t} + \sum_{j=1}^N \frac{\partial}{\partial x_j}\Big(\sum_{i=1}^m (C_i)_j q_i\Big) = 0; \ (e.1)z \mid_{t=0} = v_1 + \dots + v_m. \quad (13.14)$$

Without the condition on the C_i, all the interesting information goes to infinity, and if this condition is not satisfied, it tells at what velocity $\frac{C^*}{\varepsilon}$ one must travel in order to follow the interesting effects. In order to identify the functions q_i one multiplies the ith equation by η and, using $M\, e = 0$, one writes it as

$$\eta\frac{\partial u_i^\eta}{\partial t} + \sum_{j=1}^N \frac{\partial\big((C_i)_j u_i^\eta\big)}{\partial x_j} + \frac{a}{\eta}\sum_{k=1}^m M_{i,k}\Big(u_k^\eta - \frac{(u^\eta.1)}{(e.1)}e_k\Big) = 0, \qquad (13.15)$$

giving at the limit $\eta \to 0$

$$\sum_{j=1}^N \frac{\partial\big((C_i)_j e_i z\big)}{\partial x_j} + a\sum_{k=1}^m M_{i,k} q_k = 0. \qquad (13.16)$$

This equation has a solution if and only if the vector R with components $R_i = \sum_j \frac{\partial[(C_i)_j e_i z]}{\partial x_j}$ belongs to the range of M, i.e. is orthogonal to $\mathbf{1}$, because $\mathbf{1}$ generates the nullspace of M^T; this is indeed true as $\sum_{i=1}^m (C_i)_j e_i = 0$ for $j = 1, \ldots, N$. By construction the vector q with components q_i is orthogonal to $\mathbf{1}$ and because the nullspace of M is generated by e which has positive components (and is then not orthogonal to $\mathbf{1}$), there is a unique solution q orthogonal to $\mathbf{1}$ (one may also use the inequality $\sum_{i,j} \frac{1}{e_i} M_{i,j} \xi_i \xi_j \geq \gamma |\xi|^2$ if $(\xi.\mathbf{1}) = 0$ for constructing the solution). The final equation is of the form

$$(e.1) \frac{\partial z}{\partial t} - \sum_{i,j=1}^N \frac{\partial}{\partial x_i} \left(\frac{D_{i,j}}{a} \frac{\partial z}{\partial x_j} \right) = 0, \qquad (13.17)$$

where D is a nonnegative matrix,[1] but D is not necessarily proportional to I and the solution z may not have all its derivative in L^2; for example, if there is an index $j \in \{1, \ldots, N\}$ such that $(C_i)_j = 0$ for $i = 1, \ldots, m$, one obtains no information on $\frac{\partial z}{\partial x_j}$.

The preceding examples, and the exposition of the Perron–Frobenius theory are useful for various reasons. One reason is to think about the origin of diffusion in space, not from resulting from a random walk with jumps in position, which is not a physically realistic scenario (although it is one of many different mathematical approaches), but from jumps in velocity, after one lets a characteristic velocity tend to ∞ (and in some examples this velocity is the velocity of light c). However, the models used have the defect of postulating some scattering effects with precise probabilities of transition, independent of the state of the system, so that the equation obtained is linear. In the following lecture, I shall start describing a different type of interaction, which creates semi-linear equations with quadratic nonlinearities, because it models interaction of particles of one type with particles of the same or another type, while a linear scattering term supposes that the particles interact with a fixed background. Another reason is to observe that the repetition of a game where probabilities of transitions appear tends to create a special pattern when time tends to ∞, with only one parameter at one's disposal, and it has some similarity with the rules of thermodynamics, where equilibria are indexed by only one parameter, the temperature; however, there is an important hypothesis for arriving at that conclusion, which is a notion of irreducibility, and in thermodynamics it corresponds to the necessity of having all the parts of a body interacting together in order to end up with a unique temperature.

Of course, I have not addressed the question of the validity of the probability assumptions yet, and I have only shown games with some probabilities built in and then I have deduced something (and even if what I have deduced is observed, it is in no way a proof that there are inherent probabilities,

---

[1] D is symmetric when M is symmetric, which is the case if one assumes that the probability of transition from state i to state j is equal to the probability of transition from state j to state i, for all i, j.

of course); in the previous games, the probabilities were said to be related to scattering, and at some time one should then wonder a little more about what scattering is.

For the linear problems already considered, nonnegative data give rise to nonnegative solutions, and because of linearity there is an order-preserving property, but for the semi-linear problems nonnegative data will still give rise to nonnegative solutions but the order-preserving property will be lost; it does exist for the *Carleman model*,[2] as was first noticed by Ignace KOLODNER,[3] but the Carleman model is not a model of kinetic theory, as there is no conservation of momentum.

Starting in the following lecture, I shall switch from the linear models studied up to now to semi-linear models, and that will create important differences in properties and some of the methods already used will lose their efficiency.

For the linear cases, there were not many difficulties working with L^1, with L^∞ or with L^2, but for semi-linear problems it will be natural to obtain bounds in L^1, because of conservation of mass, but L^∞ bounds will no longer be obvious, either because they must be proven by different methods, or sometimes because they are not true. Actually, when the maximum principle does not hold, L^1 and L^∞ are not good functional spaces for solving partial differential equations, but if $1 < p < \infty$ the L^p spaces can be used for the singular integrals which appear when one uses the Green functions for elliptic partial differential equations with constant coefficients, because of Calderón–Zygmund theory,[4,5] which extended the one-dimensional study of

---

[2] Tage Gillis Torsten CARLEMAN, Swedish mathematician, 1892–1949. He had worked in Lund and in Stockholm, Sweden.

[3] Ignacs Izaak KOLODNER, Polish-born mathematician, 1920–1996. He had worked in Albuquerque, NM, at Carnegie Tech (Carnegie Institute of Technology) and at CMU (Carnegie Mellon University), Pittsburgh, PA, where he was head of the department of mathematics from 1964 to 1971, which included the period where Carnegie Tech became CMU. I had first met him in 1974 at a meeting at Brown University in Providence, RI, long before I came to CMU in 1987.

[4] Alberto Pedro CALDERÓN, Argentine-born mathematician, 1920–1998. He received the Wolf Prize in 1989, for his groundbreaking work on singular integral operators and their application to important problems in partial differential equations, jointly with John W. MILNOR. He had worked at Buenos Aires, Argentina, at OSU (Ohio State University), Columbus, OH, at MIT (Massachusetts Institute of Technology), Cambridge, MA, and at The University of Chicago, Chicago, IL. I first heard him talk at the Lions–Schwartz seminar in the late 1960s, and I met him in Buenos Aires when I visited Argentina for two months in 1973; he kept strong ties with Argentina, as can be witnessed from the large number of mathematicians from Argentina having studied harmonic analysis, often working now in United States.

[5] Antoni Szczepan ZYGMUND, Polish-born mathematician, 1900–1992. He had worked in Warsaw, Poland and in Wilno (then in Poland, now Vilnius, Lithuania), and then at the University of Chicago, Chicago IL.

the Hilbert transform, done by M. RIESZ;[6] in the case $p = 1$, one replaces L^1 by the smaller Hardy space $\mathcal{H}^1$,[7] and in the case $p = \infty$, one replaces L^∞ by the larger space BMO (Bounded Mean Oscillation), introduced by Fritz JOHN for a question in elasticity,[8] then studied by Fritz JOHN and Louis NIRENBERG,[9] which is useful for studying the limiting case of the Sobolev embedding theorem.[10] In the late 1970s, I had thought that $BMO(\mathbb{R})$ could be a good functional space for questions of kinetic theory, for a reason unrelated to singular integrals, and I had mentioned something about that to Yves MEYER,[11] but I had found later a way to prove $L^\infty(\mathbb{R})$ bounds for the cases that I was interested in.

It is natural for a density of particles to be nonnegative and in $L^1(\mathbb{R}^N)$ if a total mass is finite, and one may wonder about using spaces of Radon measures with finite total mass, because bounded nonnegative sequences in $L^1(\mathbb{R}^N)$ may approach (in a weak $\star$ topology) any nonnegative Radon measure in $\mathcal{M}_b(\mathbb{R}^N)$ (i.e. with finite total mass), but a particular use of entropy (bounds on $\int f \, \log(f) \, dx$) precludes concentration effects; for what concerns $L^\infty(\mathbb{R}^N)$, I do not know any physical reason why if one starts with nonnegative data in $L^\infty(\mathbb{R}^N)$ the densities should stay in $L^\infty(\mathbb{R}^N)$, and one should consider that $L^\infty(\mathbb{R}^N)$ and other spaces used for proving regularity of solutions of partial differential equations are chosen for reasons of personal taste rather than for reasons related to the (expected) physical content of an equation. The reason why I thought that spaces constructed like $BMO(\mathbb{R}^N)$ could be useful for some problems in kinetic theory is that they are naturally defined by integrals. The precise definition of when a function $u \in L^1_{loc}(\mathbb{R}^N)$ belongs to $BMO(\mathbb{R}^N)$ is to take any cube $Q \subset \mathbb{R}^N$, to compute the average u_Q of u on Q, so that $u - u_Q$ is related to oscillations of u on Q, and then to consider the average of $|u - u_Q|$ on Q, which is the mean oscillation on Q; the space

---

[6] Marcel RIESZ, Hungarian-born mathematician, 1886–1969 (the younger brother of Frigyes (Frederic) RIESZ). He had worked in Stockholm and in Lund, Sweden.

[7] Godfrey Harold HARDY, English mathematician, 1877–1947. He had worked in Cambridge and in Oxford, England, holding the Savilian chair of geometry in 1920–1931, and in Cambridge again, holding the Sadleirian chair of pure mathematics in 1931–1942.

[8] Fritz JOHN, German-born mathematician, 1910–1994. He had worked in Lexington, KY, and at NYU (New York University), New York, NY.

[9] Louis NIRENBERG, Canadian-born mathematician, born in 1925. He received the Crafoord Prize in 1982. He works at NYU (New York University), New York, NY.

[10] If one observes that one has $||u - u_Q||_{L^1(Q)} \leq C_p(Q)||grad(u)||_{L^p(Q)}$ for $u \in W^{1,p}(Q)$ and $1 \leq p \leq \infty$, then for reasons of homogeneity the case $p = N$ gives $C_N(Q) = M\,|Q|$; however, the same bound in $M\,|Q|$ is true if the derivatives of u belong to the Marcinkiewicz space $L^{N,\infty}$, a particular space in the family of Lorentz spaces; this is the case for $\log(|x|)$, which then belongs to $BMO(\mathbb{R}^N)$.

[11] Yves François MEYER, French mathematician, born in 1939. He worked at Université Paris Sud, Orsay (where he was my colleague from 1975 to 1979), at École Polytechnique, Palaiseau, at Université Paris IX-Dauphine, Paris, and at ENS-Cachan (École Normale Supérieure de Cachan), Cachan, France.

$BMO(\mathbb{R}^N)$ is precisely the space of functions for which this mean oscillation is bounded by a number M independent of which cube Q one has considered, hence the choice of the name BMO; the smallest M is a semi-norm for u, and it does not change by adding a constant function to u. One could consider that if a density of particles u is constant it corresponds to an equilibrium and then $u - u_Q$ could be like a mass out of equilibrium, and describing how much mass is out of equilibrium might be useful, although it may not be the precise way that it enters the definition of $BMO(\mathbb{R}^N)$ that should be important. Of course, functions in $BMO(\mathbb{R}^N)$ are not necessarily bounded, because $\log |x| \in BMO(\mathbb{R}^N)$, but Fritz JOHN and Louis NIRENBERG have shown that for $u \in BMO(\mathbb{R}^N)$ there exists $\varepsilon > 0$ such that $e^{\varepsilon |u|} \in L^1_{loc}(\mathbb{R}^N)$, with ε depending only upon the semi-norm of u in $BMO(\mathbb{R}^N)$.

[Taught on Wednesday October 3, 2001.]

Notes on names cited in footnotes for Chapter 13, MILNOR,[12] F. RIESZ,[13] SAVILE,[14] SADLEIR,[15] CRAFOORD,[16] MARCINKIEWICZ,[17] G.G. LORENTZ,[18] and for the preceding footnotes, WAYNE.[19]

---

[12] John Willard MILNOR, American mathematician, born in 1931. He received the Wolf Prize in 1989, for ingenious and highly original discoveries in geometry, which have opened important new vistas in topology from the algebraic, combinatorial, and differentiable viewpoint, jointly with Alberto CALDERÓN. He worked at Princeton University, Princeton, NJ, and at SUNY (State University of New York) at Stony Brook, NY.

[13] Frigyes (Frederic) RIESZ, Hungarian mathematician, 1880–1956. He had worked in Kolozsvár (then in Hungary, now Cluj-Napoca, Romania), in Szeged and in Budapest, Hungary. He introduced the spaces L^p in honour of LEBESGUE and the spaces $\mathcal{H}^p$ in honour of HARDY, but no spaces are named after him, and the Riesz operators have been introduced by his younger brother Marcel RIESZ.

[14] Sir Henry SAVILE, English mathematician, 1549–1622. In 1619, he established professorships of geometry and astronomy at Oxford, England.

[15] In 1701, Lady SADLEIR established a professorship of pure mathematics in Cambridge, England.

[16] Holger CRAFOORD, Swedish industrialist and philanthropist, 1908–1982. He invented the artificial kidney, and he and his wife Anna-Greta CRAFOORD, 1914–1994, established the Crafoord Prize in 1980 by a donation to the royal Swedish academy of sciences, to reward and promote basic research in scientific disciplines that fall outside the categories of the Nobel Prize (which have included mathematics, geoscience, bioscience, astronomy, and polyarthritis).

[17] Józef MARCINKIEWICZ, Polish mathematician, 1910–1940. He had worked in Wilno (then in Poland, now Vilnius, Lithuania). He died during World War II, presumably executed by the Soviets with thousands of other Polish officers.

(footnotes 18–19 on next page)

---

[18] George Gunther LORENTZ, Russian-born mathematician, born in 1910. He worked in Toronto, Ontario, at Wayne State University, Detroit, MI, in Syracuse, NY, and in Austin, TX.

[19] Anthony WAYNE, American general, 1745–1796. Wayne State University, Detroit, MI, is named after him.

14

Discrete Velocity Models

In the classical description, a gas is made of atoms and molecules, but when MAXWELL and BOLTZMANN developed the basic ideas for the kinetic theory of gases, they imagined a gas made of particles with no *internal structure*. In celestial mechanics, all planets are assumed to have spherical symmetry so that the gravitational field created outside is the same as that of a point mass at its centre, and the gravitational forces on another planet produce only a resulting attraction of its centre and no torque, so that there is no change in the angular momentum of the planets, and one neglects them. It would be different for planets with a magnetic field, because electromagnetism would have to be taken into account, and in a close encounter planets could exchange angular momentum through electromagnetic interaction. Actually, ALFVÉN observed in the 1970s that some of what is observed in the cosmos should be explained by electromagnetic effects, but those who adhere to the dogma of gravitation cannot learn about electromagnetism, and they prefer to invent dark matter, dark energy, dark fields, and so on, in order to avoid questioning their dogma.

The same problem occurs concerning the 19th century ideas in kinetic theory, that there has been enough evidence to show that they are wrong, but most people want to stick to them. The ideas of POINCARÉ about relativity have pointed out that there are no instantaneous forces at a distance, and his reason was that one cannot define instantaneity and that interaction between particles must be transmitted by a field, at the velocity of light, but a more compelling reason has come out from quantum mechanics, despite its dogmatic errors, that at a microscopic level there are only waves and no particles, so that the classical ideas about near collisions involving only two particles feeling a force at a distance should be rejected as a naive 19th century point of view.

The classical idea is that one expects particles to collide with other particles and the number of such collisions between a particle of type 1 and a particle of type 2 is expected to be proportional to the product of the density of particles of type 1 and the density of particles of type 2, but because this does not seem to explain what is observed, some people have tried to add a correction involving three types of particles, despite the fact that, from a classical point of view, triple collisions are expected to be extremely rare events. The only way out is to accept the fact that the classical language that one has been using since the end of the 19th century is too limited to explain what is really going on, and that one cannot avoid treating particles as the waves they really are.

Before studying the *collision operator* imagined by MAXWELL and by BOLTZMANN, which appears in what one calls the Boltzmann equation, I want to discuss simpler models, where velocities can only take a finite numbers of values, the discrete velocity models. In the early 1970s, Renée GATIGNOL offered me a copy of her book [16],[1] where she attributes the idea to MAXWELL, and although velocities should belong to $\mathbb{R}^3$, I shall start by considering a problem in $\mathbb{R}^N$. In this approach, all particles are equal with the same mass.[2] Conservation of mass in a collision just means that two particles come in and two particles come out, and one must concentrate then on conservation of momentum and conservation of energy, and that means that a collision between two particles with velocities V_i and V_j may result in two particles having velocities V_k and V_ℓ if

$$V_i + V_j = V_k + V_\ell$$
$$|V_i|^2 + |V_j|^2 = |V_k|^2 + |V_\ell|^2, \tag{14.1}$$

and particles with velocities V_k and V_ℓ may result in two particles having velocities V_i and V_j after a collision, of course. Notice that no conservation of angular momentum is mentioned, because one assumes that no angular momentum is carried away by the particles (unlike for billiard balls), and that the energy has only a translational kinetic part; apart from a possible rotational kinetic energy, molecules also show an energy related to the variations in distances between the atoms forming the molecule.

Again, one should emphasize that the preceding discussion supposed that the particles react as rigid bodies do, i.e. according to the rules of classical mechanics, but one immediately abandons the framework of classical mechanics by introducing probabilities for choosing the result of a collision, and if two particles with velocities V_i and V_j collide, one assumes that there are probabilities of transforming into the various possible pairs, counting the possibility

---

[1] Renée Yvonne FLANDRIN-GATIGNOL, French mathematician. She works at Université Paris VI (Pierre et Marie Curie), Paris, France.

[2] Dry air is composed mostly of molecules of nitrogen N_2, molecules of oxygen O_2, and atoms of argon Ar with proportions 75.5%, 23.2%, 1.3% in mass, or 78.08%, 20.94%, 0.93% in volume. In the usual circumstances there is a variable amount of carbon dioxide CO_2, but air is not dry and it contains variable amounts of water H_2O as vapour, and this humidity plays an important role in the weather conditions. This shows that the hypothesis of identical particles is not always realistic, and one should take it as a first step, for example for describing a gas like argon, whose atoms are spherical and show no chemical activity.

of emerging from the collision with the same velocities as before entering it, as if there had not been a collision; one denotes by $P_{i,j;k,\ell}$ the probability that a collision with velocities V_i and V_j creates particles having velocities V_k and V_ℓ, and as particles cannot be discerned,[3] one asks for a symmetry in i and j, and also a symmetry in k and ℓ; the natural conditions on the coefficients $P_{i,j;k,\ell}$ are then

$$P_{i,j;k,\ell} \geq 0, P_{j,i;k,\ell} = P_{i,j;\ell,k} = P_{i,j;k,\ell} \text{ for all pairs } i,j;k,l$$
$$P_{i,j;k,\ell} = 0 \text{ if } V_k + V_\ell \neq V_i + V_j \text{ or } |V_k|^2 + |V_\ell|^2 \neq |V_i|^2 + |V_j|^2$$
$$\text{for all pairs } i,j;k,l \qquad (14.2)$$
$$\textstyle\sum_{k,l} P_{i,j;k,\ell} = 1 \text{ for all pairs } i,j.$$

In general one has also $P_{k,\ell;i,j} = P_{i,j;k,\ell}$ for all pairs i,j and k,ℓ. The result of this analysis is that if $u_i(x,t)$ denotes the (nonnegative) density of particles with velocity V_i, for $i = 1, \ldots, m$, then one has

$$(u_i)_t + \big(V_i.grad(u_i)\big) + \sum_{k,\ell=1}^m A_{i,k,\ell} u_k u_\ell = 0 \text{ for } i = 1, \ldots, m, \qquad (14.3)$$

where $A_{i,k,\ell} = A_{i,\ell,k}$ for all $i,k,\ell = 1, \ldots, m$, and the coefficients $A_{i,k,\ell}$ are related to the probabilities $P_{a,b;c,d}$ in the following way. For each of the pairs k, ℓ and a, b, one puts a term $K\,P_{k,\ell;a,b} u_k u_\ell$ in the equation for u_k, a term $K\,P_{k,\ell;a,b} u_k u_\ell$ in the equation for u_ℓ, a term $-K\,P_{k,\ell;a,b} u_k u_\ell$ in the equation for u_a, and a term $-K\,P_{k,\ell;a,b} u_k u_\ell$ in the equation for u_b; this expresses the fact that a collision takes away a particle with velocity V_k and a particle with velocity V_ℓ and adds a particle with velocity V_a and a particle with velocity V_b, and that this happens with the proportion $P_{k,\ell;a,b}$; as a simplification,[4] the formula for $A_{i,k,\ell}$ is then

$$A_{i,k,\ell} = K \sum_{a,b=1}^m P_{k,\ell;a,b}(\delta_{i,k} + \delta_{i,\ell} - \delta_{i,a} - \delta_{i,b}) \text{ for all } i,k,\ell = 1, \ldots, m, \quad (14.4)$$

---

[3] If one looks at small waves on the surface of the sea, one sometimes can follow one and see it interact with other waves but it is often impossible to follow where a particular wave goes during interaction; physicists use an hypothesis of indiscernability of particles, for a similar reason, that there are actually no particles.

[4] Instead of $K\,P_{k,\ell;a,b}$ one should write $K_{k,\ell} P_{k,\ell;a,b}$, as this concerns only what happens in the collisions of particles with velocity V_k against particles with velocity V_ℓ. The results concerning the signs of the coefficients $A_{i,k,\ell}$ and the relations expressing conservation of mass, conservation of momentum and conservation of energy are unchanged; it is only when deriving the entropy inequality that one uses then a supplementary information, which in that general case is $K_{k,\ell} P_{k,\ell;a,b} = K_{a,b} P_{a,b;k,\ell}$.

where K is often written as $\frac{1}{\varepsilon}$ and ε is interpreted as a mean free path between collisions,[5] and I shall discuss later the question of letting ε tend to 0. One then deduces some useful properties of the coefficients $A_{i,k,\ell}$

$$A_{i,k,\ell} \leq 0 \text{ if } i \neq k \text{ and } i \neq \ell \text{ for all } i, k, \ell = 1, \ldots, m, \tag{14.5}$$

which is related to having nonnegative solutions for nonnegative data,

$$\sum_{i=1}^{m} A_{i,k,\ell} = 0 \text{ for all } k, \ell = 1, \ldots, m, \tag{14.6}$$

which is conservation of mass, from which one deduces

$$A_{i,k,\ell} \geq 0 \text{ if } i = k \text{ or } i = \ell \text{ for all } i, k, \ell = 1, \ldots, m, \tag{14.7}$$

and the fact that $P_{k,\ell;a,b} = 0$ unless $V_k + V_\ell = V_a + V_b$ and $|V_k|^2 + |V_\ell|^2 = |V_a|^2 + |V_b|^2$ implies that when $P_{k,\ell;a,b} \neq 0$ one has $\sum_{i=1}^{m} V_i(\delta_{i,k} + \delta_{i,\ell} - \delta_{i,a} - \delta_{i,b}) = V_k + V_\ell - V_a - V_b = 0$ and $\sum_{i=1}^{m} |V_i|^2(\delta_{i,k} + \delta_{i,\ell} - \delta_{i,a} - \delta_{i,b}) = |V_k|^2 + |V_\ell|^2 - |V_a|^2 - |V_b|^2 = 0$, so that

$$\begin{aligned} \sum_{i=1}^{m} A_{i,k,\ell} V_i &= 0 \text{ for all } k, \ell = 1, \ldots, m \\ \sum_{i=1}^{m} A_{i,k,\ell} |V_i|^2 &= 0 \text{ for all } k, \ell = 1, \ldots, m, \end{aligned} \tag{14.8}$$

which express conservation of momentum and conservation of energy.

Another important property, related to the H-theorem of BOLTZMANN, is that when $u_i > 0$ one has $\sum_{i=1}^{m} \log(u_i)(\delta_{i,k} + \delta_{i,\ell} - \delta_{i,a} - \delta_{i,b}) = \log(u_k) + \log(u_\ell) - \log(u_a) - \log(u_b) = \log(u_k u_\ell) - \log(u_a u_b)$, and if one assumes now that one has

$$P_{k,\ell;i,j} = P_{i,j;k,\ell} \text{ for all pairs } i, j \text{ and } k, \ell, \tag{14.9}$$

one deduces

$$\begin{aligned} \sum_{i,k,\ell=1}^{m} A_{i,k,\ell} u_k u_\ell \log(u_i) &= K \sum_{k,\ell,a,b=1}^{m} P_{k,\ell;a,b}\big(\log(u_k u_\ell) - \log(u_a u_b)\big) u_k u_\ell \\ &= \frac{K}{2} \sum_{k,\ell,a,b=1}^{m} P_{k,\ell;a,b}\big(\log(u_k u_\ell) - \log(u_a u_b)\big)(u_k u_\ell - u_a u_b) \geq 0, \end{aligned} \tag{14.10}$$

and the inequality stays valid if one allows some u_i to vanish. To the conservation of mass, conservation of momentum and conservation of energy, which can be written as

$$\begin{aligned} \frac{\partial}{\partial t}\Big(\sum_{i=1}^{m} u_i\Big) + \sum_{j=1}^{N} \frac{\partial}{\partial x_j}\Big(\sum_{i=1}^{m}(V_i)_j u_i\Big) &= 0 \\ \frac{\partial}{\partial t}\Big(\sum_{i=1}^{m} u_i V_i\Big) + \sum_{j=1}^{N} \frac{\partial}{\partial x_j}\Big(\sum_{i=1}^{m}(V_i)_j u_i V_i\Big) &= 0 \\ \frac{\partial}{\partial t}\Big(\sum_{i=1}^{m} u_i |V_i|^2\Big) + \sum_{j=1}^{N} \frac{\partial}{\partial x_j}\Big(\sum_{i=1}^{m}(V_i)_j u_i |V_i|^2\Big) &= 0, \end{aligned} \tag{14.11}$$

---

[5] As $\int_{\mathbb{R}^N} u_i(x,t)\,dx$ is a mass, one sees that u_i has units $mass\ length^{-N}$; each velocity V_i has units $length\ time^{-1}$ so $(u_i)_t$ and $\big(V_i.grad(u_i)\big)$ has units $mass\ length^{-N}\ time^{-1}$, and each term $K u_k u_\ell$ having those units, one sees that K has units $mass^{-1}\ length^N\ time^{-2}$, so that an interpretation of the inverse of K as a length does not seem so good, and there are other quantities involved, like scattering cross-sections.

one then adds the inequality

$$\frac{\partial}{\partial t}\left(\sum_{i=1}^{m} u_i \log(u_i)\right) + \sum_{j=1}^{N}\frac{\partial}{\partial x_j}\left(\sum_{i=1}^{m}(V_i)_j u_i \log(u_i)\right) \leq 0, \tag{14.12}$$

expressing the decay of entropy[6]

$$I(t) = \int_{\mathbb{R}^N}\left(\sum_{i=1}^{m} u_i \log(u_i)\right) dx. \tag{14.13}$$

If $\frac{dI}{dt} = 0$, then one must have $\left(\log(u_k u_\ell) - \log(u_a u_b)\right)\left(u_k u_\ell - u_a u_b\right) = 0$ whenever $P_{k,\ell;a,b} \neq 0$, i.e. $u_k u_\ell = u_a u_b$; if $P_{k,\ell;i,j} = P_{i,j;k,\ell}$ for all pairs i,j and k, ℓ, one deduces that the nonlinear terms vanish.

I shall describe in more detail later some simplified versions of a two-dimensional model introduced by MAXWELL, where there are four possible velocities, so I shall call it the *four velocities model*, with $V_1 = (1,0)$, $V_2 = (-1,0)$, $V_3 = (0,1)$ and $V_4 = (0,-1)$, but it is also known as a *Broadwell model*,[7] and because all the velocities have the same norm, kinetic energy is automatically conserved; this type of model is not so interesting for modelling a gas, because there is no possible temperature (as temperature is related to variations of $|v|^2$, as will be seen later), and the equations are

$$\begin{aligned}
(u_1)_t + (u_1)_x + \mathcal{N} &= 0 \\
(u_2)_t - (u_2)_x + \mathcal{N} &= 0 \\
(u_3)_t + (u_3)_y - \mathcal{N} &= 0 \\
(u_4)_t - (u_4)_y - \mathcal{N} &= 0,
\end{aligned} \tag{14.14}$$

where the nonlinear term $\mathcal{N}$ is usually taken to be

$$\mathcal{N} = K\left(u_1 u_2 - u_3 u_4\right), \tag{14.15}$$

expressing equal transition probabilities for pairs 1, 2 and 3, 4 to transform into each other in a collision (as these are the only different pairs corresponding to the same total momentum, equal to 0), and I shall take $K = 1$ in most of the discussions, which corresponds to looking at the equations for $K\,u_j$, or the equations for $u_j(K\,x, K\,y, K\,t)$.

---

[6] Mathematicians use a different sign convention than physicists, who have the entropy increasing; an interpretation is that entropy represents disorder created by irreversible processes, and this must have been what CLAUSIUS had in mind in inventing the concept (or at least in expressing it in clearer terms, because there seems to have been some controversy about who had the original idea at the time, and nationalistic questions may have obscured the facts); BOLTZMANN then proposed that entropy is $-\int_{\mathbb{R}^3 \times \mathbb{R}^3} f(\mathbf{x}, \mathbf{v}, t) \log\left(f(\mathbf{x}, \mathbf{v}, t)\right) d\mathbf{x}\,d\mathbf{v}$.

[7] James E. BROADWELL, American engineer. He worked at Caltech (California Institute of Technology), Pasadena, CA.

For this model, as for the general model with arbitrary coefficients $A_{i,j,k}$ (i.e. without imposing sign conditions or conservation properties), one has a local existence theorem for data in L^∞, and existence can be asserted at least for a time of the order of the inverse of the L^∞ norm of the initial data. This type of result is obtained by standard techniques of ordinary differential equations, and one can prove local existence and uniqueness of solutions for perturbations of a semi-group by a locally Lipschitz nonlinearity.[8] For a general first-order system of the form

$$(u_i)_t + \big(V_i.grad(u_i)\big) = F_i(u_1,\ldots,u_m) \text{ for } x \in \mathbb{R}^N, t > 0,$$
$$u_i \mid_{t=0} = v_i \text{ for } x \in \mathbb{R}^N, i = 1,\ldots,m,$$

(14.16)

with $v_1,\ldots,v_m \in L^\infty(\mathbb{R}^N)$, one assumes that the nonlinearities satisfy the bounds

$$|z_1|,\ldots,|z_m| \le r \text{ imply } \max_i |F_i(z_1,\ldots,z_m)| \le M(r)$$
$$|z_1|,\ldots,|z_m|,|\xi_1|,\ldots,|\xi_m| \le r \text{ imply}$$
$$\max_i |F_i(z_1,\ldots,z_m) - F_i(\xi_1,\ldots,\xi_m)| \le K(r)\max_j |z_j - \xi_j|.$$

(14.17)

Lemma 14.1. *If $\rho_0 = \max_i ||v_i||_{L^\infty(\mathbb{R}^N)}$ and the solution of $\frac{d\rho}{dt} = M(\rho)$ is finite on $[0,T]$, i.e. $T < \int_{\rho_0}^\infty \frac{d\rho}{M(\rho)}$, there is a unique solution in $\mathbb{R}^N \times (0,T)$, satisfying $|u_i(x,t)| \le \rho(t)$ a.e. $(x,t) \in \mathbb{R}^N \times (0,T)$, for $i = 1,\ldots,m$. One approaches the solution by the iterative method*

$$(u_i^{(n+1)})_t + \big(V_i.grad(u_i^{(n+1)})\big) = F_i(u_1^{(n)},\ldots,u_m^{(n)}), (x,t) \in \mathbb{R}^N \times (0,T);$$
$$u_i^{(n+1)} \mid_{t=0} = v_i, x \in \mathbb{R}^N, i = 1,\ldots,m,$$

(14.18)

Proof: If the initialization functions $u_1^{(0)},\ldots,u_m^{(0)}$ are bounded (measurable) on $\mathbb{R}^N \times (0,T)$ with $|u_i^{(0)}(x,t)| \le R_0$, for $(x,t) \in \mathbb{R}^N \times (0,T)$ and $i = 1,\ldots,m$, then one has $|u_i^{(n)}(x,t)| \le R_n(t)$, for $(x,t) \in \mathbb{R}^N \times (0,T)$ and $i = 1,\ldots,m$, where $R_n(t) = r_0 + \int_0^t M\big(R_{n-1}(s)\big)\,ds$ for $n \ge 1$. Assume that $R_n(t) \le R^\infty$ for all n and $t \in (0,T)$, and let $K^\infty = K(R^\infty)$; then if $\varepsilon_n(t) = \max_i ||u_i^{(n)}(\cdot,t) - u_i^{(n-1)}(\cdot,t)||_{L^\infty(\mathbb{R}^N)}$ for $t \in (0,T)$ and $n \ge 1$, one has $F_i(u_1^{(n)},\ldots,u_m^{(n)})(\cdot,t) - F_i(u_1^{(n-1)},\ldots,u_m^{(n-1)})(\cdot,t)||_{L^\infty(\mathbb{R}^N)} \le K^\infty \varepsilon_n(t)$ for $t \in (0,T)$, $i = 1,\ldots,m$, and $n \ge 1$, and therefore $\varepsilon_{n+1}(t) \le K^\infty \int_0^t \varepsilon_n(s)\,ds$, so that one deduces by

---

[8] Some authors seem to have been lost in technical details about *mild* solutions, because the semi-group of translation $S(t)$ defined by $\big(S(t)w\big)(x) = w(x - a\,t)$ is not a strongly continous semi-group if $a \ne 0$, i.e. $S(t)w$ may not converge to w in L^∞ norm as t tends to 0, but $S(t)w \rightharpoonup w$ in L^∞ weak $\star$ anyway, and one just has to observe that the natural solution in the sense of distributions of $u_t + a.u_x = f$ in $\mathbb{R}^N \times (0,\infty)$ with $u \mid_{t=0} = v$ is indeed given by $u(\cdot,t) = S(t)v(\cdot) + \int_0^t S(t-s)f(\cdot,s)\,ds$.

induction that $\varepsilon_{n+1}(t) \leq \frac{1}{n!}(K^\infty t)^n \sup_{s \in (0,t)} \varepsilon_1(s)$ for $t \in (0,T)$, showing the uniform convergence of u_n to a limit, which is then the desired solution.

If $R_{n-1} = \rho$ on $(0,T)$, then $R_n = \rho$ on $(0,T)$, so that one way to create a bounded sequence and to prove existence is to choose the initial guess $u^{(0)}$ such that $|u_i^{(0)}(x,t)| \leq \rho(t)$, for $(x,t) \in \mathbb{R}^N \times (0,T)$ and $i = 1,\ldots,m$, for example by taking all u_i^0 equal to 0. However, it is important to prove uniqueness without imposing too precise bounds, so if a solution is bounded by R_0, one chooses $0 < S < \int_{R_0}^\infty \frac{dz}{M(z)}$, and the argument shows that it coincides with the obtained solution for $0 \leq t \leq \min\{T,S\}$, and if $T > S$ one starts the argument again with initial time S and the solution must then coincide with the obtained solution for $0 \leq t \leq \min\{T, 2S\}$, etc.   □

The preceding argument with $M(r) = C r^2$ gives $T < \frac{C}{r_0}$, and this result is valid without any sign condition on the coefficients $A_{i,j,k}$ or on the initial data.

The uniqueness property shows that if the initial data are periodic in a direction, then the solution is periodic in that direction, i.e. if there exists $h \in \mathbb{R}^N$ such that $v_i(x + h) = v_i(x)$ a.e. $x \in \mathbb{R}^N$ for $i = 1,\ldots,m$, then one has $u_i(x + h, t) = u_i(x,t)$ a.e. $(x,t) \in \mathbb{R}^N \times (0,T)$ for $i = 1,\ldots,m$, where T is chosen according to Lemma 14.1. Indeed, in all cases, if one defines $\widetilde{u}_i$ by $\widetilde{u}_i(x,t) = u_i(x + h, t)$ for $i = 1,\ldots,m$, then it satisfies the equation for initial data $\widetilde{v}_i$ defined by $\widetilde{v}_i(x) = v_i(x + h)$ for $i = 1,\ldots,m$; if then $\widetilde{v}_i = v_i$ for $i = 1,\ldots,m$, the uniqueness property implies $\widetilde{u}_i = u_i$ for $i = 1,\ldots,m$. As a consequence, if the initial data are independent of one direction, the solution is independent of that direction; for example, if in the four velocities model the initial data are independent of y, then the solution is independent of y, and one finds the one-dimensional four velocities model

$$(u_1)_t + (u_i)_x + \mathcal{N} = (u_2)_t - (u_2)_x + \mathcal{N} = (u_3)_t - \mathcal{N} = (u_4)_t - \mathcal{N} = 0, \quad (14.19)$$

with $\mathcal{N} = K(u_1 u_2 - u_3 u_4)$ and initial data depending only upon x (and belonging to $L^\infty(\mathbb{R})$); the presence of the $u_3 u_4$ term in $\mathcal{N}$ may look strange from a physical point of view, because one does not expect particles with the same velocity to interact, but one should remember that 0 is not the velocity of the particles of the third and fourth families, but the projection of their velocity on the x axis; a new symmetry arises in this model, which was not true for the initial (two-dimensional) four velocities model, that the equation becomes invariant by exchanging u_3 and u_4,[9] and in the case where $u_3 = u_4$

---

[9] It means that if one defines $\widetilde{u}_1 = u_1, \widetilde{u}_2 = u_2, \widetilde{u}_3 = u_4, \widetilde{u}_4 = u_3$ then one obtains the solution for initial data $\widetilde{v}_1 = v_1, \widetilde{v}_2 = v_2, \widetilde{v}_3 = v_4, \widetilde{v}_4 = v_3$; one deduces that if $v_3 = v_4$ in $\mathbb{R}^2$, then $u_3 = u_4$ in $\mathbb{R}^2 \times (0,T)$. A different symmetry exists for the two-dimensional four velocities model, where one exchanges u_3 and u_4 but one also changes y into $-y$ (and one can also exchange u_1 and u_2 and change x into $-x$), i.e. one defines $\widetilde{u}_1(x,y,t) = u_1(x,-y,t), \widetilde{u}_2(x,y,t) = u_2(x,-y,t), \widetilde{u}_3(x,y,t) = u_4(x,-y,t), \widetilde{u}_4(x,y,t) = u_3(x,-y,t)$ then one obtains the solution for initial

one obtains the Broadwell model,[10] where I have used $u = u_1$, $v = u_2$ and $w = u_3 = u_4$,

$$u_t + u_x + u\,v - w^2 = 0$$
$$v_t - v_x + u\,v - w^2 = 0 \qquad (14.20)$$
$$w_t - u\,v + w^2 = 0,$$

where the density of mass is $u + v + 2w$, the density of momentum in x is $u - v$ (and 0 for the density of momentum in y as $u_3 = u_4 = w$), the density of kinetic energy is proportional to mass, and the density of entropy is $u \log(u) + v \log(v) + 2w \log(w)$ for the case of nonnegative data (as the solution is nonnegative for $t > 0$).

The uniqueness property can be rendered more powerful by making the statements local instead of global, and for this one should notice an important finite speed of propagation effect.

Lemma 14.2. *If initial data belong to $L^\infty(\mathbb{R}^N)$, then for $t > 0$ the solution at (x,t) only depends upon the initial data at points $y \in \{x\} - t\,conv\{V_1, \ldots, V_m\}$, where conv A is the convex hull of A, i.e. of the form $y = x - t\sum_i \theta_i V_i$ for some $\theta_i \geq 0$ for $i = 1, \ldots, m$, with $\sum_i \theta_i = 1$.*

Proof: One initializes the iterative method with $u_i^{(0)} = 0$ for $i = 1, \ldots, m$, and one notices by induction that each $u_i^{(n)}(x,t)$ only depends upon the initial data on $\{x\} - t\,conv\{V_1, \ldots, V_m\}$. This follows from the formula $u_i^{(n)}(x,t) = v_i(x - t\,V_i) + \int_0^t F_i\big(u_1^{(n-1)}(x - s\,V_i, t - s), \ldots, u_m^{(n-1)}(x - s\,V_i, t - s)\big)\,dx$, using the fact that $x - t\,V_i \in \{x\} - t\,conv\{V_1, \ldots, V_m\}$ and that $\{x - s\,V_i\} - (t - s)\,conv\{V_1, \ldots, V_m\} \subset \{x\} - t\,conv\{V_1, \ldots, V_m\}$.   □

This result permits us to compare solutions which are not necessarily defined on a strip $\mathbb{R}^N \times (0, T)$ but on a set $A \subset \mathbb{R}^N \times (0, \infty)$ with A such that $(x,t) \in A$ implies $(y, t - s) \in A$ for all $s \in (0, t)$ and all $y \in \{x\} - s\,conv\{V_1, \ldots, V_m\}$.

It is useful then to develop criteria which are necessary, or sufficient, for the solution to be nonnegative when the initial data are nonnegative, as this corresponds to the physical property that a density of particles should be nonnegative.

---

data $\widetilde{v}_1(x, y) = v_1(x, -y), \widetilde{v}_2(x, y) = v_2(x, -y), \widetilde{v}_3(x, y) = v_4(x, -y), \widetilde{v}_4(x, y) = v_3(x, -y)$.

[10] One may start from a three-dimensional six velocities model, where one adds velocities $V_5 = (0, 0, +1)$ and $V_6 = (0, 0, -1)$, and the nonlinearity in the first and second families is $2u_1 u_2 - u_3 u_4 - u_5 u_6$ for example; if one starts with data independent of y, z, then the solution is independent of y, z and if one imposes also $v_3 = v_4 = v_5 = v_6$, then one has $u_3 = u_4 = u_5 = u_6$ for $t > 0$, and one finds the model $u_t + u_x + 2u\,v - 2w^2 = v_t - v_x + 2u\,v - 2w^2 = w_t - u\,v + w^2 = 0$, where the density of mass is $u + v + 4w$, etc.

Lemma 14.3. *If for $i = 1, \ldots, m$, the function F_i has (also) the property that $z_j \geq 0$ for $j = 1, \ldots, m$, and $z_i = 0$ imply $F_i(z_1, \ldots, z_m) \geq 0$, then for nonnegative initial data the solution is nonnegative for $t \geq 0$ (as long as it exists).*

Proof: One assumes that $0 \leq v_i(x) \leq r_0$ a.e. $x \in \mathbb{R}^N$ for $i = 1, \ldots, m$, and one chooses $T < \int_{r_0}^{\infty} \frac{dr}{M(r)}$, so that the solution of $\frac{d\rho}{dt} = M(\rho)$ with $\rho(0) = r_0$ is well defined on $[0, T]$; one chooses $\lambda \geq K(\rho(T))$, and one uses a different iterative technique, where one starts from $u_i^{(0)} = 0$ for $i = 1, \ldots, m$, but for $n \geq 1$ one defines $u_i^{(n)}$ by

$$
(u_i^{(n)})_t + (V_i.grad(u_i^{(n)})) + \lambda u_i^{(n)} = \lambda u_i^{(n-1)} + F_i(u_1^{(n-1)}, \ldots, u_m^{(n-1)}),
$$
$$
u_i \mid_{t=0} = v_i, \text{ for } i = 1, \ldots, m.
$$

$$(14.21)$$

If $0 \leq u_j^{(n-1)}(x, t) \leq \rho(t)$ a.e. $x \in \mathbb{R}^N, t \in (0, T)$ for $j = 1, \ldots, m$, then one has $0 \leq \lambda u_i^{(n-1)} + F_i(u_1^{(n-1)}, \ldots, u_m^{(n-1)}) \leq \lambda \rho(t) + M(\rho(t))$ because of the choice of λ and the definition of M; this implies $0 \leq u_i^{(n)} \leq r(t)$, where r is the solution of $\frac{dr}{dt} + \lambda r = \lambda \rho + M(\rho)$ with $r(0) = r_0$, which gives $r(t) = \rho(t)$ for $t \in (0, T)$. Having uniform bounds for all $u_i^{(n)}$, one estimates the differences $u_i^{(n)} - u_i^{(n-1)}$ in L^{∞} norm as was done before, and $u^{(n)}$ then converges uniformly to a fixed point, which is the solution. □

The conditions imposed on the functions F_i for $i = 1, \ldots, m$, are then sufficient to obtain nonnegative solutions for $t > 0$ when the initial data are nonnegative, but they are actually also necessary conditions, and a more general result is proven first in the case of ordinary differential equations.

Definition 14.4. *A closed set $C \subset \mathbb{R}^m$ is* forward invariant *for the differential equation $\frac{dz}{dt} = F(z)$ (with F locally Lipschitz), if $z(0) \in C$ implies $z(t) \in C$ for $t \geq 0$ as long as the solution exists.*

Lemma 14.5. *If F is a locally Lipschitz mapping and C is a closed set of $\mathbb{R}^m$, then it is forward invariant for the differential equation $\frac{dz}{dt} = F(z)$ if and only if C satisfies the condition*

$$
dist(c + \varepsilon F(c); C) = o(\varepsilon) \text{ for } \varepsilon > 0 \text{ small, for all } c \in C
$$
$$
\text{(or equivalently, for all } c \in \partial C).
$$

$$(14.22)$$

Proof: If $z(0) = c_0 \in C$, then one has $z(\varepsilon) = c_0 + \varepsilon F(c_0) + o(\varepsilon)$; if C is forward invariant one has $z(\varepsilon) \in C$ and so the distance from $c_0 + \varepsilon F(c_0)$ to C is less than or equal to the distance from $c_0 + \varepsilon F(c_0)$ to $z(\varepsilon)$, which is $o(\varepsilon)$.

Conversely, assume that C satisfies the condition; to simplify the argument assume that F is globally Lipschitz continuous (with constant K); then one has $|z_1(t) - z_2(t)| \leq e^{Kt}|z_1(0) - z_2(0)|$ for $t > 0$ for any two solutions of the differential equation. Let $z(t)$ be any solution of the differential equation, and for some time t_0 choose a projection ξ_0 of $z(t_0)$ onto

C, and let ξ be the solution of the differential equation with $\xi(t_0) = \xi_0$; then for $\varepsilon > 0$ one has $|z(t_0 + \varepsilon) - \xi(t_0 + \varepsilon)| \leq e^{K\varepsilon}|z(t_0) - \xi_0|$, and as $\xi(t_0 + \varepsilon) = \xi_0 + \varepsilon\,F(\xi_0) + o(\varepsilon)$ one has $dist(\xi(t_0 + \varepsilon); C) = o(\varepsilon)$ and therefore $dist(z(t_0 + \varepsilon); C) \leq |z(t_0 + \varepsilon) - \xi(t_0 + \varepsilon)| + dist(\xi(t_0 + \varepsilon); C) \leq e^{K\varepsilon}dist(z(t_0); C) + o(\varepsilon) = dist(z(t_0); C) + K\varepsilon\,dist(z(t_0); C) + o(\varepsilon)$, showing that $\frac{d[dist(z;C)]}{dt}\big|_{t=t_0} \leq K\,dist(z(t_0); C)$ (where the derivative is a right derivative), and as this holds for all t_0 one deduces that $dist(z(s); C) \leq e^{K s}dist(z(0); C)$ for all $s \geq 0$, showing that $z(0) \in C$ implies $z(s) \in C$ for all $s \geq 0$. $\square$

Definition 14.6. *For a system* $(u_i)_t + (V_i.grad(u_i)) = F_i(u_1, \ldots, u_m)$, $i = 1, \ldots, m$, *with* F_i *locally Lipschitz for* $i = 1, \ldots, m$, *a closed subset* C *of* $\mathbb{R}^m$ *is* forward invariant *if when the initial data satisfy* $v(x) \in C$ *a.e.* $x \in \mathbb{R}^N$ *then the solution satisfies* $u(x, t) \in C$ *a.e.* $x \in \mathbb{R}^N, t \in (0, T)$ *(as long as the solution exists).*

Using initial data independent of x, u must solve the differential equation $\frac{du}{dt} = F(u)$, and C must then be forward invariant for the differential equation, so that it must satisfy the condition $dist(c + \varepsilon\,F(c); C) = o(\varepsilon)$ for $\varepsilon > 0$ small and all $c \in C$. The presence of the transport part in the equation imposes supplementary conditions on C: taking as an example the Broadwell model, which has three distinct velocities, let (u_1, v_1, w_1) and (u_2, v_2, w_2) be two points in C, and consider the initial data

$$u_0(x) = \begin{cases} u_1 \text{ for } x < 0 \\ u_2 \text{ for } x > 0 \end{cases}, v_0(x) = \begin{cases} v_1 \text{ for } x < 0 \\ v_2 \text{ for } x > 0 \end{cases}, w_0(x) = \begin{cases} w_1 \text{ for } x < 0 \\ w_2 \text{ for } x > 0 \end{cases}, \tag{14.23}$$

then one has for small $t > 0$

$$u(x, t) = \begin{cases} u_1 + O(t) \text{ for } x < t \\ u_2 + O(t) \text{ for } x > t \end{cases},$$
$$v(x, t) = \begin{cases} v_1 + O(t) \text{ for } x < -t \\ v_2 + O(t) \text{ for } x > -t \end{cases}, \tag{14.24}$$
$$w(x, t) = \begin{cases} w_1 + O(t) \text{ for } x < 0 \\ w_2 + O(t) \text{ for } x > 0 \end{cases},$$

and this shows that a forward invariant set C must be a product, because in the region $-t < x < 0$ one has points of C of the form $(u_1, v_2, w_1) + O(t)$, and in the region $0 < x < t$ one has points of C of the form $(u_1, v_2, w_2) + O(t)$, and as C is closed one finds that $(u_1, v_2, w_1) \in C$ and $(u_1, v_2, w_2) \in C$, so that C has the form $C_u \times C_v \times C_w$; one extends easily this condition to an arbitrary system.

For the Broadwell model, one can deduce easily all the forward invariant sets: they are the points corresponding to constant solutions (C_u, C_v, C_w reduced to one point, satisfying $u\,v = w^2$), the degenerate solutions ($C_v = $

$C_w = \{0\}$, corresponding to $v = w = 0$ and u satisfying $u_t + u_x = 0$) or ($C_u = C_w = \{0\}$, corresponding to $u = w = 0$ and v satisfying $v_t - v_x = 0$), the nonnegative quadrant corresponding to nonnegative solutions ($C_u = C_v = C_w = [0, \infty)$), and the whole space ($C_u = C_v = C_w = R$).

It is natural then to look for bounds depending upon t, and for initial data satisfying $0 \leq u_0 \leq a_0, 0 \leq v_0 \leq b_0, 0 \leq w_0 \leq c_0$ a.e. $x \in R$, one wants to deduce that one has $0 \leq u(x,t) \leq a(t), 0 \leq v(x,t) \leq b(t), 0 \leq w(x,t) \leq c(t)$ a.e. $x \in R, t > 0$, for the best possible bounds a, b, c, but apart from the trivial necessary condition that $\frac{da}{dt}(0) \geq c_0^2, \frac{db}{dt}(0) \geq c_0^2, \frac{dc}{dt}(0) \geq a_0 b_0$, one does not know how to derive other necessary conditions.[11]

One may want to impose a stronger condition, that the rectangular box used, i.e. $C(t) = [0, a(t)] \times [0, b(t)] \times [0, c(t)]$ is such that if the initial data takes its values in $C(s)$ then the solution at time t takes its values in $C(s+t)$. One can write necessary conditions for this condition to hold: because one may use points where $u_0(x) = a(s), v_0(x) = 0, w_0(x) = c(s)$, one finds that one must have $\frac{da}{dt}(s) \geq c^2(s)$, and similarly one must have $\frac{db}{dt}(s) \geq c^2(s)$, and $\frac{dc}{dt}(s) \geq a(s)b(s)$, but these differential inequalities do not have global solutions if a_0 or b_0 or c_0 is > 0.[12]

There are only special cases in kinetic theory where one can find bounded forward invariant regions, giving easily global L^∞ bounds, and one such example, which I learnt from Henri CABANNES,[13] consists in starting from the (two-dimensional) four velocities model, and uses data depending only upon $x+y$, so that the solutions are functions of $x+y$ and t and the system obtained is then

$$(u_1)_t + (u_1)_z + \mathcal{N} = (u_2)_t - (u_2)_z + \mathcal{N} = (u_3)_t + (u_3)_z - \mathcal{N}$$
$$= (u_4)_t - (u_4)_z - \mathcal{N} = 0 \text{ with } \mathcal{N} = u_1 u_2 - u_3 u_4, \tag{14.25}$$

and one deduces that

$$(u_1 + u_3)_t + (u_1 + u_3)_z = (u_2 + u_4)_t - (u_2 + u_4)_z = 0, \tag{14.26}$$

and therefore for any $A, B > 0$ one has the forward invariant region

$$0 \leq u_1, u_2, u_3, u_4, u_1 + u_3 \leq A, u_2 + u_4 \leq B. \tag{14.27}$$

---

[11] As will be shown in the next lecture, if $a_0, b_0, c_0 < \infty$, then the solution exists globally, as I proved with Michael CRANDALL in 1975, extending an idea of Takaaki NISHIDA and MIMURA. However, the best possible bounds in L^∞ are not known.

[12] The system of equations $\frac{da}{dt} = \frac{db}{dt} = c^2$ and $\frac{dc}{dt} = a\,b$ gives lower bounds and it may be solved by quadratures, because $a - b$ and $a^3 - 3a^2 b + 2c^3$ are constants, but as soon as a, b, c are all positive a lower bound which blows up in finite time is easily obtained, as $\varphi = \min\{a, b, c\}$ satisfies $\frac{d\varphi}{dt} \geq \varphi^2$, and the time of existence is $\leq \frac{1}{\varphi(0)}$.

[13] Henri CABANNES, French mathematician, born in 1923. He worked at Université Paris VI (Pierre et Marie Curie), Paris, France.

One has a system with order-preserving property if $\frac{\partial F_i}{\partial u_j} \geq 0$ for $i \neq j$, and this does not occur for systems from kinetic theory, but it occurs for the Carleman model,[14]

$$u_t + u_x + u^2 - v^2 = 0$$
$$v_t - v_x - u^2 + v^2 = 0, \tag{14.28}$$

which has the bounded invariant regions

$$0 \leq u, v \leq M, \tag{14.29}$$

in which the solution is order preserving, as was noticed by Ignace KOLODNER in the early 1960s. Not knowing about his work, I had rediscovered that property in the early 1970s, but I knew that such a property is not shared by models from kinetic theory, and I decided to work on obtaining L^∞ bounds for the Broadwell model, and during the year 1974–1975 that I spent at the University of Wisconsin in Madison, I discussed that question with Michael CRANDALL. There is an L^1 contraction property for the Carleman model, which had been noticed by Thomas LIGGETT,[15] but I had noticed that the models for which an L^1 contraction property was known (including also the Burgers equation and the porous medium equation) were also order preserving, and I had shown Michael CRANDALL why it was necessary because of a conservation of integral, and he had shown the converse, and I have given our result at Lemma 4.1, but there was no objection for an L^1 contraction property based on a distance on $\mathbb{R}^m$ different from the Euclidean distance. We found a distance adapted to the nonlinear terms, but it is not compatible with the transport terms; after three months we had not found much, when we were given an article from Takaaki NISHIDA and MIMURA which opened a new approach.[16]

Although I am discussing at length mathematical questions about discrete velocity models, one should be aware of the limitations of such models; actually, when I first met Clifford TRUESDELL,[17] he told me that these models are not a good replacement for the Boltzmann equation, because they are not invariant by rotation, but at the time I could not see why that should be so

---

[14] The model is found in an appendix of a book by CARLEMAN [1], but as the work of CARLEMAN was not finished when he died, the book was edited by Lennart CARLESON and FROSTMAN, who completed some proofs; if the model was found in CARLEMAN's papers, they must have known that it is not a good model of kinetic theory, as momentum is not conserved, but it has an entropy inequality.

[15] Thomas Milton LIGGETT, American mathematician, born in 1944. He works at UCLA (University of California Los Angeles), Los Angeles, CA.

[16] Masayasu MIMURA, Japanese mathematician. He works in Hiroshima, Japan.

[17] Clifford Ambrose TRUESDELL III, American mathematician, 1919–2000. He had worked at Indiana University, Bloomington, IN, and at Johns Hopkins University, Baltimore, MD, where I first met him, in the spring of 1975.

important. He may have thought of the problem posed by the formal limit $\varepsilon \to 0$, i.e. the Hilbert expansion in the case of the Boltzmann equation, which is formal (despite having been proposed by HILBERT, so it should have been called the Hilbert conjecture, but it may have been known to BOLTZMANN),[18] and the first term is the Euler equation for an ideal gas, which is isotropic. Conversely, the formal expansion for (14.14)–(14.15) when $K \to +\infty$ consists in postulating that

$$u_j = U_j + K^{-1}U_j^1 + K^{-2}U_j^2 + \ldots \quad \text{for } j = 1, 2, 3, 4, \tag{14.30}$$

so that

one conjectures that $u_j \rightharpoonup U_j$ weakly for $j = 1, 2, 3, 4,$ $\qquad$ (14.31)

with

$$
\begin{aligned}
(U_1 + U_2 + U_3 + U_4)_t + (U_1 - U_2)_x + (U_3 - U_4)_y &= 0 \\
(U_1 - U_2)_t + (U_1 + U_2)_x &= 0 \\
(U_3 - U_4)_t + (U_3 + U_4)_y &= 0 \\
U_1 U_2 - U_3 U_4 &= 0,
\end{aligned}
\tag{14.32}
$$

and the first equation of (14.32) corresponds to conservation of mass

$$\varrho_t + (q_1)_x + (q_2)_y = 0, \tag{14.33}$$

with density of mass ϱ, density of linear momentum $q = \varrho V$, and macroscopic velocity V given by

$$\varrho = U_1 + U_2 + U_3 + U_4, \ q_1 = \varrho V_1 = U_1 - U_2, \ q_2 = \varrho V_2 = U_3 - U_4, \tag{14.34}$$

which imply

$$\varrho \geq 0, \ |V_1| + |V_2| \leq 1, \ \text{almost everywhere.} \tag{14.35}$$

The last equation of (14.32) then serves in expressing each U_j in terms of $\varrho, q_1, q_2,$

$$
\begin{aligned}
U_1 &= \tfrac{\varrho}{4} + \tfrac{q_1^2 - q_2^2}{4\varrho} + \tfrac{q_1}{2} \\
U_2 &= \tfrac{\varrho}{4} + \tfrac{q_1^2 - q_2^2}{4\varrho} - \tfrac{q_1}{2} \\
U_3 &= \tfrac{\varrho}{4} + \tfrac{q_2^2 - q_1^2}{4\varrho} + \tfrac{q_2}{2} \\
U_4 &= \tfrac{\varrho}{4} + \tfrac{q_2^2 - q_1^2}{4\varrho} - \tfrac{q_2}{2},
\end{aligned}
\tag{14.36}
$$

---

[18] I have been told that BOLTZMANN expected that some macroscopic properties of a fluid, like viscosity and heat conductivity, would be very dependent upon which law describes the forces at a distance between particles, so that he could deduce what kind of forces exist at a microscopic level from macroscopic measurements. The formal expansion shattered his belief, as the first term is the Euler equation for an ideal gas whatever the law about forces is, so that either the formal expansion is wrong, or the Boltzmann equation is not a good model for describing real gases! For different reasons, one knows now that the Boltzmann equation is not a good model for describing real gases, but no one knows if the conjecture about the expansion, which is a question in mathematics, is true or not.

but the second and third equations of (14.32) then give

$$(q_1)_t + \left(\frac{q_1^2 - q_2^2}{2\varrho} + \frac{\varrho}{2}\right)_x = 0$$
$$(q_2)_t + \left(\frac{q_2^2 - q_1^2}{2\varrho} + \frac{\varrho}{2}\right)_y = 0,$$

(14.37)

which do not resemble the corresponding equations for the balance of linear momentum given by Newton's law,[19] and in consequence (14.32) cannot be interpreted as describing the motion of a fluid, because the directions of the axes play an important role so that there is no isotropy, and there is no Galilean invariance, and I guess that this was a reason behind Clifford TRUESDELL's remark.

In 1989, I thought of a way that could help avoid the angular cut-off which has been used in the Boltzmann equation since the work of Harold GRAD,[20] and I mentioned it to Pierre-Louis LIONS.[21] Back in 1983, I had understood why the Boltzmann equation is not a good physical model, and I had told him, maybe in too cryptic terms, as I had said that there are only mathematical problems on the Boltzmann equation; he may not have understood that I meant that it is a bad physical model, and I had added that they were the question of removing the angular cut-off and the question of letting ε tend to 0. In 1989, I had been playing with a different problem, for which the fact that the Fourier transform of the uniform measure on the unit circle decays at infinity (and it decays in $1/\sqrt{r}$) was quite useful, and I could observe that memory is a curious phenomenon because I had the feeling that I had seen something like that before, but I could not remember where, and after a while I thought of checking an article of Charles FEFFERMAN,[22] in the proceedings of ICM (International Congress of Mathematicians) 1974 in Vancouver, and there it was, about his work and the work of STEIN,[23] on restrictions of Fourier transform on spheres. Because of invariance by rotations, the computation of

---

[19] The corresponding equations for Euler equation are $(q_1)_t + \left(\frac{q_1^2}{\varrho} + p\right)_x + \left(\frac{q_1 q_2}{\varrho}\right)_y = 0$ and $(q_2)_t + \left(\frac{q_1 q_2}{\varrho}\right)_x + \left(\frac{q_2^2}{\varrho} + p\right)_y = 0$.

[20] Harold GRAD, American physicist, 1923–1987. He had worked at NYU (New York University), New York, NY.

[21] Pierre-Louis LIONS, French mathematician, born in 1956. He received the Fields medal in 1994. He worked at Université Paris IX-Dauphine, Paris, France, and he holds now a chair (équations aux dérivées partielles et applications, 2002–) at Collège de France, Paris.

[22] Charles Louis FEFFERMAN, American mathematician, born in 1949. He received the Fields Medal in 1978. He worked at the University of Chicago, Chicago, IL, and he works now at Princeton University, Princeton, NJ.

[23] Elias M. STEIN, Belgian-born mathematician, born in 1931. He received the Wolf Prize in 1999, for his contributions to classical and "Euclidean" Fourier analysis and for his exceptional impact on a new generation of analysts through his eloquent teaching and writing, jointly with Laszlo LOVASZ. He worked at the University of Chicago, Chicago, IL, and at Princeton University, Princeton, NJ.

the bilinear term $Q(f,f)$ for the Boltzmann equation (in $\mathbb{R}^3$) contains the evaluation of integrals on circles, of bilinear quantities similar to convolution products, and when I told Pierre-Louis LIONS about theorems of restriction on circles I expected him to be interested in collaborating with me on that question; he was obviously not interested, but later he seemed to rediscover a property of smoothing by convolution with uniform measures on spheres, and I assume that it is different from what I had in mind, which I never checked in detail. I also thought that Clifford TRUESDELL's remark could have been about this kind of effect, which cannot be seen in discrete velocity models.

The basic estimate that I thought useful for estimates on the collision kernel is to show that $f \star g$ has a restriction on the circle $\mathbb{S}^1$,[24] for f, g in suitable Lorentz spaces $L^{p,q}(\mathbb{R}^2)$; the uniform measure $d\theta$ on $\mathbb{S}^1$ has $\mathcal{F}d\theta \in L^\infty(\mathbb{R}^2) \cap L^{4,\infty}(\mathbb{R}^2)$ because it decays in $|\xi|^{-1/2}$, and for f, g smooth one has $\int_{\mathbb{S}^1} f \star g \, d\theta = \int_{\mathbb{R}^2} \mathcal{F}f \, \mathcal{F}g \, \mathcal{F}d\theta$, so that (using estimates for the absolute values of f and g) one bounds the $L^1(\mathbb{S}^1)$ norm of $(f \star g)\big|_{\mathbb{S}^1}$ by bounding $\mathcal{F}f \, \mathcal{F}g$ in $L^{4/3,1}(\mathbb{R}^2)$, giving

$$(f \star g)\big|_{\mathbb{S}^1} \in L^1(\mathbb{S}^1) \text{ if } f \in L^2(\mathbb{R}^2) \text{ and } g \in L^{4/3,2}(\mathbb{R}^2),$$
$$\text{or } f \in L^{4/3,2}(\mathbb{R}^2) \text{ and } g \in L^2(\mathbb{R}^2)$$
$$\text{or } f \in L^{p,r}(\mathbb{R}^2), g \in L^{q,r'}(\mathbb{R}^2) \text{ with } \tfrac{4}{3} < p, q < 2, \tfrac{1}{p} + \tfrac{1}{q} = \tfrac{5}{4}, 1 \leq r \leq \infty$$
$$\tag{14.38}$$

and that uses the Lions–Peetre interpolation theory.[25],[26]

[Taught on Friday October 5, 2001.]

Notes on names cited in footnotes for Chapter 14, CARLESON,[27] FROSTMAN,[28] LOVASZ,[29] and for the preceding footnotes, THOMPSON,[30] YALE,[31] EÖTVÖS.[32]

---

[24] The curvature of the circle plays a role, and $f \star g$ may not have traces on segments.

[25] Of course, it was Jacques-Louis LIONS who developed the theory of interpolation spaces, and not his son Pierre-Louis.

[26] Jaak PEETRE, Estonian-born mathematician, born in 1935. He worked at Lund University, Sweden.

[27] Lennart CARLESON, Swedish mathematician, born in 1928. He received the Wolf Prize in 1992, for his fundamental contributions to Fourier analysis, complex analysis, quasi-conformal mappings and dynamical systems, jointly with John G. THOMPSON. He worked in Uppsala and in Stockholm, Sweden.

[28] Otto FROSTMAN, Swedish mathematician, 1907–1977. He had worked in Stockholm, Sweden.

(footnotes 29–32 on next page)

---

[29] Laszlo LOVASZ, Hungarian-born mathematician, born in 1948. He received the Wolf Prize in 1999, for his outstanding contributions to combinatorics, theoretical computer science and combinatorial optimization, jointly with Elias M. STEIN. He works at Yale University, New Haven, CT, and Eötvös University, Budapest, Hungary.

[30] John Griggs THOMPSON, American-born mathematician, born in 1932. He received the Wolf Prize in 1992, for his profound contributions to all aspects of finite group theory and connections with other branches of mathematics, jointly with Lennart CARLESON. He worked in Cambridge, England.

[31] Elihu YALE, American-born English philanthropist, Governor of Fort St George, Madras, India, 1649–1721. Yale University, New Haven, CT, is named after him.

[32] Baron Loránd EÖTVÖS, Hungarian physicist, 1848–1919. Eötvös University, Budapest, Hungary, is named after him.

15

The Mimura–Nishida and the Crandall–Tartar Existence Theorems

A different idea was needed for discrete velocity models, like the Broadwell model, a simplified version of the Maxwell four velocities model, and it came from the strong Japanese school specialized in questions of kinetic theory and fluid dynamics, by a result of MIMURA and Takaaki NISHIDA, who mixed L^1 and L^∞ estimates in the following way.

Lemma 15.1. *(Mimura–Nishida) For the Broadwell model, for every $k > 1$ there exists $\varepsilon(k) > 0$ such that if the initial data satisfy*

$$
\begin{aligned}
&u_0, v_0, w_0 \in L^\infty(\mathbb{R}) \cap L^1(\mathbb{R}) \\
&0 \leq u_0, v_0, w_0 \leq M_0 \ \text{a.e. in } \mathbb{R} \\
&\int_{\mathbb{R}} (u_0 + v_0 + 2w_0) \, dx \leq \varepsilon(k),
\end{aligned}
\tag{15.1}
$$

then the solution exists for all $t > 0$ and satisfies

$$
0 \leq u(x,t), v(x,t), w(x,t) \leq k \, M_0 \ \text{in } \mathbb{R} \times (0, \infty). \square
\tag{15.2}
$$

I shall show their proof in a moment, but in explaining the general result that I derived after that with Michael CRANDALL, one does not need to know the method of proof of the result, and the argument extends to general systems if one can prove a preliminary result of the type obtained by MIMURA and Takaaki NISHIDA (without needing the refinement that k can be taken as near to 1 as one likes), and one deduces a global existence theorem for nonnegative bounded data by using the finite speed of propagation property and an entropy inequality.

Proposition 15.2. *(Crandall–Tartar) Let a general system satisfy $A_{i,j,k} \leq 0$ if $i \neq j$ and $i \neq k$, so that nonnegative data give rise to nonnegative solutions, and satisfy an entropy inequality, i.e. such that $\sum_{i,j,k} A_{i,j,k} w_j w_k \log(w_i) \geq 0$ for all $w \in \mathbb{R}^m$ such that $w_i \geq 0$ for all i, so that $\int_{\mathbb{R}^N} \left(\sum_i u_i \log(u_i) \right) dx$ is nonincreasing for nonnegative data with compact support.*

Assume that a Mimura–Nishida estimate holds, i.e. that there exists $k_0 \geq 1$ and $\varepsilon_0 > 0$ such that for initial data such that

$$0 \leq v_i \leq M_0, i = 1, \ldots, m \text{ a.e. in } \mathbb{R}^N$$
$$\int_{\mathbb{R}^N} \left(\sum_{i=1}^m v_i \right) dx \leq \varepsilon_0, \tag{15.3}$$

the solution exists globally for $t > 0$ and satisfies

$$0 \leq u_i \leq k_0 M_0, i = 1, \ldots, m \text{ a.e. in } \mathbb{R}^N \times (0, \infty). \tag{15.4}$$

Then, there exists a function $F(t, M)$ (depending upon ε_0, k_0, m, N but not on the precise values of the coefficients $A_{i,j,k}$ for example) such that for any bounded nonnegative data the solution exists for all $t > 0$, and satisfies

$$0 \leq v_i \leq M_0, i = 1, \ldots, m \text{ a.e. in } \mathbb{R}^N \text{ implies}$$
$$0 \leq u_i(x, t) \leq F(t, M_0), i = 1, \ldots, m \text{ a.e. in } \mathbb{R}^N \times (0, \infty). \square \tag{15.5}$$

Before showing the proof, a few remarks are in order. In principle, the proposition applies to all dimensions N, but in practice one only knows how to use it in one dimension, and it seems unlikely (but not impossible) that there are genuine cases of N-dimensional problems where it applies. The reason is scaling,[1] as all the models with linear transport and quadratic nonlinearities are invariant if one changes $u(x, t)$ into $\lambda u(\lambda x, \lambda t)$, and such transformations leave the $L^N(\mathbb{R}^N)$ norm invariant, and this corresponds to mass in dimension 1, but in dimension $N \geq 2$, the total mass can be made arbitrarily small by a suitable scaling. If such a Mimura–Nishida estimate was valid, then the L^∞ norm would just be multiplied by k_0 for initial data with compact support, and because of the finite speed of propagation property it would also be true without imposing that the total mass be finite; that does happen in cases with forward invariant sets which are bounded, but there is no hint that this could be true in a general case, although it does not seem to contradict any known result; of course, if k_0 could be taken arbitrarily near 1, then in dimension N it would imply that the L^∞ norm does not increase, and that is not realistic (and one should remember that the nonlinearities appearing in the Carleman model are not consistent with principles of kinetic theory).

The critical part of the proof is to use an entropy inequality for deducing an equi-integrability property, and it is related to classical results in functional

---

[1] It is natural to change the units of length and time in the same way, so that the characteristic velocities which enter into the equation do not change, but it may look a little strange that the unknown u scales as the inverse of a length, which leaves the mass invariant in one dimension but not in dimension $N \geq 2$; one reason is that I have neglected factors K in front of the nonlinearities, which have a dimension, and which incorporate some effective scattering cross-sections (so that the particle "collides" with the other particles present in a cylinder around its path).

analysis, the Dunford–Pettis theorem,[2,3] and the De La Vallée Poussin criterion.[4] It is important that mass is conserved, but it is equally important in kinetic theory that there cannot be concentrations of mass on sets of small measure, and this results from the entropy inequality, for which Lemma 15.3 is the key.

Lemma 15.3. *If $u_1, \ldots, u_m$ are nonnegative in $\mathbb{R}^N$, have a support with finite measure, and*

$$\int_{\mathbb{R}^N} \left(\sum_{i=1}^m u_i \log(u_i) \right) dx \leq I < \infty, \tag{15.6}$$

then for every $\varepsilon > 0$, there exists $\delta > 0$ such that

$$\int_\omega \left(\sum_{i=1}^m u_i \right) dx \leq \varepsilon \text{ for all measurable subsets } \omega \text{ having measure } \leq \delta, \tag{15.7}$$

and one can choose $\delta = \frac{\varepsilon}{2m} e^{-2J/\varepsilon}$, with $J = I + \frac{1}{e} \sum_{i=1}^m meas(support(u_i))$.

Proof: The function $s \log(s)$ is negative for $0 < s < 1$ and it attains its minimum at $s = \frac{1}{e}$ and the minimum is $-\frac{1}{e}$; one deduces that $s \log_+(s) \leq s \log(s) + \frac{1}{e}$ for $s \geq 0$, where $\log_+(s)$ is the nonnegative part of the logarithm, equal to 0 on $[0,1]$ and $\log(s)$ on $[1,\infty)$. One then deduces that $\int_{\mathbb{R}^N} \left(\sum_{i=1}^m u_i \log_+(u_i) \right) dx \leq J = I + \frac{1}{e} \sum_{i=1}^m meas(support(u_i))$, although it is only the measure of the points where $0 < u_i < 1$ that should be added, but in practice one bounds this measure by an upper bound of the measure of the support. For any measurable set ω of finite measure, and for $i = 1, \ldots, m$, one decomposes ω into the part where $0 \leq u_i \leq M$ and the part where $u_i \geq M$, and one has $\int_\omega u_i \, dx \leq M \, meas(\omega) + \frac{1}{\log_+(M)} \int_\omega u_i \log_+(u_i) \, dx$, where $M > 1$ has to be chosen; summing in i, one obtains $\int_\omega \left(\sum_{i=1}^m u_i \right) dx \leq m M \, meas(\omega) + \frac{J}{\log_+(M)}$, so one chooses $M = e^{-2J/\varepsilon}$, which makes the second term $\frac{\varepsilon}{2}$, and one chooses then $meas(\omega)$ so that $m M \, meas(\omega) \leq \frac{\varepsilon}{2}$.   □

Proof of Proposition 15.2: Let $V = \max_i |V_i|$. In order to find an upper bound of $u_i(x_0, t_0)$ for some $t_0 \in [0, T]$, one only needs to know the initial data $v_j, j = 1, \ldots, m$, at points y such that $|y - x_0| \leq V \, t_0$; so one changes the initial data by putting all $v_j(y)$ equal to 0 if $|y - x_0| > V \, T$, and keeping the former value if $|y - x_0| \leq V \, T$, and the bounds that one will be able to prove for the solution will apply for bounding each $u_i(x_0, t_0)$. With the new initial data, the initial entropy $\int_{\mathbb{R}^N} \left(\sum_{i=1}^m v_i \log(v_i) \right) dx$ is $\leq C_N V^N T^N m M_0 \log_+(M_0)$, where

[2] Nelson DUNFORD, American mathematician, 1906–1986. He had worked at Yale University, New Haven, CT.

[3] Billy James PETTIS, American mathematician. He worked at Tulane University, New Orleans, LA, and University of North Carolina, Chapel Hill, NC.

[4] Charles Jean Gustave Nicolas DE LA VALLÉE POUSSIN, Belgian mathematician, 1866–1962. He was made Baron in 1928. He had worked in Louvain, Belgium.

C_N is the volume of the unit ball of $\mathbb{R}^N$; by hypothesis, as long as the solution exists in $(L^\infty(\mathbb{R}^N))^m$, the entropy will be bounded by that quantity; to apply Lemma 15.3, one needs to estimate the measure of the support of the solution, but by the finite speed of propagation property the support at time 0 is in a ball of radius VT and grows at most at speed V, so for $0 \le t \le T$ it is included in a ball of radius $2VT$, and the coefficient J of Lemma 15.3 may be taken to be $C_N V^N T^N m \left(M_0 \log_+(M_0) + \frac{4^N}{e}\right)$. For the value ε_0, Lemma 15.3 gives a value δ, and one defines ρ by $C_N \rho^N = \delta$, so that as long as the solution exists and for any time $t \in [0, T]$, the total mass in a ball of radius ρ is less or equal than the critical value for the Mimura–Nishida estimate.

Then one applies the estimate to prove that if the solution exists up to time t_1, with $t_1 \le T$, then it exists up to time $t_1 + \frac{\rho}{V}$ and its L^∞ norm between t_1 and $t_1 + \frac{\rho}{V}$ is at most multiplied by k_0. Indeed, one performs a second type of truncation, just for the purpose of estimating the norm of the solution: one restricts the data at time t_1 inside a ball centered at a point x_1 and with radius ρ, and one replaces the data at time t_1 by 0 outside this ball, and the hypothesis of the Mimura–Nishida estimate is valid, and the solution exists for $t > t_1$ with a norm in $L^\infty(\mathbb{R}^N)$ multiplied at most by a factor k_0; of course, this last solution only coincides with ours in a small cone, namely the set of points (x, t) with $t_1 \le t \le t_1 + \frac{\rho}{V}$ and $|x - x_1| \le \rho - V(t - t_1)$, but by moving the point x_1 this small cone sweeps the entire strip $\mathbb{R}^N \times (t_1, t_1 + \frac{\rho}{V})$, and therefore the norm of the solution is at most multiplied by k_0 in this strip. Then one repeats the process, and because ρ has been estimated uniformly, one attains the time T starting from time 0 in a finite number of operations (bounded by $1 + \frac{VT}{\rho}$), and the L^∞ estimate is obtained. $\square$

Proof of Lemma 15.1: For the Broadwell model, one has the conservation of mass and the conservation of momentum, written as $(u + w)_t + u_x = 0$ and $(v + w)_t - v_x = 0$. One starts from nonnegative bounded initial data u_0, v_0, w_0 such that $\int_{\mathbb{R}}(u_0 + v_0 + 2w_0)\, dx \le \varepsilon_0$, and a precise value of ε_0 will be obtained. For $x_0 \in \mathbb{R}$ and $t_0 > 0$, one integrates $(u + w)_t + u_x = 0$ on a triangle with vertices $(x_0, 0), (x_0 + t_0, 0), (x_0 + t_0, t_0)$, and one obtains a boundary integral

$$\int_0^t w(x_0 + s, s)\, ds + \int_0^t u(x_0 + t_0, s)\, ds = \int_{x_0}^{x_0+t_0}(u_0 + w_0)\, dx, \qquad (15.8)$$

and similarly, if one integrates $(v + w)_t - v_x = 0$ on a triangle with vertices $(x_0 - t_0, 0), (x_0, 0), (x_0 - t_0, t_0)$, one obtains

$$\int_0^t w(x_0 - s, s)\, ds + \int_0^t u(x_0 - t_0, s)\, ds = \int_{x_0-t_0}^{x_0}(v_0 + w_0)\, dx, \qquad (15.9)$$

so that one deduces that

$$\int_0^\infty u(x+s, s)\, ds \le \varepsilon_0, \int_0^\infty v(x-s, s)\, ds \le \varepsilon_0, \int_0^\infty w(x, s)\, ds \le \varepsilon_0 \text{ a.e. } x \in \mathbb{R},$$
$$(15.10)$$

where the upper bounds ∞ are only used after one has shown global existence, but until then are restricted to the finite time of existence. Then, as long as the solution exists one defines $M(t)$ as the smallest number such that

$$u(x,s), v(x,s), w(x,s) \leq M(t) \text{ a.e. } x \in \mathbb{R}, s \in (0,t), \tag{15.11}$$

so that $M(0)$ is $\max\{\|u_0\|_{L^\infty(\mathbb{R})}, \|v_0\|_{L^\infty(\mathbb{R})}, \|w_0\|_{L^\infty(\mathbb{R})}\}$. For almost all $x_0 \in \mathbb{R}$, one can work on the characteristic line parametrized by $(x_0 + s, s)$, and one has $\frac{d}{dt}u(x_0 + t, t) \leq w^2(x_0 + t, t) \leq M(t)w(x_0 + t, t)$, giving after integration $u(x_0 + t, t) \leq u_0(x_0) + \int_0^t M(s)w(x_0 + s, s)\,ds \leq M(0) + M(t)\int_0^t w(x_0 + s, s)\,ds \leq M(0) + \varepsilon_0 M(t)$; therefore one finds that the essential supremum of $u(x,s)$ for $x \in \mathbb{R}$ and $s \in (0,t)$ is $\leq M(0) + \varepsilon_0 M(t)$. Similarly, working on the characteristic line parametrized by $(x_0 - s, s)$ for v one finds that the essential supremum of $v(x,s)$ for $x \in \mathbb{R}$ and $s \in (0,t)$ is $\leq M(0) + \varepsilon_0 M(t)$, and working on the characteristic line parametrized by (x_0, s) for w one finds that the essential supremum of $w(x,s)$ for $x \in \mathbb{R}$ and $s \in (0,t)$ is $\leq M(0) + \varepsilon_0 M(t)$. These three bounds, and the definition of $M(t)$ as the smallest number for some inequality to hold, shows that one has

$$M(t) \leq M(0) + \varepsilon_0 M(t), \tag{15.12}$$

as long as the solution exists; therefore one finds global existence if $\varepsilon_0 < 1$, and by choosing $\varepsilon(k) = 1 - \frac{1}{k}$, one finds $M(t) \leq \frac{M(0)}{1-\varepsilon(k)} = k\,M(0)$.   □

Henri CABANNES has checked that a Mimura–Nishida estimate holds for various classical discrete velocity models. I shall describe in the next lecture a different way to obtain similar estimates, which is not based on using conservations, like in the proof of MIMURA and Takaaki NISHIDA.

[Taught on Monday October 8, 2001.]

Notes on names cited in footnotes for Chapter 15, TULANE.[5]

---

[5] Paul TULANE, American philanthropist, 1801–1887. Tulane University, New Orleans, LA, is named after him.

16

Systems Satisfying My Condition (S)

In this lecture, I shall describe some local and global existence results for a special class (S) of semi-linear systems in only one space variable, that I introduced in 1979, those which have the form

$$(u_i)_t + C_i(u_i)_x + \sum_{j,k} A_{i,j,k} u_j u_k = 0, \, i = 1, \ldots, m, \tag{16.1}$$

with the *condition (S)*

$$(S) \qquad C_j = C_k \text{ implies } A_{i,j,k} = 0 \text{ for all } i. \tag{16.2}$$

The reason why I had first looked at this class of system was that it has a property of stability with respect to weak convergence, which I shall describe in the next lecture, a simple consequence of the *div-curl lemma* (a first example of *compensated compactness*), that I had proven a few years before with François MURAT.

As often happens when doing research in mathematics, when one looks for something and one has found it, one may overlook another interesting property that one was not really looking for; one must stay alert and pay attention to details, and one may discover some unexpected result. I was checking a new proof of that particular application of the div-curl lemma, different from the one using the Fourier transform, which was our initial approach in 1974, or the one using the framework of differential forms and Hodge decomposition,[1] which was shown to me in 1975 by Joel ROBBIN,[2] and it seemed to apply to a more general setting using L^p spaces for $p < 2$; by looking to the best possible value of p, I finally found a simple reason why if $u_1, u_2 \in L^1(\mathbb{R}^2)$ with $(u_1)_{x_1}, (u_2)_{x_2} \in L^1(\mathbb{R}^2)$, then one has $u_1 u_2 \in L^1(\mathbb{R}^2)$; I now call that

---

[1] William Vallance Douglas HODGE, Scottish mathematician, 1903–1975. He had worked in Bristol and in Cambridge, England.

[2] Joel William ROBBIN, American mathematician, born in 1941. He works at University of Wisconsin, Madison, WI.

type of result *compensated integrability* [19], because it should not be confused with compensated compactness, and it is useful for proving the existence of solutions for systems in the class (S), with initial data in $L^1(\mathbb{R})$, and it appears to be a completely different approach from the semi-group point of view (which had not really succeeded, apart from cases where an L^1 contraction property holds).

Definition 16.1. *For $c \in \mathbb{R}$, V_c is the space of functions u such that $u_t + c u_x = f \in L^1(\mathbb{R} \times \mathbb{R})$ and $u\mid_{t=0} = g \in L^1(\mathbb{R})$, with the norm $\|u\|_{V_c} = \|f\|_{L^1(\mathbb{R}^2)} + \|g\|_{L^1(\mathbb{R})}$; W_c is the space of functions u such that there exists $h \in L^1(\mathbb{R})$ with $|u(x,t)| \leq h(x - ct)$ a.e. in $\mathbb{R}^2$, with the norm $\|u\|_{W_c} = \inf_h \|h\|_{L^1(\mathbb{R})}$.*

Notice that the time t is not restricted to be ≥ 0, because one makes no hypothesis on the sign of the coefficients $A_{i,j,k}$; taking $t \in \mathbb{R}$ will be possible for small L^1 data, but for large L^1 data one will only obtain local existence, and the definition of the spaces V_c and W_c can be restricted to functions defined in $\mathbb{R} \times (-\alpha, \beta)$ for positive α, β (so the interval in time contains 0), and even some other sets, as will be seen in some proofs.

Lemma 16.2. *For every $c \in \mathbb{R}$, one has $V_c \subset W_c$, with $\|u\|_{W_c} \leq \|u\|_{V_c}$ for all $u \in V_c$.*
For $c_1 \neq c_2$, one has $u_1 u_2 \in L^1(\mathbb{R}^2)$ whenever $u_1 \in W_{c_1}, u_2 \in W_{c_2}$, with

$$\int_{\mathbb{R} \times \mathbb{R}} |u_1| \, |u_2| \, dx \, dt \leq \frac{1}{|c_1 - c_2|} \|u_1\|_{W_{c_1}} \|u_2\|_{W_{c_2}} \text{ for all } u_1 \in W_{c_1}, u_2 \in W_{c_2}.$$
$$(16.3)$$

Proof: Using the Fubini theorem, one has $u(x,t) = g(x - ct) + \int_0^t f(x - cs, t - s) \, ds$ for almost every x, t, where $f = u_t + c u_x$ and $g = u\mid_{t=0}$, and this gives $|u(x,t)| \leq h(x - ct)$ with $h(y) = |g(y)| + \int_{\mathbb{R}} |f(y + cs, s)| \, ds$ and $\|h\|_{L^1(\mathbb{R})} = \|u\|_{V_c}$.
For $i = 1, 2$, one has $|u_i(x,t)| \leq h_i(x - c_i t)$ a.e. in $\mathbb{R}^2$, with $\|h_i\|_{L^1(\mathbb{R})} \leq \|u_i\|_{W_{c_i}} + \varepsilon$, so that $\int_{\mathbb{R} \times \mathbb{R}} |u_1| \, |u_2| \, dx \, dt \leq \int_{\mathbb{R} \times \mathbb{R}} |h_1(x - c_1 t)| \, |h_2(x - c_2 t)| \, dx \, dt$; one changes variables in the last integral, defining $y_1 = x - c_1 t, y_2 = x - c_2 t$, which is a good change of variables because $c_1 \neq c_2$, and one has $dx \, dt = \frac{1}{|c_1 - c_2|} dy_1 \, dy_2$ and the integral is equal to $\frac{1}{|c_1 - c_2|} \int_{\mathbb{R}} h_1(y_1) \, dy_1 \int_{\mathbb{R}} h_2(y_2) \, dy_2$; then, one lets ε tend to 0. $\square$

The technical adantage of using the functional space V_c instead of W_c is that functions in V_c have a trace at $t = 0$, and that their behaviour at infinity is also well described; indeed, if one moves with speed c and one writes $u(x,t) = U(x - ct, t)$ then $U \in V_0$, and if one denotes $E = L^1(\mathbb{R})$, a function $U \in V_0$ is such that $U(\cdot, 0) \in E$ and $U_t \in L^1(\mathbb{R}; E)$, so U is absolutely continuous in t with values in E; apart from having well defined values at every time $U(\cdot, t) \in L^1(\mathbb{R})$, it implies that U also has well defined limits $U_+ \in L^1(\mathbb{R})$ as $t \to +\infty$ and $U_- \in L^1(\mathbb{R})$ as $t \to -\infty$.

Because condition (S) is assumed, it is possible to define solutions with initial data in $L^1(\mathbb{R})$ in a unique way, globally for small L^1 data, locally in time for large L^1 data, and it is the adopted choice of functional spaces which permits that, and the solution is sought such that $u_i \in V_{C_i}$ for $i = 1, \ldots, m$, and thanks to Lemma 16.2 all the products $u_j u_k$ appearing with a nonzero coefficient $A_{i,j,k}$ belong to $L^1(\mathbb{R} \times \mathbb{R})$.

Proposition 16.3. *Assuming condition (S), there exists $\varepsilon_0 > 0$ (depending upon the coefficients $A_{i,j,k}$ and the velocities C_i in an explicit way) such that if the initial data $v_i, i = 1, \ldots, m$, satisfy*

$$\int_{\mathbb{R}} \left(\sum_{i=1}^m |v_i| \right) dx < \varepsilon_0, \tag{16.4}$$

there is a unique solution $u = (u_1, \ldots, u_m) \in V_{C_1} \times \ldots \times V_{C_m}$. $\tag{16.5}$

Proof: One proves the existence of a solution with small norm by applying a fixed point argument for a strict contraction; uniqueness can be proven without assuming that the solution has a small norm. One looks for a fixed point of the map Φ, defined for $u \in W_{C_1} \times \ldots \times W_{C_m}$, with $\Phi(u) = U \in V_{C_1} \times \ldots \times V_{C_m}$ solution of

$$(U_i)_t + C_i(U_i)_x + \sum_{j,k} A_{i,j,k} u_j u_k = 0, U_i \mid_{t=0} = v_i, i = 1, \ldots, m. \tag{16.6}$$

Because of Lemma 16.2 and condition (S), all the terms $A_{i,j,k} u_j u_k$ belong to $L^1(\mathbb{R} \times \mathbb{R})$, so that $U_i \in V_{C_i}$ for $i = 1, \ldots, m$. The spaces used are Banach spaces, and one looks for a closed set which is mapped into itself, and then that it is a strict contraction. For

$$\alpha_i = ||v_i||_{L^1(\mathbb{R})}, i = 1, \ldots, m, \tag{16.7}$$

$$||u_i||_{W_{C_i}} \leq \xi_i, i = 1, \ldots, m \text{ implies}$$
$$||U_i||_{V_{C_i}} \leq \eta_i = \alpha_i + \sum_{j,k}' \frac{|A_{i,j,k}|}{|C_j - C_k|} \xi_j \xi_k, i = 1, \ldots, m, \tag{16.8}$$

where $'$ means that one avoids indices for which $C_j = C_k$ (which correspond to $A_{i,j,k} = 0$). For

$$\beta = \max_{\{j,k | C_j \neq C_k\}} \sum_{i=1}^m \frac{|A_{i,j,k}|}{|C_j - C_k|}, \tag{16.9}$$

$$\sum_{i=1}^m \eta_i \leq \sum_{i=1}^m \alpha_i + \beta \sum_{j,k}' \xi_j \xi_k \leq \varepsilon_0 + \beta \left(\sum_{j=1}^m \xi_j \right)^2, \tag{16.10}$$

and one checks immediately that

$$\varepsilon_0 \le \frac{1}{4\beta} \text{ and } \sum_{i=1}^{m} \xi_i \le 2\varepsilon_0 \text{ imply } \sum_{i=1}^{m} \eta_i \le 2\varepsilon_0. \tag{16.11}$$

If one imposes $\varepsilon_0 \le \frac{1}{4\beta}$, then one has found a closed set mapped into itself, defined by $\sum_{i=1}^{m} ||u_i||_{W_{C_i}} \le 2\varepsilon_0$; in order to check if Φ is a strict contraction on this set, one takes u, u' in the set and one estimates the norm of $U_i - U_i'$, using the usual decomposition $u_j u_k - u_j' u_k' = u_j(u_k - u_k') + (u_j - u_j')u_k'$, and one finds

$$||U_i - U_i'||_{V_{C_i}} \le \sum_{j,k}' \frac{|A_{i,j,k}|}{|C_j - C_k|}(\xi_j ||u_k - u_k'||_{W_{C_k}} + ||u_j - u_j'||_{W_{C_j}} \xi_j'), i = 1, \ldots, m, \tag{16.12}$$

so that if $\sum_{i=1}^{m} ||u_i||_{W_{C_i}} \le 2\varepsilon_0$ and $\sum_{i=1}^{m} ||u_i'||_{W_{C_i}} \le 2\varepsilon_0$ one has

$$\sum_{i=1}^{m} ||U_i - U_i'||_{V_{C_i}} \le 4\varepsilon_0 \beta \sum_{i=1}^{m} ||u_i - u_i'||_{W_{C_i}}, \tag{16.13}$$

and a strict contraction property follows from the choice $\varepsilon_0 < \frac{1}{4\beta}$. Uniqueness of the solution is true without assuming that the second solution has a small norm, as will be shown later.   □

With a simple adaptation, one can obtain a local existence theorem for arbitrary data in L^1, but the time of existence is not just a function of the norm of the initial data in L^1, like for ordinary differential equations (once again one observes the limitations of the point of view of the theory of semigroups for this kind of problem), and that will be seen by considering the example

$$u_t + u_x = u\,v; \ u(\cdot,0) = u_0 \tag{16.14}$$
$$v_t - v_x = u\,v; \ v(\cdot,0) = v_0,$$

for which the preceding proposition applies with $\beta = 1$, and gives global existence if $\int_{\mathbb{R}}(|u_0| + |v_0|)\,dx < 1$. For $a > 0$ and $L > \frac{1}{a}$, one chooses the initial data u_0 and v_0 equal to a in $(-L, +L)$ and 0 outside; in the domain of dependence $\{(x,t) \mid |x| \le L - |t|\}$, the solution then solves the ordinary differential system $u_t = v_t = u\,v, u(0) = v(0) = a$, whose solution is $u(t) = v(t) = \frac{a}{1-a\,t}$ and it blows up at time $t_c = \frac{1}{a} < L$; the initial data satisfy $\int_{\mathbb{R}}(|u_0| + |v_0|)\,dx = 4L\,a$, which can be any number > 4, but the time of existence is $\frac{1}{a}$ (because one has $0 \le u(x,t), v(x,t) \le \frac{a}{1-a\,t}$), which can be arbitrarily small by taking a large. This explains that the time of local existence requires a more precise analysis than the evaluation of a few global norms (but it can be seen on the nondecreasing rearrangements of the initial data).

Proposition 16.4. *Let $v_1, \ldots, v_m \in L^1(\mathbb{R})$ and let $r_0 > 0$ be such that*

$$\int_{z-r_0}^{z+r_0} \left(\sum_{j=1}^{m} |v_j(x)| \right) dx \le \varepsilon_0 \text{ for all } z \in \mathbb{R}, \tag{16.15}$$

with ε_0 as in Proposition 16.3, for example $\varepsilon_0 < \frac{1}{4\beta}$. Then there is a unique solution for $|t| \leq \frac{r_0}{\max_i |C_i|}$.

For a given $z \in \mathbb{R}$, one takes as new initial data the functions v_i in the interval $(z - r_0, z + r_0)$ and 0 outside, and for these new initial data the solution exists globally, but it may only coincide with the desired solution in the domain of dependence $\{(x,t) \mid x - C_i t \in (z - r_0, z + r_0)$ for $i = 1, \ldots, m\}$.

However, one must prove that two solutions starting from two intervals coincide on the intersection of the domains of dependence. This comes from using the uniqueness of small solutions as in Proposition 16.3, but observing that it applies to domains of the form $D_J = \{(x,t) \mid x - C_i t \in J$ for $i = 1, \ldots, m\}$ for any interval J given at time 0, where the L^1 norm of the initial data has to be small, and this is because given $g_i \in L^1(J)$ and $f_i \in L^1(D_J)$ there is one solution of $(U_i)_t + C_i (U_i)_x = f_i$ in D_J and $U|_{t=0} = g_i$ on J, belonging to a space V_{C_i} defined in an obvious way, for $i = 1, \ldots, m$.

If a solution has large norm in L^1 of a subset of $\mathbb{R} \times \mathbb{R}$ or $\mathbb{R}$, then one also uses the fact that for every $\varepsilon > 0$ there exists $\delta > 0$ such that the integral on any subset of measure at most δ is bounded by ε, a property which has been used in asserting the existence of r_0 in (16.15).   □

The fixed point property is used on a set of $W_{C_1} \times \ldots \times W_{C_m}$ but the fixed point is necessarily in the range of Φ, in $V_{C_1} \times \ldots \times V_{C_m}$. In the case of small data in L^1, where the solution exists for all time, I have already pointed out that the solution u_i belonging to V_{C_i} gives information on the asymptotic behaviour $t \to \pm\infty$, i.e. there exists $u_i^-, u_i^+ \in L^1(\mathbb{R})$ such that

$$\int_{\mathbb{R}} |u_i(x,t) - u_i^{\pm}(x - C_i t)| \, dx \to 0 \text{ as } t \to \pm\infty. \tag{16.16}$$

I have also obtained results of scattering, i.e. about the map $u^- \mapsto u^+$, which I shall not describe.

The global existence result for small data, and the counter-example showing that the time of existence for large data is not only a function of the L^1 norm, shows an important effect due to the transport term. The model of the counter-example is similar to that of a chemical chain reaction which creates an explosion in finite time, but the reaction needs the two constituents to be present for sustaining itself, and if the constituents move at a different velocities, the reaction started at one point must be sustained by molecules coming from elsewhere, and there is a problem of timing and it is important that there is a sufficient amount to sustain the reaction to its end. This interpretation explains one defect of using only global norms of functional spaces like L^p, which give no clue about where the information is located, and one could think of using rearrangement methods, as started by HARDY and LITTLEWOOD,[3]

---

[3] John Edensor LITTLEWOOD, English mathematician, 1885–1977. He had worked in Manchester and in Cambridge, England, where he held the newly founded Rouse Ball professorship (1928–1950).

but I do not know of any efficient way to do that for models of kinetic theory; techniques of maximal functions, also started by HARDY and LITTLEWOOD and extended by WIENER should be more adapted, probably in the way used by Lars HEDBERG,[4] and he traced his idea to some earlier work of Lennart CARLESON, and of STEIN.

Before generalizing the preceding results to the Broadwell model, which violates condition (S) because of the presence of the w^2 terms, one should observe that the method also permits one to give bounds in L^∞, and the argument is analogous to that of MIMURA and Takaaki NISHIDA, but relies on the bounds in the V_{C_i} spaces instead of a conservation property.

Proposition 16.5. *Assuming condition (S), for every $k > 1$ there exists $\varepsilon(k) > 0$ such that*

$$v_i \in L^1(\mathbb{R}) \cap L^\infty(\mathbb{R}), i = 1, \ldots, m, \ and \ \int_{\mathbb{R}} \left(\sum_{i=1}^m |v_i| \right) dx \le \varepsilon(k) \ imply$$
$$|u_i(x,t)| \le k \max_j ||v_j||_{L^\infty(\mathbb{R})}, i = 1, \ldots, m \ a.e. \ (x,t) \in \mathbb{R} \times \mathbb{R}.$$
$$(16.17)$$

Proof: Taking $\varepsilon(k) \le \varepsilon_0$ one has a global solution satisfying $|u_i(x,t)| \le h_i(x - C_i t)$ with $\sum_i ||h_i||_{L^1(\mathbb{R})} \le 2\varepsilon_0$. One has local existence in L^∞, and one must find a bound for the L^∞ norm. Integrating along a characteristic line with velocity C_i one bounds each of the terms $|A_{i,j,k} u_j u_k|$ by replacing $|u_j|$ by $h_j(x - C_j t)$ and $|u_k|$ by $M(t)$ if $C_i \ne C_j$, or $|u_k|$ by $h_k(x - C_k t)$ and $|u_j|$ by $M(t)$ if $C_i \ne C_k$, the case $C_i = C_j = C_k$ being of no consequence as $A_{i,j,k} = 0$ in that case; by integrating one finds that $|u_i(x,t)| \le |v_i(x - C_i t)| + K\varepsilon_0 M(t)$ with K depending only on the coefficients $A_{i,j,k}$ and the velocities C_i, and this gives an estimate $M(t) \le M(0) + K\varepsilon_0 M(t)$, and one chooses $\varepsilon(k) \le \frac{1}{K}\left(1 - \frac{1}{k}\right)$. $\square$

In order to treat the Broadwell model, I introduced a slightly more general framework, but there the nonnegative character of solutions is crucial, and $t \ge 0$. The main idea is that an estimate for the integral of w^2 is now obtained by integration of the third equation, which gives $\int_{\mathbb{R} \times (0,T)} w^2 \, dx \, dt + \int_{\mathbb{R}} w(x,T) \, dx = \int_{\mathbb{R} \times (0,T)} u v \, dx \, dt + \int_{\mathbb{R}} w_0 \, dx$, and because of $u_0, v_0, w_0 \ge 0$ one has $w \ge 0$ and therefore the missing bound is replaced by $\int_{\mathbb{R} \times (0,T)} w^2 \, dx \, dt \le \int_{\mathbb{R} \times (0,T)} u v \, dx \, dt + ||w_0||_{L^1(\mathbb{R})} \le \frac{1}{2}||u||_{W_1} ||v||_{W_2} + \varepsilon_0$; one concludes as before that $||u||_{V_1} \le ||u_0||_{L^1(\mathbb{R})} + \int_{\mathbb{R} \times (0,T)} w^2 \, dx \, dt$ and $||v||_{V_{-1}} \le ||v_0||_{L^1(\mathbb{R})} + \int_{\mathbb{R} \times (0,T)} w^2 \, dx \, dt$.[5] The solution obtained is such that for large t one has

---

[4] Lars Inge HEDBERG, Swedish mathematician, 1935–2005. He had worked in Linköping, Sweden.

[5] I had not wanted to repeat the same procedure as before in my report, and that may be why Reinhard ILLNER interpreted that I had not really proven the existence for data in L^1 for the Broadwell model.

$u \approx u^+(x-t)$, $v \approx v^+(x+t)$ and $w \approx w^+(x)$, but as pointed out by Russell CAFLISCH,[6] one has $w^+ = 0$ because $w \in L^2(\mathbb{R} \times (0, \infty))$. The conservation of mass and the conservation of momentum show that $\int_{\mathbb{R}} u^+ \, dx = \int_{\mathbb{R}} (u_0 + w_0) \, dx$ and $\int_{\mathbb{R}} uv + dx = \int_{\mathbb{R}} (v_0 + w_0) \, dx$.

[Taught on Wednesday October 10, 2001.]

Notes on names cited in footnotes for Chapter 16, R. BALL,[7] ILLNER.[8]

---

[6] Russell Edward CAFLISCH, American mathematician. He has worked at NYU (New York University), New York, NY, and at UCLA (University of California Los Angeles), Los Angeles, CA.

[7] Walter William Rouse BALL, English mathematician, 1850–1925. He had worked in Cambridge, England.

[8] Reinhard ILLNER, German-born mathematician. He has worked in Kaiserslautern, Germany, at Duke University, Durham, NC, and at University of Victoria, British Columbia.

17

Asymptotic Estimates for the Broadwell and the Carleman Models

For nonnegative initial data with a small total mass $\int_{\mathbb{R}} (u_0 + v_0 + 2w_0)\, dx$, the asymptotic behaviour is that of a free streaming solution, i.e. without the nonlinear interaction terms, but with a particularity that w tends to 0, and that is due to the presence of the w^2 term in the equations. Actually, using a remark of Raghu VARADHAN,[1] which simplified a result of Thomas BEALE,[2] which I shall discuss later, the result is true for all nonnegative data with a finite total mass.

In principle, models of kinetic theory have no interaction between particles travelling at the same velocity, but one should remember that the four velocities model has a term $u_3 u_4$ corresponding to particles going in opposite directions and parallel to the y axis, and it is only because the initial data have been assumed independent of y that the velocities seem to be the same, equal to 0, but 0 is just the projection of the velocity onto the x axis; actually the conservation of kinetic energy is concerned with $u + v + 2w$ and not with $u + v$ as it would have been if the particles with density w had a zero velocity.

The presence of the w^2 term acts as a destruction mechanism, and indeed the particles from the third and fourth families eventually all transform into particles of the first and second families by collisions. One may wonder why the process is not symmetric and why the collisions of particles of the first and second families do not produce enough particles of the third and fourth family. My analysis, which went further than the result of MIMURA and Takaaki NISHIDA, explained that the hypothesis of finite mass, together with the fact that the particles created in the first and second families are taken away (because they have different velocities) puts a severe limitation on the production of particles of the third and fourth families, which are not replaced, so these families die out.

---

[1] Sathamangalam Raghu Srinivasa VARADHAN, Indian-born mathematician, born in 1940. He works at NYU (New York University), New York, NY.

[2] James Thomas BEALE, American mathematician. He works at Duke University, Durham, NC.

Once one knows that $u \in V_1, v \in V_{-1}, w \in V_0 \cap L^2$, one deduces that there exist $u_\infty, v_\infty \in L^1(\mathbb{R})$ such that, as t tends to ∞, one has

$$\begin{array}{l} \int_\mathbb{R} |u(x,t) - u_\infty(x - t)| \, dx \to 0 \\ \int_\mathbb{R} |v(x,t) - v_\infty(x + t)| \, dx \to 0 \\ \int_\mathbb{R} |w(x,t)| \, dx \to 0. \end{array} \tag{17.1}$$

The conservation of mass $\int_\mathbb{R} (u + v + 2w) \, dx$ and the conservation of momentum $\int_\mathbb{R} (u - v) \, dx$ imply that

$$\begin{array}{l} \int_\mathbb{R} (u_\infty + v_\infty) \, dx = \int_\mathbb{R} (u_0 + v_0 + 2w_0) \, dx \\ \int_\mathbb{R} (u_\infty - v_\infty) \, dx = \int_\mathbb{R} (u_0 - v_0) \, dx, \end{array} \tag{17.2}$$

and solving this system gives

$$\begin{array}{l} \int_\mathbb{R} u_\infty \, dx = \int_\mathbb{R} (u_0 + w_0) \, dx \\ \int_\mathbb{R} v_\infty \, dx = \int_\mathbb{R} (v_0 + w_0) \, dx, \end{array} \tag{17.3}$$

and therefore it is more natural to express the conservation of mass and the conservation of momentum as

$$\begin{array}{l} (u + w)_t + u_x = 0 \\ (v + w)_t - v_x = 0, \end{array} \tag{17.4}$$

emphasizing $u + w$ as the mass which will eventually go to infinity on the right side and $v + w$ as the mass which will eventually go to infinity on the left side.

In the two-dimensional four velocities model, one may tag particles, and this point of view might be classical, but I only noticed it a few years ago, while trying with Chun LIU to prove global existence results for the two-dimensional four velocities model;[3] I had been led to integrating along the direction $(1, -1)$ by a remark of Robert PESZEK,[4] but such integrals had already appeared before in a method of Shuichi KAWASHIMA,[5] and I may have been just finding an intuitive explanation for his estimates, which I had not read, but asked my student Kamel HAMDACHE to read and generalize,[6]

[3] Chun LIU, Chinese-born mathematician. He was a post doctoral associate of CNA (Center for Nonlinear Analysis) at CMU (Carnegie Mellon University), Pittsburgh, PA, and he works now at Penn State (Pennsylvania State University), University Park, PA.

[4] Robert W. PESZEK, Polish-born mathematician. He was a post doctoral associate of CNA (Center for Nonlinear Analysis) at CMU (Carnegie Mellon University), Pittsburgh, PA, and he works now at MTU (Michigan Technological University), Houghton, MI.

[5] Shuichi KAWASHIMA, Japanese mathematician. He works at Kyushu University, Fukuoka, Japan.

[6] Kamel HAMDACHE, French mathematician, born in 1948. He has worked in Algiers, Algeria, and then in various laboratories of CNRS (Centre National de la

which he did. When two particles from the first and second families collide, one may decide that it is the particle from the first family which switches to the third family, and when two particles from third and fourth families collide that it is the particle from the third family which switches to the first family. Conservation of mass is $(u_1 + u_2 + u_3 + u_4)_t + (u_1 - u_2)_x + (u_3 - u_4)_y = 0$, conservation of momentum in x is $(u_1 - u_2)_t + (u_1 + u_2)_x = 0$ and conservation of momentum in y is $(u_3 - u_4)_t + (u_3 + u_4)_y = 0$, from which one deduces $(u_1 + u_3)_t + (u_1)_x + (u_3)_y = 0$, and it is natural to integrate along parallels to the direction $(1, -1)$, because for a particle of the first or third family one cannot predict the position $(\xi(t), \eta(t))$ that it will occupy at time t, but one can predict what $\xi(t) + \eta(t)$ will be; indeed, one has $\xi'(t) + \eta'(t) = 1$, because one has $\xi'(t) = 1$ and $\eta'(t) = 0$ while the particle is in the first family and $\xi'(t) = 0$ and $\eta'(t) = 1$ while the particle is in the third family. If one defines

$$M_{13}(x, y, t) = \int_{\mathbb{R}} (u_1 + u_3)(x + z, y - z, t)\, dz, \text{ so that}$$
$$(M_{13})_x = (M_{13})_y, \text{ or } M_{13}(x, y, t) = N_{13}(x + y, t), \tag{17.5}$$

one obtains

$$(M_{13})_t + (M_{13})_x = 0, \tag{17.6}$$

so that one can compute M_{13} directly from the initial data,

$$M_{13}(x, y, t) = \int_{\mathbb{R}} (v_1 + v_3)(x - t + z, y - z)\, dz. \tag{17.7}$$

If one has proven that the asymptotic behaviour is that u_1, u_2, u_3, u_4 look eventually like $u_1^\infty(x - t, y)$, $u_2^\infty(x + t, y)$, $u_3^\infty(x, y - t)$, $u_4^\infty(x, y + t)$ (which is true for nonnegative data with small L^2 norm), and if m_i is the integral of u_i^∞, then one has $m_1 + m_3 = \int_{\mathbb{R}^2} (v_1 + v_3)\, dx\, dy$; similarly, $m_1 + m_4 = \int_{\mathbb{R}^2} (v_1 + v_4)\, dx\, dy$, $m_2 + m_3 = \int_{\mathbb{R}^2} (v_2 + v_3)\, dx\, dy$, and $m_2 + m_4 = \int_{\mathbb{R}^2} (v_2 + v_4)\, dx\, dy$, but it is not clear if there are simple formulas giving separately m_1, m_2, m_3, m_4.

Coming back to the asymptotic behaviour for the Broadwell model, it is important to realize that there is no precise shape for the limiting functions u_∞ and v_∞, and that they can be arbitrary nonnegative functions with compact support for example, if one accepts to translate them. Indeed let φ, ψ be two nonnegative functions with compact support, and consider the initial data $u_0(x) = \varphi(x - a)$, $v_0(x) = \psi(x - b)$, $w_0(x) = 0$, with $a, b \in \mathbb{R}$ chosen in such a way that the support of u_0 is entirely to the right of the support of v_0; in that case the explicit solution will be $u(x, t) = \varphi(x - a - t)$, $v(x, t) = \psi(x - b + t)$, $w(x, t) = 0$, because these formulas imply $u\, v - w^2 = 0$. However, if one wants exactly $u_\infty = \varphi$, and $v_\infty = \psi$, and the support does not satisfy the condition,

Recherche Scientifique), at ENSTA (École Normale Supérieure des Techniques Avancées), Palaiseau, at ENS (École Normale Supérieure) Cachan, in Bordeaux, at Université Paris-Nord, Villetaneuse, and at École Polytechnique, Palaiseau, France. He did his thesis (doctorat d'état, 1986) under my supervision.

one has a more technical problem of scattering (which I have only studied for systems satisfying condition (S) for small data).

The condition (S) (or its generalization) is not satisfied by the Carleman model, and I was not trying to include it in my analysis (as it is not a model from kinetic theory, and global L^∞ bounds were known for that model), but I then learnt of an estimate by Reinhard ILLNER and Michael REED,[7] that the solution of the Carleman model with nonnegative data with finite total mass, i.e. $\int_\mathbb{R}(u_0 + v_0)\,dx = m < \infty$ satisfies a uniform estimate

$$0 \le u(x,t), v(x,t) \le \frac{C(m)}{t}, \tag{17.8}$$

and that shows that the asymptotic behaviour is quite different than for the Broadwell model; I included a simplified proof of their result in the appendix of my 1980 report, but the estimates for $C(m)$ were much too large. Their result suggested me to look at self-similar solutions of the Carleman model, which I shall describe in a moment, and from that study I conjectured a bound $C(m) = O(m^2 + 1)$ in the decay estimate, which I proved a few years after, and I shall describe that later, in connection with the method of generalized invariant regions. One cannot hope for a faster decay, because of conservation of mass, i.e. $\int_\mathbb{R}(u(x,t) + v(x,t))\,dx = m$, and if one starts with initial data having their support in an interval of length L, then the support at time t is included in an interval of length $L + 2t$, and therefore one must have $m \le (L+2t)(||u(\cdot,t)||_{L^\infty(\mathbb{R})} + ||v(\cdot,t)||_{L^\infty(\mathbb{R})})$; by letting L tend to 0, it shows that $C(m) \ge \frac{m}{4}$. One has $C(m) \ge 1$ for all $m > 0$, because if $u(x_0) > 0$ then $f(t) = u(x_0 + t, t)$ satisfies the differential inequality $f' + f^2 \ge 0$ with $f(0) > 0$, and the solution satisfies $f(t) \ge \frac{f(0)}{1+f(0)t}$ for all $t > 0$; by letting $f(0)$ tend to ∞, it gives $C(m) \ge 1$.

For any of our semi-linear hyperbolic systems with a quadratic nonlinearity, if $u_i, i = 1,\dots,m$, is a solution, then $\tilde{u}_i, i = 1,\dots,m$, is a solution if one defines it by $\tilde{u}_i(x,t) = \lambda\,u_i(\lambda x, \lambda t), i = 1,\dots,m$, where $\lambda > 0$ is arbitrary. It is then natural to look for solutions such that $\tilde{u}_i = u_i$ for $i = 1,\dots,m$, independently of λ, and that means that $u_i(x,t) = \frac{1}{t}U_i(\frac{x}{t})$ (by choosing $\lambda = \frac{1}{t}$), and such solutions are called self-similar.

One looks for a self-similar nonnegative solution of the Carleman model, $u(x,t) = \frac{1}{t}U(\frac{x}{t}), v(x,t) = \frac{1}{t}V(\frac{x}{t})$, with finite total mass m, which must then be $\int_\mathbb{R}(U + V)\,d\sigma$. One finds that $t^2(u_t + u_x + u^2 - v^2) = -U - \sigma U' + U' + U^2 - V^2 = 0$, and that $t^2(v_t - v_x - u^2 + v^2) = -V - \sigma V' - V' - U^2 + V^2 = 0$, where $\sigma = \frac{x}{t}$ and $' = \frac{d}{d\sigma}$, i.e.

$$\begin{aligned} ((1-\sigma)U)' + U^2 - V^2 = 0 \\ ((1+\sigma)V)' + U^2 - V^2 = 0, \end{aligned} \tag{17.9}$$

---

[7] Michael Charles REED, American mathematician, born in 1942. He works at Duke University, Durham, NC.

from which one deduces by subtracting the two equations

$$(1 - \sigma)U - (1 + \sigma)V = C_0, \tag{17.10}$$

and one only considers the case $C_0 = 0$; the reason is that when t tends to 0, the data for U and for V converge to $\alpha \, \delta_0$ and to $\beta \, \delta_0$, and one then expects that U and V vanish if $|x - t| > t$, because of the finite speed of propagation, which means that $U(\sigma)$ and $V(\sigma)$ vanish for $|\sigma| > 1$. Using $U = (1 + \sigma)Z$ and $V = (1 - \sigma)Z$, one finds that $[(1 - \sigma^2)Z]' + 4\sigma \, Z^2 = 0$, which after division by Z^2 gives a linear equation $(\sigma^2 - 1)\left(\frac{1}{Z}\right)' - \frac{2\sigma}{Z} + 4\sigma = 0$, which has the particular solution $\frac{1}{Z} = 2$, and general solution $2 + C(1 - \sigma^2)$; this gives

$$\begin{aligned} U(\sigma) &= \frac{1+\sigma}{2+C(1-\sigma^2)} \text{ in } [-1, 1), \ 0 \text{ outside},\\ V(\sigma) &= \frac{1-\sigma}{2+C(1-\sigma^2)} \text{ in } (-1, 1], \ 0 \text{ outside}, \end{aligned} \tag{17.11}$$

for which one must have $C > -2$. One sees that $V(\sigma) = U(-\sigma)$, so that both U and V have the same integral, and the relation between C and m is

$$m = \int_{-1}^{+1} \frac{2}{2 + C(1 - \sigma^2)} \, d\sigma, \tag{17.12}$$

which can be computed explicitly (and m tends to 0 as C tends to $+\infty$ and tends to 0 as C tends to -2), and I shall come back to this computation later.

For the Broadwell model, the self-similar solutions do not have finite mass, but I have suggested that they could be useful for another question, discussed later.

[Taught on Friday October 12, 2001.]

18

Oscillating Solutions; the 2-D Broadwell Model

There are various reasons why it is useful to consider what happens for sequences of solutions of evolution equations when one starts from a sequence of initial data which converges only weakly. My motivation in the mid 1970s was that topologies like weak convergence and more general topologies of weak type, like those appearing in homogenization, are a good way to express the relations between different scales, the finest scale being called microscopic (or mesoscopic for those who are rigid enough to consider that the term microscopic only applies to the level of atoms) and the coarsest scale being called macroscopic; I had initiated that philosophy in the early 1970s, influenced by some work of Évariste SANCHEZ-PALENCIA.

Homogenization is understood in the general context that I had developed with François MURAT in the early 1970s, i.e. related to the H-convergence approach that we had introduced, which is a little more general than the G-convergence approach that Sergio SPAGNOLO had developed in the late 1960s, with the help of Ennio DE GIORGI.[1] I had borrowed the term homogenization from Ivo BABUŠKA, but I applied it in general situations, without a restriction to a periodically modulated framework, which I had first seen in the work of Évariste SANCHEZ-PALENCIA, and the way I used that term certainly conformed to the spirit of what Ivo BABUŠKA had meant when he introduced it. However, among those who often use the mathematical tools that I had developed for the general framework, many limit themselves to the periodically modulated case, for one reason or another, but project their limitations on their students by not emphasizing that the method that they use had been developed for a general framework; probably for some other reason, they rarely mention the pioneering work that had been done by Évariste

---

[1] Ennio DE GIORGI, Italian mathematician, 1928–1996. He received the Wolf Prize in 1990, for his innovating ideas and fundamental achievements in partial differential equations and calculus of variations, jointly with Ilya PIATETSKI-SHAPIRO. He had worked at Scuola Normale Superiore, Pisa, Italy.

SANCHEZ-PALENCIA in the early 1970s, precisely on the periodic framework that they want to limit themselves to.

According to my philosophy, the weak convergence is adapted to some quantities, which are usually coefficients of differential forms (as it appeared after discussions with Joel ROBBIN), for example it applies to the density of mass ρ, and to the linear momentum q, which both appear in the equation expressing conservation of mass, $\frac{\partial \rho}{\partial t} + div(q) = 0$, but if one wants to define an effective velocity u for transport of mass by writing $q = \rho u$, then the weak convergence may not be adapted to u itself.[2] If some physical phenomenon occurs at a microscopic level, and a model pretends to describe the relevant physical quantities observed at a macroscopic level, then the macroscopic equations should be stable when using the right type of weak convergence which describes the passage from the microscopic or mesoscopic level to the macroscopic level (and one should then be careful about identifying what the right convergence should be); if they are not stable it means that the effective equations have not been identified correctly.

I was wondering if equations used in kinetic theory are stable with respect to weak convergence, which is well adapted for densities of particles, but I observed that most discrete velocity models are not stable. This negative fact is in itself difficult to use, because these models are not believed to be exact but are considered as simplifications; this mathematical exercise should then be considered in its right context, that it may help in developing better mathematical tools that one needs for studying more complicated models, believed to describe accurately a part of the physical reality.

My first step was to show that a simple model like the Carleman model (which is not a model from kinetic theory as it does not conserve momentum), is not stable with respect to weak convergence. In order to see that, I considered sequences a_n, b_n satisfying $0 \leq a_n, b_n \leq M$ in $\mathbb{R}$, and I used the solutions u_n, v_n satisfying

$$
\begin{aligned}
(u_n)_t + (u_n)_x + (u_n)^2 - (v_n)^2 = 0 \text{ in } \mathbb{R} \times (0, \infty); \ u_n(\cdot, 0) = a_n \text{ in } \mathbb{R}, \\
(v_n)_t - (v_n)_x - (u_n)^2 + (v_n)^2 = 0 \text{ in } \mathbb{R} \times (0, \infty); \ v_n(\cdot, 0) = b_n \text{ in } \mathbb{R},
\end{aligned} \tag{18.1}
$$

and one has $0 \leq u_n, v_n \leq M$. If $a_n \rightharpoonup a_\infty$ and $b_n \rightharpoonup b_\infty$ in $L^\infty(\mathbb{R})$ weak $\star$, then I wanted to show that it is not always true that u_n and v_n converge in $L^\infty(\mathbb{R} \times (0, \infty))$ weak $\star$ to the solutions u_∞ and v_∞ corresponding to the initial data a_∞ and b_∞. I later characterized the oscillations created in the sequence, and I shall describe that in another lecture, but at that time I did

---

[2] For charged particles, one denotes by ϱ the density of electric charge and by j the density of electric current, and conservation of charge takes the form $\frac{\partial \varrho}{\partial t} + div(j) = 0$, but one usually does not introduce an effective velocity for transport of charge defined by $j = \varrho u$, because the "particles" carrying the charges have very different masses and velocities, as they are light electrons or heavy ions, and an average velocity would be useless.

not need to be as precise, and integrating along characteristic lines, I first observed that

$$u_n(x,t) = a_n(x - t) + O(t); \quad v_n(x,t) = b_n(x + t) + O(t). \tag{18.2}$$

Taking $a_n = 1 + \sin(n\,x)$ and $b_n = 1$ (so that $M = 2$) gives $a_\infty = b_\infty = 1$ but $(a_n)^2 \rightharpoonup \frac{3}{2}$, and as $(u_n)_t + (u_n)_x = -(a_n(x - t) + O(t))^2 + (1 + O(t))^2$ for which a subsequence converges weakly to $-\frac{1}{2} + O(t)$, one finds that the weak $\star$ limit u_* of a subsequence u_m is $1 - \frac{t}{2} + O(t^2)$, different from $u_\infty = 1$ (and the weak $\star$ limit v_* of a subsequence v_m is $1 + \frac{t}{2} + O(t^2)$, different from $v_\infty = 1$).

The same type of negative result applies to the Broadwell model, and using $u_n(\cdot,0) = v_n(\cdot,0) = 1$ and $w_n(\cdot,0) = 1 + \sin(n\,\cdot)$ (which imply an estimate $0 \le u_n, v_n, w_n \le M(t)$ by Proposition 15.2), one obtains $u_n = 1 + O(t)$, $v_n = 1 + O(t)$ and $w_n(x,t) = \sin(n\,x) + O(t)$ and then weak limits of subsequences are of the form $u_* = 1 + \frac{t}{2} + O(t^2)$, $v_* = 1 + \frac{t}{2} + O(t^2)$, $w_* = 1 - \frac{t}{2} + O(t^2)$, different from the solution $u_\infty = v_\infty = w_\infty = 1$.

I know a class of semi-linear hyperbolic systems which has the property of being stable by weak convergence, which is precisely the class satisfying condition (S) that I have already described, and that property follows from a simple application of the div-curl lemma that I had proven in 1974 with François MURAT,[3] which we generalized a few years later to a more general theory, called compensated compactness.[4] The initial form of the div-curl lemma is as follows.

Lemma 18.1. *If Ω is an open subset of $\mathbb{R}^N$ and*

$$E^{(n)} \rightharpoonup E^{(\infty)} \text{ in } L^2(\Omega; \mathbb{R}^N) \text{ weak, and}$$
$$\frac{\partial E_i^{(n)}}{\partial x_j} - \frac{\partial E_j^{(n)}}{\partial x_i} \text{ is bounded in } L^2(\Omega) \text{ for all } i,j = 1, \ldots, N, \tag{18.3}$$

and

$$D^{(n)} \rightharpoonup D^{(\infty)} \text{ in } L^2(\Omega; \mathbb{R}^N) \text{ weak, and}$$
$$\sum_{i=1}^{N} \frac{\partial D_i^{(n)}}{\partial x_i} \text{ is bounded in } L^2(\Omega), \tag{18.4}$$

---

[3] As I have already mentioned, Joel ROBBIN had provided a different proof in 1975, using differential forms and Hodge decomposition.

[4] The name was coined by Jacques-Louis LIONS, who had asked François MURAT to generalize the div-curl lemma as part of the work for his thesis, and he had given him an article of SCHULENBERGER & WILCOX, which he thought related; François MURAT proved a result of sequential weak continuity for a more general quadratic setting, using a condition of constant rank, choosing a slightly different method than the one that we had followed for proving the div-curl lemma. I generalized the framework for predicting the weak limits of all quadratic forms (without imposing a rank condition on the differential constraints used), the germ of the theory of H-measures which I developed ten years after.

then $(E^{(n)}.D^{(n)})$ converges to $(E^{(\infty)}.D^{(\infty)})$ weakly $\star$ in the sense of Radon measures,[5] i.e.

$$\int_\Omega (E^{(n)}.D^{(n)})\varphi\, dx \to \int_\Omega (E^{(\infty)}.D^{(\infty)})\varphi\, dx \ \text{for all } \varphi \in C_c(\Omega). \qquad (18.5)$$

Proof: The initial proof that we followed, which I used later for the more general compensated compactness theory, is a simple adaptation of a proof by Lars HÖRMANDER of the compactness of the injection of $H_0^1(\Omega)$ into $L^2(\Omega)$ for Ω bounded (and even for Ω having finite Lebesgue measure), and it uses the Fourier transform; it differs from the other proof that we had learnt from our advisor, Jacques-Louis LIONS, based on a characterization of compact sets in L^p, due to FRÉCHET and/or KOLMOGOROV. Choosing $\psi \in C_c(\Omega)$ equal to 1 on the support of φ, one replaces $E^{(n)}$ by $\varphi E^{(n)}$ and $D^{(n)}$ by $\psi D^{(n)}$, which satisfy similar hypotheses, and one must show that with the added hypothesis that $E^{(n)}$ and $D^{(n)}$ have their support in a fixed compact set of $\mathbb{R}^N$ (and one extends them by 0 outside Ω), one has $\int_{\mathbb{R}^N}(E^{(n)}.D^{(n)})\,dx \to \int_{\mathbb{R}^N}(E^{(\infty)}.D^{(\infty)})\,dx$, which one checks by applying the Plancherel theorem, i.e. one proves that $\int_{\mathbb{R}^N}(\mathcal{F}E^{(n)}.\overline{\mathcal{F}D^{(n)}})\,d\xi \to \int_{\mathbb{R}^N}(\mathcal{F}E^{(\infty)}.\overline{\mathcal{F}D^{(\infty)}})\,d\xi$. Because $\mathcal{F}E^{(n)}$ converges pointwise to $\mathcal{F}E^{(\infty)}$ and is uniformly bounded, it converges in $L^2_{loc}(\mathbb{R}^N;\mathbb{R}^N)$ strong by the Lebesgue dominated convergence theorem, and the only technical point is to show that $(\mathcal{F}E^{(n)}.\overline{\mathcal{F}D^{(n)}})$ is small at infinity. This follows from decomposing the two vectors $\mathcal{F}E^{(n)}(\xi)$ and $\mathcal{F}D^{(n)}(\xi)$ on the subspaces parallel to ξ or perpendicular to ξ; the Lagrange identity $|\xi|^2|a|^2 = |\sum_i \xi_i a_i|^2 + \sum_{i<j}|\xi_i a_j - \xi_j a_i|^2$ for all $a \in C^N$ and all $\xi \in \mathbb{R}^N$ permits us to estimate the component $a_\parallel$ on the subspace parallel to ξ by $|\xi|^2|a_\parallel|^2 = \sum_{i<j}|\xi_i a_j - \xi_j a_i|^2$ and the component $a_\perp$ on the subspace perpendicular to ξ by $|\xi|^2|a_\perp|^2 = |\sum_i \xi_i \overline{a_i}|^2$, and this implies that $|\xi|^2|\mathcal{F}E_\perp^{(n)}(\xi)|^2 = \sum_{i<j}|\xi_i \mathcal{F}E_j^{(n)}(\xi) - \xi_j \mathcal{F}E_i^{(n)}(\xi)|^2 \in L^1(\mathbb{R}^N)$ and $|\xi|^2|\mathcal{F}D_\parallel^{(n)}(\xi)|^2 = |\sum_i \xi_i \mathcal{F}D_i^{(n)}(\xi)|^2 \in L^1(\mathbb{R}^N)$, so that $|\xi|(\mathcal{F}E^{(n)}.\overline{\mathcal{F}D^{(n)}})$ is bounded in $L^1(\mathbb{R}^N)$. □

The way the div-curl lemma is used for proving the stability with respect to weak convergence of semi-linear systems in one space variable satisfying condition (S) is the following.

Lemma 18.2. *If $\omega \subset \mathbb{R}^2$ and $u_n \rightharpoonup u_\infty, v_n \rightharpoonup v_\infty$ in $L^2(\omega)$ weak and*

$$\begin{aligned}
\frac{\partial u_n}{\partial t} + c_1 \frac{\partial u_n}{\partial x} \text{ bounded in } L^2(\omega) \\
\frac{\partial v_n}{\partial t} + c_2 \frac{\partial v_n}{\partial x} \text{ bounded in } L^2(\omega),
\end{aligned} \qquad (18.6)$$

then if $c_1 \neq c_2$ one has $u_n v_n \rightharpoonup u_\infty v_\infty$ weakly $\star$ in the sense of Radon measures.

---

[5] Under the preceding hypotheses, I have shown that one cannot always take φ to be the characteristic function of a smooth set with closure in Ω, so the convergence does not hold in general in $L^1_{loc}(\Omega)$ weak.

Proof: Using $x_1 = x$ and $x_2 = t$, one applies the div-curl lemma with $E^{(n)} = (u_n, -c_1 u_n)$ and $D^{(n)} = (c_2 v_n, v_n)$, and one deduces that $(c_2 - c_1) u_n v_n$ converges to $(c_2 - c_1) u_\infty v_\infty$ weakly $\star$ in the sense of Radon measures.   □

I had introduced the particular class (S) of first-order semi-linear hyperbolic systems with quadratic nonlinearities because I knew that class to be stable by weak convergence, a simple example of compensated compactness, but the existence theorem that I proved after that is something of a different nature, which I later called *compensated integrability*, and I also coined the term *compensated regularity* for another type of result that I had obtained, after it had been improved by Raphaël COIFMAN,[6] Pierre-Louis LIONS, Yves MEYER and Steven SEMMES,[7] using Hardy space $\mathcal{H}^1$, because they had created some confusion by wrongly claiming that they had improved a result of compensated compactness,[8] and as I consider that the worst sin of a teacher is to mislead students and researchers, I coined the new terms (compensated integrability, compensated regularity) precisely for explaining the differences.

I had checked that the class (S) is almost the right one,[9] by considering a general system of the form

$$\frac{\partial u_i}{\partial t} + C^i.grad(u_i) = F_i(u_1, \ldots, u_m) \text{ in } \mathbb{R}^N \times (0, T), i = 1, \ldots, m$$
$$u_i(\cdot, 0) = v_i \text{ in } \mathbb{R}^N, i = 1, \ldots, m, \quad (18.7)$$

with $C^1, \ldots, C^m \in \mathbb{R}^N$, and $F_1, \ldots, F_m$ locally Lipschitz functions on $\mathbb{R}^m$, so that for bounded initial data in $L^\infty(\mathbb{R}^N)$ the solution exists on an interval $(0, T)$, with T depending eventually upon the L^∞ norm of the initial data, and

---

[6] Ronald Raphaël COIFMAN, Israeli-born mathematician, born in 1941. He worked at Washington University, St Louis, MO, and at Yale University, New Haven, CT.

[7] Stephen William SEMMES, American mathematician, born in 1962. He works at Rice University, Houston, TX.

[8] What they had done could hardly be called an improvement of the div-curl lemma anyway, because with more hypotheses (i.e. $curl(E^{(n)}) = 0$ and $div(D^{(n)}) = 0$) they did not even prove the convergence of the whole sequence $(E^n.D^n)$ to $(E^\infty.D^\infty)$, which I had shown by a simple integration by parts in the particular case where $curl(E^{(n)}) = 0$, because if $E^{(n)} = grad(u_n)$ then $(E^{(n)}.D^{(n)}) = \sum_i \frac{\partial(u_n D_i^{(n)})}{\partial x_i} - u_n div(D^{(n)})$ and u_n converges strongly. Using more hypotheses, they had proven that $(E^{(n)}.D^{(n)})$ is bounded in $\mathcal{H}^1$, which is the dual of VMO, so that a subsequence converges in $\mathcal{H}^1$ weak $\star$, but they did not identify limits, so their result is not about compensated compactness.

[9] I know that this kind of sentence which I like to use is not so good from a grammatical point of view, but it is my way of recalling that mathematical truths are not subject to change with time, i.e. if a mathematical result has been proven in the past, then it is still right in the present, and it will still be right in the future: I had proven something in the past, and my result is true. If I had written that it was true, some readers may wrongly interpret that it is like some "truths" which evolve with time, like the statements which physicists often make, which depend upon being believed by a majority, until they are shown to be wrong!

I looked for those nonlinearities which make the system weakly $\star$ stable, i.e. for all sequences of initial data converging in $L^\infty(\mathbb{R}^N)$ weak $\star$ the corresponding solutions converge in $L^\infty(\mathbb{R}^N \times (0,T))$ weak $\star$ to the solution corresponding to the limit. I proved then that it is true if and only if either all the F_i are affine, or if $N = 1$ and each F_i has the form

$$F_i(u_1, \ldots, u_m) = \sum_{j,k} a_{i,j,k} u_j u_k + affine(u_1, \ldots, u_m), \qquad (18.8)$$

where the coefficients $a_{i,j,k}$ are such that $a_{i,j,k} = a_{i,k,j}$ for all $i, j, k = 1, \ldots, m$, and satisfy condition (S):

$$C^j = C^k \text{ implies } a_{i,j,k} = 0 \text{ for all } i, \qquad (18.9)$$

so that, apart from the added affine parts (which are not so natural in kinetic theory, except for using Galilean invariance), the condition that I had found is indeed the more general one.

Of course, the fact that a system is not stable by weak convergence does not mean that one cannot prove the existence and uniqueness of solutions for it, and a way to see the difference between compensated compactness and compensated integrability is to consider a classical remark of Emilio GAGLIARDO,[10] and of Louis NIRENBERG, in their independent proofs of the Sobolev embedding theorem, which states in the case $N = 3$ that

$$\int_{\mathbb{R}^3} |u_1(x_2, x_3)|\, |u_2(x_1, x_3)|\, |u_3(x_1, x_2)|\, dx_1\, dx_2\, dx_3 \\ \leq ||u_1||_{L^2(\mathbb{R}^2)} ||u_2||_{L^2(\mathbb{R}^2)} ||u_3||_{L^2(\mathbb{R}^2)}, \qquad (18.10)$$

and is a simple consequence of the Cauchy–Schwarz inequality, because the function v_3 defined by $v_3(x_1, x_2) = \int_{\mathbb{R}} |u_1(x_2, x_3)|\, |u_2(x_1, x_3)|\, dx_3$ satisfies $|v_3(x_1, x_2)|^2 \leq \left(\int_{\mathbb{R}} |u_1(x_2, x_3)|^2\, dx_3\right)\left(\int_{\mathbb{R}} |u_2(x_1, x_3)|^2\, dx_3\right)$, which one then integrates in (x_1, x_2) to obtain $||v_3||_{L^2(\mathbb{R}^2)} \leq ||u_1||_{L^2(\mathbb{R}^2)} ||u_2||_{L^2(\mathbb{R}^2)}$. However, there is no analogous result of compensated compactness, i.e. if $\frac{\partial u_i^n}{\partial x_i} = 0$ and $u_i^n \rightharpoonup u_i^\infty$ in L^∞ weak $\star$ for $i = 1, 2, 3$, then in general $u_1^n u_2^n u_3^n$ does not converge to $u_1^\infty u_2^\infty u_3^\infty$ in L^∞ weak $\star$; actually, the compensated compactness theory shows that only affine functions F have the property that one can deduce that $F(u_1^n, u_2^n, u_3^n)$ converges to $F(u_1^\infty, u_2^\infty, u_3^\infty)$ in L^∞ weak $\star$.

The preceding type of estimate is useful for proving uniform L^2 estimates for the (two-dimensional) four velocities model, for nonnegative initial data with a small L^2 norm. In 1985, Takaaki NISHIDA had mentioned having proven an existence theorem for small data in L^2, after I had mentioned my computations to him; I realize now that my computations did not prove existence, and were just a first step.

---

[10] Emilio GAGLIARDO, Italian mathematician, born in 1930. He worked at Università di Pavia, Pavia, Italy.

For nonnegative initial data v_1, v_2, v_3, v_4 belonging to $L^2(\mathbb{R}^2)$ and having small norms, one looks for bounds for the (nonnegative) solutions of the (two-dimensional) four velocities model of the form

$$
\begin{aligned}
0 &\leq u_1(x, y, t) \leq U_1(x - t, y) \\
0 &\leq u_2(x, y, t) \leq U_2(x + t, y) \\
0 &\leq u_3(x, y, t) \leq U_3(x, y - t) \\
0 &\leq u_4(x, y, t) \leq U_4(x, y + t),
\end{aligned}
\tag{18.11}
$$

with $U_1, U_2, U_3, U_4 \in L^2(\mathbb{R}^2)$. From $(u_1)_t + (u_1)_x + u_1 u_2 = u_3 u_4$, and $u_2 \geq 0$, one finds that $0 \leq u_1(x, y, t) \leq v_1(x - t, y) + \int_0^t f(x - s, y, t - s)\, ds$, with $f(\xi, \eta, \tau) = U_3(\xi, \eta - \tau) U_4(\xi, \eta + \tau)$, so that $f(x - s, y, t - s) = U_3(x - s, y - t + s) U_4(x - s, y + t - s)$ and $\int_0^t f(x - s, y, t - s)\, ds = \int_0^t U_3(x - t + \sigma, y - \sigma) U_4(x - t + \sigma, y + \sigma)\, d\sigma$, and therefore

$$
\begin{aligned}
&0 \leq u_1(x, y, t) \leq \widetilde{U}_1(x - t, y), \text{ with} \\
&\widetilde{U}_1(\xi, \eta) = v_1(\xi, \eta) + \int_0^\infty U_3(\xi + \sigma, \eta - \sigma) U_4(\xi + \sigma, \eta + \sigma)\, d\sigma.
\end{aligned}
\tag{18.12}
$$

Similarly, one has

$$
\begin{aligned}
&0 \leq u_2(x, y, t) \leq \widetilde{U}_2(x + t, y), \text{ with} \\
&\widetilde{U}_2(\xi, \eta) = v_2(\xi, \eta) + \int_0^\infty U_3(\xi - \sigma, \eta - \sigma) U_4(\xi - \sigma, \eta + \sigma)\, d\sigma \\
&0 \leq u_3(x, y, t) \leq \widetilde{U}_3(x, y - t), \text{ with} \\
&\widetilde{U}_3(\xi, \eta) = v_3(\xi, \eta) + \int_0^\infty U_1(\xi - \sigma, \eta + \sigma) U_2(\xi + \sigma, \eta + \sigma)\, d\sigma \\
&0 \leq u_4(x, y, t) \leq \widetilde{U}_4(x, y + t), \text{ with} \\
&\widetilde{U}_4(\xi, \eta) = v_4(\xi, \eta) + \int_0^\infty U_1(\xi - \sigma, \eta - \sigma) U_2(\xi + \sigma, \eta - \sigma)\, d\sigma.
\end{aligned}
\tag{18.13}
$$

One wants a fixed point of the mapping $(U_1, U_2, U_3, U_4) \mapsto (\widetilde{U}_1, \widetilde{U}_2, \widetilde{U}_3, \widetilde{U}_4)$; it is a well-defined mapping from $\left(L^2(\mathbb{R}^2)\right)^4$ into itself, because it is like the Gagliardo–Nirenberg remark (18.10), using different directions; writing $U_j(a, b) = F_j(a + b, a - b)$ for $j = 3, 4$, one has

$$
\begin{aligned}
|\widetilde{U}_1(\xi, \eta) - v_1(\xi, \eta)|^2 &= |\int_0^\infty F_3(\xi + \eta, \xi - \eta + 2\sigma) F_4(\xi + \eta + 2\sigma, \xi - \eta)\, d\sigma|^2 \\
&\leq \tfrac{1}{4}(\int_0^\infty |F_3(\xi + \eta, \tau)|^2\, d\tau)(\int_0^\infty |F_4(\tau, \xi - \eta)|^2\, d\tau)
\end{aligned}
\tag{18.14}
$$

by the Cauchy–Schwarz inequality, and using $d\xi\, d\eta = \tfrac{1}{2} d(\xi + \eta)\, d(\xi - \eta)$ one then finds that

$$
\begin{aligned}
\int_{\mathbb{R}^2} |\widetilde{U}_1(\xi, \eta) - v_1(\xi, \eta)|^2\, d\xi\, d\eta &\leq \tfrac{1}{8}(\int_{\mathbb{R}^2} |F_3|^2\, d\xi\, d\eta)(\int_{\mathbb{R}^2} |F_4|^2\, d\xi\, d\eta) \\
&= \tfrac{1}{2}(\int_{\mathbb{R}^2} |U_3|^2\, d\xi\, d\eta)(\int_{\mathbb{R}^2} |U_4|^2\, d\xi\, d\eta),
\end{aligned}
\tag{18.15}
$$

and therefore

$$
\|\widetilde{U}_1\|_{L^2(\mathbb{R}^2)} \leq \|v_1\|_{L^2(\mathbb{R}^2)} + \frac{1}{\sqrt{2}} \|U_3\|_{L^2(\mathbb{R}^2)} \|U_4\|_{L^2(\mathbb{R}^2)},
\tag{18.16}
$$

and similarly

$$\|\widetilde{U_2}\|_{L^2(\mathbb{R}^2)} \leq \|v_2\|_{L^2(\mathbb{R}^2)} + \tfrac{1}{\sqrt{2}}\|U_3\|_{L^2(\mathbb{R}^2)}\|U_4\|_{L^2(\mathbb{R}^2)}$$
$$\|\widetilde{U_3}\|_{L^2(\mathbb{R}^2)} \leq \|v_3\|_{L^2(\mathbb{R}^2)} + \tfrac{1}{\sqrt{2}}\|U_1\|_{L^2(\mathbb{R}^2)}\|U_2\|_{L^2(\mathbb{R}^2)} \qquad (18.17)$$
$$\|\widetilde{U_4}\|_{L^2(\mathbb{R}^2)} \leq \|v_4\|_{L^2(\mathbb{R}^2)} + \tfrac{1}{\sqrt{2}}\|U_1\|_{L^2(\mathbb{R}^2)}\|U_2\|_{L^2(\mathbb{R}^2)}.$$

If $\max_i \|v_i\|_{L^2(\mathbb{R}^2)} = \varepsilon < \frac{1}{2\sqrt{2}}$, then choosing $\max_i \|U_i\|_{L^2(\mathbb{R}^2)} \leq 2\varepsilon$ implies $\max_i \|\widetilde{U_i}\|_{L^2(\mathbb{R}^2)} \leq 2\varepsilon$, and the mapping is a strict contraction on this set, with constant $\theta \leq 2\sqrt{2}\,\varepsilon < 1$; the mapping has a (unique) fixed point, and as long as the solution exists it satisfies the bounds with the functions U_1, U_2, U_3, U_4 found.

[Taught on Monday October 15, 2001.]

Notes on names cited in footnotes for Chapter 18, PIATETSKI-SHAPIRO,[11] SCHULENBERGER,[12] WILCOX,[13] WASHINGTON,[14] W. RICE.[15]

---

[11] Ilya PIATETSKI-SHAPIRO, Russian-born mathematician, born in 1929. He received the Wolf Prize in 1990, for his fundamental contributions in the fields of homogeneous complex domains, discrete groups, representation theory and automorphic forms, jointly with Ennio DE GIORGI. He worked in Tel Aviv, Israel.

[12] John R. SCHULENBERGER, American mathematician. He worked in Denver, CO, at University of Utah, Salt Lake City, UT and at Texas Tech University, Lubbock, TX.

[13] Calvin Hayden WILCOX, American mathematician. He worked at University of Wisconsin, Madison, WI, and at University of Utah, Salt Lake City, UT.

[14] George WASHINGTON, American general, 1732–1799. First President of the United States.

[15] William Marsh RICE, American financier and philanthropist, 1816–1900. Rice University, Houston, TX, is named after him.

19

Oscillating Solutions: the Carleman Model

After showing that the Carleman model is not stable by weak convergence, I did not try immediately to characterize the oscillations. My philosophy that good physical models should be stable with respect to some adapted convergence did not apply to that model, as it is not a model of kinetic theory, and one reason why I was led to study oscillations for this model was related to studying the asymptotic behaviour (i.e. as t tends to ∞) of the solution. The question of looking at the asymptotic behaviour often has no physical interest, as most models have lost their validity long before time has become large enough,[1] but discrete velocity models are not very good physics, and the Carleman model is not about physics at all, and I was interested by the mathematical result of decay in $\frac{C(m)}{t}$ for nonnegative solutions with finite total mass m, that Reinhard ILLNER and Michael REED had obtained. It is easy to understand that the nonlinearities describe some kind of self-destructive process, and I wanted to understand more about what was going on. If one starts from initial data with compact support in an interval of length L, the support at time t is included in an interval of length $L+2t$, and the solution being $O\left(\frac{1}{t}\right)$ it is then natural to rescale the x variable and the u and v functions in opposite ways, and one is led to consider the sequences u_n and v_n defined by

---

[1] One example is the kind of nonsense that one often hears from some people who pretend to work on turbulence, as letting time tend to infinity has hardly anything to do with turbulence. Large time could be of importance if one is working in an infinite domain and one rescales the equations in an appropriate way, but those who advocate this question are usually working in a box, often with periodic conditions, and any resemblance to turbulence in these conditions could only be a lucky accident. Of course, turbulent flows show complicated behaviour, and it has been known since POINCARÉ that ordinary differential equations may show strange effects as time tends to ∞, and those who have coined the word chaos have certainly decided to translate what POINCARÉ did into a more recent language, but making people believe that the two problems are related is pure political propaganda.

$$u_n(x,t) = n\,u(n\,x, n\,t), \; v_n(x,t) = n\,v(n\,x, n\,t). \tag{19.1}$$

They stay bounded in $L^\infty(\mathbb{R}^2)$ by the Illner–Reed estimate, and as they satisfy the same Carleman model, I found it natural to start by investigating what happens for general bounded sequences of solutions, which I generated by considering bounded sequences of (nonnegative) initial data.[2]

I started with a sequence

$$0 \le a_n, b_n \le M \text{ in } \mathbb{R}, \tag{19.2}$$

and I considered the Carleman model

$$(u_n)_t + (u_n)_x + (u_n)^2 - (v_n)^2 = 0 \text{ in } \mathbb{R} \times (0, \infty); \; u_n(\cdot, 0) = a_n \text{ in } \mathbb{R} \atop (v_n)_t - (v_n)_x - (u_n)^2 + (v_n)^2 = 0 \text{ in } \mathbb{R} \times (0, \infty); \; v_n(\cdot, 0) = b_n \text{ in } \mathbb{R}, \tag{19.3}$$

for which one has the uniform estimate

$$0 \le u_n, v_n \le M \text{ in } \mathbb{R} \times (0, \infty). \tag{19.4}$$

I then wondered if the knowledge of oscillations in the sequence (a_n, b_n), for example if the *Young measure* for the sequence (a_n, b_n),[3] which describes the one-point statistics for the data by identifying all the weak $\star$ limits of $f(a_n, b_n)$ for all continuous functions f, is sufficient for deducing the Young measure for the sequence of solutions (u_n, v_n). Indeed, I found that the Young measure for (u_n, v_n), which is a tensor product, is actually determined by the sole knowledge of the Young measure for a_n and the Young measure for b_n (which is less information than the Young measure for (a_n, b_n), of course), and this property is actually valid for all systems of two equations of the form

$$(u_n)_t + C_1(u_n)_x = F(u_n, v_n) \text{ in } \mathbb{R} \times (0, T); \; u_n(\cdot, 0) = a_n \text{ in } \mathbb{R} \atop (v_n)_t + C_2(v_n)_x = G(u_n, v_n) \text{ in } \mathbb{R} \times (0, T); \; v_n(\cdot, 0) = b_n \text{ in } \mathbb{R}, \tag{19.5}$$

if $C_1 \ne C_2$, if F, G are locally Lipschitz continuous and if the solutions stay bounded for the time interval considered (the analysis being a little more

---

[2] It is not necessary to consider nonnegative data, but without this condition one must assume that the solutions stay bounded on some interval $[0, T]$ independent of n.

[3] Laurence Chisholm YOUNG, English-born mathematician, 1905–2000. He had worked in Cape Town, South Africa, and at University of Wisconsin, Madison, WI. I had met Laurence YOUNG in Madison in 1971 during my first visit to United States, and as my English was not so good I had conversed with him in French, which he spoke without accent, and he might have learnt it when his father (W.H. YOUNG) was teaching in Lausanne, Switzerland. I only learnt much later about his work in the calculus of variations, and when I pioneered the introduction of Young measures in the partial differential equations of continuum mechanics in the late 1970s, I used the term *parametrized measures* instead, that I had heard in seminars on "control theory" in Paris, France.

technical in this general case). One should notice that even for the case of two equations, the same results would not hold if there was more than one space variable (except for affine functions F, G, of course).

I shall also show later that the same result does not always hold for three equations, by investigating the case of the Broadwell model.

I assume that

$$(a_n)^k \rightharpoonup A_k \text{ in } L^\infty(\mathbb{R}) \text{ weak } \star, k = 1, \ldots$$
$$(b_n)^k \rightharpoonup B_k \text{ in } L^\infty(\mathbb{R}) \text{ weak } \star, k = 1, \ldots, \qquad (19.6)$$

and this is equivalent to using the Young measure for the sequence a_n and the Young measure for the sequence b_n.[4] I extract a subsequence (u_m, v_m) such that

$$(u_m)^k \rightharpoonup U_k \text{ in } L^\infty(\mathbb{R} \times (0, \infty)) \text{ weak } \star, k = 1, \ldots$$
$$(v_m)^k \rightharpoonup V_k \text{ in } L^\infty(\mathbb{R} \times (0, \infty)) \text{ weak } \star, k = 1, \ldots, \qquad (19.7)$$

and I shall identify the list of all U_i and all V_j in terms of the list of all A_k and all B_ℓ, which shows that it is not necessary to extract a subsequence. There is something special here, that the Young measure of (u_m, v_m) is a tensor product, so that (19.7) implies

$$(u_m)^j (v_m)^k \rightharpoonup U_j V_k \text{ in } L^\infty(\mathbb{R} \times (0, \infty)) \text{ weak } \star, j, k = 1, \ldots, \qquad (19.8)$$

which is equivalent to the Young measure being a tensor product, and this is a consequence of the div-curl lemma, if one notices that

$$\left((u_m)^j\right)_t + \left((u_m)^j\right)_x = j(u_m)^{j-1}\left((v_m)^2 - (u_m)^2\right)$$
$$\text{is bounded in } L^\infty(\mathbb{R} \times (0, \infty))$$
$$\left((v_m)^k\right)_t - \left((v_m)^k\right)_x = k(v_m)^{k-1}\left((u_m)^2 - (v_m)^2\right) \qquad (19.9)$$
$$\text{is bounded in } L^\infty(\mathbb{R} \times (0, \infty)),$$

so that $(u_m)^j (v_m)^k \rightharpoonup U_j V_k$ in $L^\infty(\mathbb{R} \times (0, \infty))$ weak $\star$, for all integers $j, k \geq 0$ by Lemma 18.2 (and one takes $U_0 = V_0 = 1$, of course); one deduces easily that the weak $\star$ limit of $f(u_m)g(v_m)$ is the product of the weak $\star$ limit of $f(u_m)$ by the weak $\star$ limit of $g(v_m)$, for all continuous functions f, g. I then deduced the equations satisfied by the list of all U_j and all V_k by passing to the limit in the equations for $(u_m)^j$ and the equations for $(v_m)^k$, and I found

---

[4] Because the sequence a_n is bounded, each continuous function f can be approximated uniformly by polynomials on the closed bounded interval where the a_n take their values, by the Weierstrass theorem, and for a polynomial P, the limit of $P(a_n)$ is a finite combination of the A_k, and this permits one to identify the limit of $f(a_n)$. Although the list of all A_k is equivalent to the knowledge of the Young measure, it is not easy to extract the information, but after George PAPANICOLAOU suggested using a particular class of oscillating initial data, which I shall describe later, it appeared that there is a simple way to present the computations, where the Young measure becomes explicit.

$$(U_j)_t + (U_j)_x = j\,U_{j-1}V_2 - j\,U_{j+1} \text{ in } \mathbb{R} \times (0,\infty); \; U_j(\cdot,0) = A_j, j = 1,\ldots$$
$$(V_k)_t - (V_k)_x = k\,V_{k-1}U_2 - k\,V_{k+1} \text{ in } \mathbb{R} \times (0,\infty); \; V_k(\cdot,0) = B_k, k = 1,\ldots.$$
$$\tag{19.10}$$

To prove the uniqueness of the solution (so that the extraction of a subsequence is not necessary), one must use the bounds

$$0 \leq U_k, V_k \leq M^k \text{ in } \mathbb{R} \times (0,\infty), k = 1,\ldots, \tag{19.11}$$

which follow from the uniform bound on u_m and v_m. For two solutions satisfying this infinite system, I denoted by δU_k and δV_k the differences of the corresponding solutions, then by subtracting the corresponding inequalities I obtained

$$|(\delta U_k)_t + (\delta U_k)_x| \leq k\,|\delta U_{k+1}| + k\,M^2|\delta U_{k-1}| + k\,M^{k-1}|\delta V_2| \text{ in } \mathbb{R} \times (0,\infty);$$
$$\delta U_k(\cdot,0) = 0 \text{ in } \mathbb{R}, k = 1,\ldots$$
$$|(\delta V_k)_t - (\delta V_k)_x| \leq k\,|\delta V_{k+1}| + k\,M^2|\delta V_{k-1}| + k\,M^{k-1}|\delta U_2| \text{ in } \mathbb{R} \times (0,\infty);$$
$$\delta V_k(\cdot,0) = 0 \text{ in } \mathbb{R}, k = 1,\ldots,$$
$$\tag{19.12}$$

and I improved the initial bounds $|\delta U_k(x,t)|, |\delta V_k(x,t)| \leq 2M^k$ in $\mathbb{R} \times (0,\infty)$ by integrating the preceding inequalities in t, and I obtained

$$|\delta U_k(x,t)|, |\delta V_k(x,t)| \leq 2.3.k\,M^{k+1}t \text{ in } \mathbb{R} \times (0,\infty), k = 1,\ldots, \tag{19.13}$$

and then I used these new bounds instead of $2M^k$, and I repeated this procedure, and that gave

$$|\delta U_k(x,t)|, |\delta V_k(x,t)| \leq 2.3^2 k(k+1)M^{k+2}\tfrac{t^2}{2!} \text{ in } \mathbb{R} \times (0,\infty), k \geq 1$$
$$|\delta U_k(x,t)|, |\delta V_k(x,t)| \leq \ldots$$
$$|\delta U_k(x,t)|, |\delta V_k(x,t)| \leq 2.3^{r+1}k(k+1)\ldots(k+r)M^{k+r}\tfrac{t^{r+1}}{r!}$$
$$\text{in } \mathbb{R} \times (0,\infty), k \geq 1, r \geq 2,$$
$$\tag{19.14}$$

and letting r tend to ∞, I deduced that $\delta U_k = \delta V_k = 0$ in $\mathbb{R} \times (0,T)$ if $3M\,T < 1$, proving then uniqueness on $\left(0, \frac{1}{3K}\right)$; a reiteration of the argument gives then uniqueness for all $t \geq 0$.

I deduced an important effect from the sole knowledge of the equations for U_1 and U_2, by introducing the quantity

$$\sigma_u(x,t) = \sqrt{U_2(x,t) - \big(U_1(x,t)\big)^2} \text{ in } \mathbb{R} \times (0,\infty), \tag{19.15}$$

which measures the strength of the oscillations in the sequence u_n. Of course, one has $U_2 \geq (U_1)^2$ a.e. in $\mathbb{R} \times (0,\infty)$, and $U_2 - (U_1)^2$ is a quantity similar to what the internal energy is for a gas, measuring the amount of kinetic energy that cannot be described in terms of the macroscopic velocity; the computations shown here are then like deriving information for the internal energy by using a part of the equation describing the complete phenomena, and the analogy with questions of kinetic theory may become more apparent once the equation for *Young measures* is described in more detail.

From $(U_2)_t + (U_2)_x + 2U_3 - 2U_1 V_2 = 0$, I subtracted the equation $(U_1)_t + (U_1)_x + U_2 - V_2 = 0$ multiplied by $2U_1$, and I obtained

$$\left(U_2 - (U_1)^2\right)_t + \left(U_2 - (U_1)^2\right)_x + 2(U_3 - U_1 U_2) = 0 \text{ in } \mathbb{R} \times (0, \infty);$$
$$\left(U_2 - (U_1)^2\right)|_{t=0} = A_2 - (A_1)^2 \text{ in } \mathbb{R}. \tag{19.16}$$

Because u_n is bounded there are inequalities that U_3 must satisfy once U_1 and U_2 are known, and because $u_n \geq 0$ one of these inequalities has a very simple form:[5] developing $u_n(u_n - U_1)^2$,[6] one finds $(u_n)^3 - 2U_1(u_n)^2 + (U_1)^2 u_n \geq 0$, giving at the limit $U_3 \geq 2U_1 U_2 - (U_1)^3$, or $U_3 - U_1 U_2 \geq U_1 U_2 - (U_1)^3 = U_1(U_2 - (U_1)^2)$. Formally writing $\left((\sigma_u)^2\right)_t = 2\sigma_u (\sigma_u)_t$ for example and simplifying by σ_u,[7] one obtains

$$(\sigma_u)_t + (\sigma_u)_x + U_1 \sigma_u \leq 0 \text{ in } \mathbb{R} \times (0, \infty), \tag{19.17}$$

and a similar analysis for the oscillations in the sequence v_n gives

$$(\sigma_v)_t - (\sigma_v)_x + V_1 \sigma_v \leq 0 \text{ in } \mathbb{R} \times (0, \infty). \tag{19.18}$$

I learnt from these inequalities that, independently of the detail of the oscillations in the sequence v_n, the strength of the oscillations in the sequence u_n tends to decrease along the natural characteristic lines, and the local average of u_n can be seen as a factor for making the strength decrease, in accordance with considering the process described by the equation as a self-destruction mechanism, but one should observe that the equation is not an exact one, and U_1 may be replaced by the larger quantity $\frac{U_2}{U_1}$ in the decay term.

I also learnt an important property, that the oscillations can only be created at initial time, and this can be deduced from a weaker form of the inequality $(\sigma_u)_t + (\sigma_u)_x \leq 0$, because if one has $A_2 = (a_1)^2$ on a measurable subset ω of the real line, so that σ_u starts equal to 0 on ω, then σ_u is 0 almost everywhere on the points (x, t) with $x - t \in \omega$; indeed, σ is nonincreasing along the characteristic lines and as it starts 0 and cannot become negative, it must stay 0 there. This property will be used for studying the asymptotic behaviour of solutions.

The impossibility of creating oscillations is shared by all systems with only two equations, but it is not always true for some systems of three equations, as I shall show for the Broadwell model.

[Taught on Wednesday October 17, 2001.]

---

[5] The analysis of oscillations can be carried out for initial data changing sign, but one must restrict attention to an interval in time where a bound exists.

[6] A better inequality can be obtained by developing $u_n(u_n - w)^2$, giving $U_3 - 2w U_2 + w^2 U_1 \geq 0$ for all w, and therefore $(U_2)^2 \leq U_1 U_3$; this implies $U_3 - U_1 U_2 \geq \frac{(U_2)^2}{U_1} - U_1 U_2 = \frac{U_2}{U_1}(U_2 - (U_1)^2)$, and one has $\frac{U_2}{U_1} \geq U_1$, of course.

[7] A natural procedure for proving such a statement, which I first learnt as a student from a work of Olga OLEINIK, consists in writing the equation for $\sqrt{\varepsilon^2 + \sigma_u^2}$ for $\varepsilon > 0$, and then letting ε tend to 0.

Notes on names cited in footnotes for Chapter 19, W.H. YOUNG,[8] PAPANICO-LAOU,[9] and for the preceding footnotes, CHISHOLM-YOUNG,[10] HARDINGE.[11]

---

[8] William Henry YOUNG, English mathematician, 1863–1942. There are many results attributed to him which may be joint work with his wife, Grace, as they collaborated extensively. He had worked in Liverpool, England, in Calcutta, India, holding the first Hardinge professorship (1913-1917), in Aberystwyth, Wales, and in Lausanne, Switzerland.

[9] George C. PAPANICOLAOU, Greek-born mathematician, born in 1943. He has worked at NYU (New York University), New York, NY, and at Stanford University, Stanford, CA.

[10] Grace Emily CHISHOLM-YOUNG, English mathematician, 1868–1944.

[11] Sir Charles HARDINGE, first Baron HARDINGE of Penshurst, English diplomat, 1858–1944. He was Viceroy and Governor-General of India (1910–1916).

20

The Carleman Model: Asymptotic Behaviour

I apply now what I have found about oscillating solutions for the Carleman model to the study of the asymptotic behaviour, as t tends to ∞, of the solution of the system for fixed nonnegative initial data with finite total mass, i.e.

$$u_t + u_x + u^2 - v^2 = 0 \text{ in } \mathbb{R} \times (0, \infty); \quad u(\cdot, 0) = a \text{ in } \mathbb{R}$$
$$v_t - v_x - u^2 + v^2 = 0 \text{ in } \mathbb{R} \times (0, \infty); \quad v(\cdot, 0) = b \text{ in } \mathbb{R}, \tag{20.1}$$

with

$$a, b \in L^\infty(\mathbb{R}) \cap L^1(\mathbb{R}), a, b \geq 0 \text{ a.e. in } \mathbb{R}, \int_{\mathbb{R}} (a + b) \, dx = m < \infty. \tag{20.2}$$

Of course, my analysis uses the uniform Illner–Reed bound

$$0 \leq u(x, t), v(x, t) \leq \frac{C(m)}{t} \text{ a.e. in } \mathbb{R} \times (0, \infty), \tag{20.3}$$

but a good estimate of $C(m)$ is not necessary. In order to analyse what is going on for large t, I consider the sequence (u_n, v_n) defined by

$$u_n(x, t) = n\, u(n\, x, n\, t), v_n(x, t) = n\, v(n\, x, n\, t) \text{ in } \mathbb{R} \times (0, \infty), \tag{20.4}$$

and because one has $0 \leq u_n, v_n \leq \frac{C(m)}{t}$, the solutions are uniformly bounded for $t \geq \varepsilon > 0$. The first result relies on the finite speed of propagation.

Lemma 20.1. *For every $\varepsilon, \eta > 0$, the sequences u_n and v_n converge to 0 in L^∞ weak $\star$ and L^p_{loc} strong for $1 \leq p < \infty$ on the subsets $\{(x, t) \mid x \leq -t - \eta, t \geq \varepsilon\}$ and $\{(x, t) \mid x \geq t + \eta, t \geq \varepsilon\}$.*

Proof: Because $(u_n + v_n)_t + (u_n - v_n)_x = 0$, which expresses conservation of mass, one sees by integrating on the subset $\{(x, s) \mid x \leq x_0 - s, 0 \leq s \leq t\}$ that

$$\int_{-\infty}^{x_0-t} (u_n + v_n)(\cdot, t)\, dx + 2 \int_0^t u_n(x_0 - s, s)\, ds = \int_{-\infty}^{x_0} (u_n + v_n)(\cdot, 0)\, dx, \quad (20.5)$$

and therefore, using $u_n \geq 0$, one has

$$\int_{-\infty}^{x_0-t} (u_n + v_n)(\cdot, t) \leq \int_{-\infty}^{x_0} (a_n + b_n)\, dx, \quad (20.6)$$

where $a_n(x) = n\, a(n\, x)$ and $b_n(x) = n\, b(n\, x)$ on $\mathbb{R}$; similarly, one has

$$\int_{x_1+t}^{+\infty} (u_n + v_n)(\cdot, t) \leq \int_{x_1}^{+\infty} (a_n + b_n)\, dx, \quad (20.7)$$

and this is valid for $x_0 = -\eta$ and $x_1 = +\eta$, and because all the mass of a_n and b_n concentrates at 0 and eventually enters the interval $(-\eta, +\eta)$, one deduces that $u_n + v_n$ converges to 0 in $L^1_{loc}(\Omega)$ strong, where Ω is either $\{(x, t) \mid x \leq -t - \eta\}$ or $\{(x, t) \mid x \geq t + \eta\}$. Adding the constraint $t \geq \varepsilon > 0$ permits one to use the uniform L^∞ estimate and the bound in L^∞ together with the convergence in L^1_{loc} strong implies the convergence in L^p_{loc} strong for every $p \in [1, \infty)$ (by using Hölder inequality),[1] and in L^∞ weak $\star$.   □

Lemma 20.2. *Some subsequence $(u_{n'}, v_{n'})$ converges, and any limit (u_*, v_*) of a subsequence is automatically a solution of the Carleman model for $t > 0$, having support in $\{(x, t) \mid -t \leq x \leq t\}$, and having total mass m.*

Proof: Because of the uniform L^∞ bound for $t \geq \varepsilon > 0$, and using the diagonal argument of CANTOR,[2] one may extract a subsequence such that every power of $u_{n'}$ or $v_{n'}$ converges in L^∞ weak $\star$ for any set $\{(x, t) \mid t \geq \varepsilon\}$ (and one only needs that $u_{n'}, v_{n'}, (u_{n'})^2, (v_{n'})^2$ converge in L^∞ weak $\star$). Then one observes that $\sigma_u = 0$ for $x < -t$ (and for $x > t$) by applying Lemma 20.1, and the inequality $(\sigma_u)_t + (\sigma_u)_x \leq 0$ implies that $\sigma_u = 0$ for $x < t$, and therefore $\sigma_u = 0$ almost everywhere; this implies strong convergence of $u_{n'}$ in L^2_{loc}, and therefore strong convergence in L^p_{loc} for $1 \leq p < \infty$ because of the uniform bound in L^∞. A similar argument applies to σ_v, which is 0 in $x > t$ by applying Lemma 20.1, and satisfies $(\sigma_v)_t - (\sigma_v)_x \leq 0$ so that $\sigma_v = 0$ for $x > -t$. Because of strong convergence, one may take the limit of the equation for $t \geq \varepsilon$ for every $\varepsilon > 0$, and because $\int_{\mathbb{R}} (u_n + v_n)\, dx = m$ for every $t > 0$ one obtains $\int_{\mathbb{R}} (u_* + v_*)\, dx = m$ for all $t > 0$.   □

If one knew that all the sequence converges then the limit would automatically be self-similar, and (u_*, v_*) would be the self similar solution of total mass m; however, one has extracted a subsequence, and $k\, u_*(k\, x)$, is the limit

[1] Otto Ludwig HÖLDER, German mathematician, 1859–1937. He had worked in Leipzig, Germany.

[2] Georg Ferdinand Ludwig Philipp CANTOR, Russian-born mathematician, 1845–1918. He had worked in Halle, Germany.

of $k\, u_{n'}(k\, x) = (k\, n')u(k\, n'\, x)$, and one cannot conclude because $k\, n'$ may not be a part of the subsequence, in which case the limit would be $u_*(x)$. In 1980, I had derived a complicated proof that any solution with support in $|x| \le t$ must be self-similar, but I had not written it down and I have forgotten some of the details (I remember that I had used in an essential way the L^1 contraction property for the Carleman model, which had been noticed by Thomas LIGGETT). As for every unwritten proof, it might be that it was not complete, and one may prefer to consider this result as a conjecture.

Before trying to apply the same ideas to the Broadwell model, where important differences will appear, it is useful to mention another reason why this type of study may be useful, and it is related to what is usually described as letting the mean free path tend to 0, but after discussing the principles used for the derivation of the Boltzmann equation, it will be apparent that it is only reasonable for rarefied gases, and that it does not make any sense to use it for dense gases, and to pretend that it explains the behaviour of fluids.

For $\varepsilon > 0$ (believed to represent a mean free path between collisions), one considers the system

$$
\begin{aligned}
u_t^\varepsilon + u_x^\varepsilon + \tfrac{1}{\varepsilon}\big(u^\varepsilon\, v^\varepsilon - (w^\varepsilon)^2\big) &= 0 \text{ in } \mathbb{R} \times (0, \infty),\ u^\varepsilon(\cdot, 0) = a \text{ in } \mathbb{R} \\
v_t^\varepsilon - v_x^\varepsilon + \tfrac{1}{\varepsilon}\big(u^\varepsilon\, v^\varepsilon - (w^\varepsilon)^2\big) &= 0 \text{ in } \mathbb{R} \times (0, \infty),\ v^\varepsilon(\cdot, 0) = b \text{ in } \mathbb{R} \\
w_t^\varepsilon - \tfrac{1}{\varepsilon}\big(u^\varepsilon\, v^\varepsilon - (w^\varepsilon)^2\big) &= 0 \text{ in } \mathbb{R} \times (0, \infty),\ w^\varepsilon(\cdot, 0) = c \text{ in } \mathbb{R}.
\end{aligned}
\tag{20.8}
$$

For initial data which are nonnegative and bounded, the solution exists for all $t > 0$ by applying Proposition 15.2, because one may apply the estimate for the case $\varepsilon = 1$, to $\frac{u^\varepsilon}{\varepsilon}, \frac{v^\varepsilon}{\varepsilon}, \frac{w^\varepsilon}{\varepsilon}$; the bound obtained in L^∞ is unfortunately much too large as ε tends to 0. However, the conservation of mass shows that for $t > 0$, the functions $u^\varepsilon(\cdot, t), v^\varepsilon(\cdot, t), w^\varepsilon(\cdot, t)$ are uniformly bounded in $L^1(\mathbb{R})$. In the case of initial conditions with compact support, which is not a big restriction because of the finite speed of propagation, the entropy inequality gives a bound independent of ε for the integral of $u^\varepsilon \log(u^\varepsilon), v^\varepsilon \log(v^\varepsilon), w^\varepsilon \log(w^\varepsilon)$, which implies that $u^\varepsilon(\cdot, t), v^\varepsilon(\cdot, t), w^\varepsilon(\cdot, t)$ stay in a weakly compact set of $L^1(\mathbb{R})$, and one may extract a subsequence such that $u^\varepsilon, v^\varepsilon, w^\varepsilon$ converge in L^1_{loc} weak to u_*, v_*, w_*. One then defines the density of mass ϱ and the density of momentum q by

$$
\begin{aligned}
\varrho &= u_* + v_* + 2w_* \\
q &= u_* - v_*,
\end{aligned}
\tag{20.9}
$$

and one finds the equation

$$
\varrho_t + q_x = 0
\tag{20.10}
$$

for conservation of mass, and the equation

$$
q_t + (u_* + v_*)_x = 0
\tag{20.11}
$$

for conservation of momentum.

A natural problem is then to express $u_* + v_*$ in terms of ϱ and q, and the computations are now purely formal, and not much is known about the validity of the procedure. For the Boltzmann equation, this procedure gives the Euler equation for ideal fluids (i.e. with no viscosity),[3] but it is purely formal and has not been proven to be valid (despite the name of HILBERT being attached to that formal expansion!). Another formal derivation, attributed to CHAPMAN and ENSKOG,[4,5] the *Chapman–Enskog procedure*, makes the Navier–Stokes equation appear (with a small viscosity). In the context of the Broadwell model, the formal idea is that $u^\varepsilon\, v^\varepsilon - (w^\varepsilon)^2$ must be small and therefore one postulates that $u_*v_* - (w_*)^2 = 0$; under this postulate one has $\varrho^2 = u^2 + v^2 + 6w^2 + 4u\,w + 4v\,w = q^2 + 4u\,w + 4v\,w + 8w^2 = q^2 + 4\varrho\,w$, giving w as a function of ϱ, and showing that

$$u_*v_* - (w_*)^2 = 0 \text{ implies } u_* + v_* = \frac{\varrho^2 - q^2}{2\varrho}. \tag{20.12}$$

The system in (ϱ, q) becomes then a quasi-linear hyperbolic system of conservation laws.

However, using the inequality $\big(\log(a) - \log(b)\big)(a^2 - b^2) \geq 2(a - b)^2$ for all $a, b > 0$,[6] one finds that $\sqrt{u^\varepsilon v^\varepsilon} - w^\varepsilon$ tends to 0 in L^2_{loc} strong, and it is indeed true, as I shall show later, that both $\sqrt{u^\varepsilon v^\varepsilon}$ and w^ε are bounded in L^2, but both could be oscillating and if this was the case, the formal derivation would be wrong.[7] My analysis does not address directly this question, but considers the case $\varepsilon = 1$ and studies how oscillations will propagate if one puts oscillations in the initial data.

Before studying oscillations for the Broadwell model, it is useful to observe that letting the mean free path go to 0 for the Carleman model is a much easier question, without much interest.

Lemma 20.3. *For $a, b \in L^\infty(\mathbb{R})$ with $a, b \geq 0$ in $\mathbb{R}$, the solutions $(u^\varepsilon, v^\varepsilon)$ of*

---

[3] In his lectures about physics [14], FEYNMAN wrote that the Euler equation describes "dry water" and that the Navier–Stokes equation describes "wet water".

[4] Sydney CHAPMAN, English mathematician, 1888–1970. He had worked in Cambridge, in Manchester, in London and in Oxford, England, where he held the Sedleian chair of natural philosophy.

[5] David ENSKOG, Swedish mathematician, 1884–1947. He had worked in Stockholm, Sweden.

[6] The inequality is invariant if one replaces a, b by $t\,a, t\,b$ for $t > 0$, and it is enough to take $a = 1 + x$ and $b = 1$, and the inequality is then $\log(1 + x) \geq \frac{2x}{x+2}$ for $x \geq 0$; one has equality for $x = 0$ and the right inequality between the derivatives, $\frac{1}{1+x} \geq \frac{4}{(x+2)^2}$ for $x \geq 0$.

[7] Although I pointed out this possibility many years ago, most people do not seem to believe in the possibility of oscillations, and some people prove theorems saying that if some function of the solution converges strongly then another function of the solution converges strongly, and although these results could be valid, they do not rule out the possibility that there could be oscillations and that none of these particular functions of the solution would converge strongly.

$$(u^\varepsilon)_t + (u^\varepsilon)_x + \tfrac{1}{\varepsilon}\big((u^\varepsilon)^2 - (v^\varepsilon)^2\big) = 0 \ in \ \mathbb{R} \times (0,\infty); \ u^\varepsilon(\cdot,0) = a \ in \ \mathbb{R}$$
$$(v^\varepsilon)_t - (v^\varepsilon)_x - \tfrac{1}{\varepsilon}\big((u^\varepsilon)^2 - (v^\varepsilon)^2\big) = 0 \ in \ \mathbb{R} \times (0,\infty); \ v^\varepsilon(\cdot,0) = b \ in \ \mathbb{R}$$
$$(20.13)$$

converge to

$$u_* = v_* = \frac{a+b}{2} \ in \ \mathbb{R} \times (0,\infty). \tag{20.14}$$

Proof: If $0 \le a,b \le M$, then one has $0 \le u^\varepsilon, v^\varepsilon \le M$, and one can extract a subsequence such that $u^\varepsilon \rightharpoonup u_*, v^\varepsilon \rightharpoonup v_*$ in $L^\infty(\mathbb{R} \times (0,\infty))$ weak $\star$. Integrating $\big(u^\varepsilon \log(u^\varepsilon) + v^\varepsilon \log(v^\varepsilon)\big)_t + \big(u^\varepsilon \log(u^\varepsilon) - v^\varepsilon \log(v^\varepsilon)\big)_x + \tfrac{1}{\varepsilon}\big((u^\varepsilon)^2 - (v^\varepsilon)^2\big)\big(\log(u^\varepsilon) - \log(v^\varepsilon)\big) = 0$, and using the inequality $\tfrac{1}{\varepsilon}\big((u^\varepsilon)^2 - (v^\varepsilon)^2\big)\big(\log(u^\varepsilon) - \log(v^\varepsilon)\big) \ge \tfrac{2}{\varepsilon}(u^\varepsilon - v^\varepsilon)^2$ shows that $u^\varepsilon - v^\varepsilon$ tends to 0 in L^2_{loc} strong, and therefore $u_* = v_*$, and because $(u_* + v_*)_t + (u_* - v_*)_x = 0$ and $(u_* + v_*)\,|_{t=0} = a + b$ by taking the limit of $(u^\varepsilon + v^\varepsilon)_t + (u^\varepsilon - v^\varepsilon)_x = 0$, one finds that $(u_*)_t = 0$ and $2u_* \,|_{t=0} = a + b$. $\qquad\square$

A different scaling for the Carleman model creates a more technical problem,

$$(u^\varepsilon)_t + \tfrac{1}{\varepsilon}(u^\varepsilon)_x + \tfrac{1}{\varepsilon^2}\big((u^\varepsilon)^2 - (v^\varepsilon)^2\big) = 0 \ in \ \mathbb{R} \times (0,\infty); \ u^\varepsilon(\cdot,0) = a \ in \ \mathbb{R}$$
$$(v^\varepsilon)_t - \tfrac{1}{\varepsilon}(v^\varepsilon)_x - \tfrac{1}{\varepsilon^2}\big((u^\varepsilon)^2 - (v^\varepsilon)^2\big) = 0 \ in \ \mathbb{R} \times (0,\infty); \ v^\varepsilon(\cdot,0) = b \ in \ \mathbb{R},$$
$$(20.15)$$

which was studied by Tom KURTZ.[8] Like for the linear case, it creates a diffusion equation at the limit, but of a nonlinear degenerate type. One extracts a subsequence such that $u^\varepsilon \rightharpoonup u_*, v^\varepsilon \rightharpoonup v_*$ in $L^\infty(\mathbb{R} \times (0,\infty))$ weak $\star$, and also $\tfrac{1}{\varepsilon}(u^\varepsilon - v^\varepsilon) \rightharpoonup q$ in L^2_{loc} weak, and that last inequality assumes also that $a, b \in L^1(\mathbb{R})$ (because one cannot use the finite speed of propagation anymore as it tends to ∞); of course, a consequence is that $u_* = v_*$, and taking the limit of $(u^\varepsilon + v^\varepsilon)_t + \tfrac{1}{\varepsilon}(u^\varepsilon - v^\varepsilon)_x = 0$ one obtains

$$2(u_*)_t + q_x = 0 \ in \ \mathbb{R} \times (0,\infty); \ u_* \,|_{t=0} = \frac{a+b}{2} \ in \ \mathbb{R}. \tag{20.16}$$

In order to find a relation between q and u and u_x, one subtracts the two equations and one multiplies by ε, so that $\varepsilon(u^\varepsilon - v^\varepsilon)_t + (u^\varepsilon + v^\varepsilon)_x + \tfrac{2}{\varepsilon}\big((u^\varepsilon)^2 - (v^\varepsilon)^2\big) = 0$, and formally one postulates that $\tfrac{2}{\varepsilon}\big((u^\varepsilon)^2 - (v^\varepsilon)^2\big) = \tfrac{2}{\varepsilon}(u^\varepsilon - v^\varepsilon)(u^\varepsilon + v^\varepsilon)$ converges to $4q\,u_*$, so that the guess is

$$(u_*)_x + 2q\,u_* = 0 \ in \ \mathbb{R} \times (0,\infty), \tag{20.17}$$

showing that u_* satisfies

$$(u_*)_t - \left(\frac{(u_*)_x}{4u_*}\right)_x = 0 \ in \ \mathbb{R} \times (0,\infty). \tag{20.18}$$

---

[8] Thomas Gordon KURTZ, American mathematician. He works at University of Wisconsin, Madison, WI.

In order to prove that this is the right equation, Tom KURTZ used techniques of contraction semi-groups in L^1, and constructed enough solutions of the limiting equation. I have a different method, which requires my improvement of the Illner–Reed bound, namely $C(m) = O(1 + m^2)$, and I shall prove it later; the reason is that if one writes $u^\varepsilon(x, t) = \varepsilon^2 U^\varepsilon(\varepsilon x, t)$ and $v^\varepsilon(x, t) = \varepsilon^2 V^\varepsilon(\varepsilon x, t)$, then U^ε and V^ε satisfy the usual Carleman model but for a sequence of initial data of total norm $\frac{m}{\varepsilon}$, and therefore the bound that one obtains is $0 \leq u^\varepsilon, v^\varepsilon \leq \varepsilon^2 C\left(\frac{m}{\varepsilon}\right)\frac{1}{t}$, which is $\leq \frac{K}{t}$ if one has shown that $C(m) = O(1 + m^2)$; then an application of the div-curl lemma shows that $(u^\varepsilon + v^\varepsilon)^2$ converges weakly to $(u_* + v_*)^2$ for $t \geq \eta > 0$, and therefore $u^\varepsilon + v^\varepsilon$ converges strongly to $2u_*$ and the preceding formal computation is proven.

If one considers a sequence of solutions of the Broadwell model with a sequence of nonnegative bounded data,

$$u_t + u_x + u\,v - w^2 = 0 \text{ in } \mathbb{R} \times (0, \infty); \ u_n(\cdot, 0) = a_n \text{ in } \mathbb{R}$$
$$v_t - v_x + u\,v - w^2 = 0 \text{ in } \mathbb{R} \times (0, \infty); \ v_n(\cdot, 0) = b_n \text{ in } \mathbb{R} \qquad (20.19)$$
$$w_t - u\,v + w^2 = 0 \text{ in } \mathbb{R} \times (0, \infty); \ w_n(\cdot, 0) = c_n \text{ in } \mathbb{R},$$

with $0 \leq a_n, b_n, c_n \leq M$, then one obtains a sequence of solutions satisfying $0 \leq u_n, v_n, w_n \leq F(M, t)$, by Proposition 15.2. One extracts a subsequence (for which one keeps the index n for simplification) such that the sequence of initial data corresponds to a Young measure, and for example

$$(a_n)^i (b_n)^j (c_n)^k \rightharpoonup D_{i,j,k} \text{ in } L^\infty(\mathbb{R}) \text{ weak } \star, i, j, k = 0, \dots, \qquad (20.20)$$

with the notation $A_i = D_{i00}, B_j = D_{0j0}, C_k = D_{00k}$, for $i, j, k = 0, \dots$, and one wonders if the sequence of solutions corresponds to a Young measure, i.e. if one can identify all the following weak $\star$ limits:

$$(u_n)^i (v_n)^j (w_n)^k \rightharpoonup X_{i,j,k} \text{ in } L^\infty(\mathbb{R}) \text{ weak } \star, i, j, k = 0, \dots, \qquad (20.21)$$

with the notation

$$U_i = X_{i,0,0}, V_j = X_{0,j,0}, W_k = X_{0,0,k}, \text{ for } i, j, k = 0, \dots. \qquad (20.22)$$

The equations for $(u_n)^i, (v_n)^j, (w_n)^k$ and the div-curl lemma show that

$$X_{i,j,0} = U_i V_j, X_{0,j,k} = V_j W_k, X_{i,0,k} = U_i W_k \text{ in } \mathbb{R} \times (0, \infty), i, j, k = 0, \dots, \qquad (20.23)$$

but one needs at least to identify $X_{1,1,1}$, and I shall show that it is not always equal to $U_1 V_1 W_1$.

[Taught on Wednesday October 24, 2001 (Friday October 19 and Monday October 22 were mid-semester break).]

Notes on names cited in footnotes for Chapter 20, SEDLEY.[9]

---

[9] Sir William SEDLEY, English philanthropist, 1558–1618. He endowed a chair of natural philosophy at Oxford, England.

21

Oscillating Solutions: the Broadwell Model

For the sequence of solutions of the Broadwell model, one can write equations for powers

$$
\begin{aligned}
&\left((u_n)^i\right)_t + \left((u_n)^i\right)_x + i(u_n)^i v_n - i(u_n)^{i-1}(w_n)^2 = 0 \\
&\quad \text{in } \mathbb{R} \times (0, \infty), i = 1, \ldots \\
&\left((v_n)^j\right)_t - \left((v_n)^j\right)_x + j\, u_n(v_n)^j - j(v_n)^{j-1}(w_n)^2 = 0 \\
&\quad \text{in } \mathbb{R} \times (0, \infty), j = 1, \ldots \\
&\left((w_n)^k\right)_t - k\, u_n v_n (w_n)^{k-1} + k(w_n)^{k+1} = 0 \\
&\quad \text{in } \mathbb{R} \times (0, \infty), k = 1, \ldots
\end{aligned}
\tag{21.1}
$$

and one observes an important difference between the equations for powers of u_n or v_n on one side, and the equations for powers of w_n on the other side. In the equations for powers of u_n or v_n, there only appear products $(u_n)^i v_n, (u_n)^{i-1}(w_n)^2, u_n(v_n)^j, (v_n)^{j-1}(w_n)^2$ whose limits can be expressed in terms of the list of all U_i, V_j, W_k, and one deduces

$$
\begin{aligned}
&(U_i)_t + (U_i)_x + i\, U_i V_1 - i\, U_{i-1} W_2 = 0 \text{ in } \mathbb{R} \times (0, \infty); \\
&U_i(\cdot, 0) = A_i \text{ in } \mathbb{R}, i = 1, \ldots \\
&(V_j)_t - (V_j)_x + j\, U_1 V_j - j\, V_{j-1} W_2 = 0 \text{ in } \mathbb{R} \times (0, \infty); \\
&V_j(\cdot, 0) = B_j \text{ in } \mathbb{R}, j = 1, \ldots,
\end{aligned}
\tag{21.2}
$$

while in the equation for powers of w_n, there is a term $u_n v_n (w_n)^{k-1}$ whose limit in the case $k \geq 2$ cannot be determined in the same way.[1] Taking the limit of the equations for u_n and for $(u_n)^2$ gives

---

[1] In my 1978 lectures at Heriot–Watt University, I had already used that idea for finding more necessary conditions for sequential weak continuity under differential constraints. The basic example, which I had thought of in connection with the Broadwell model, was that in $\mathbb{R}^2$ if one has bounds on $(f_n)_x$, $(g_n)_y$, and $(h_n)_x + (h_n)_y$, then one cannot always pass to the limit in the product $f_n g_n h_n$, although the product $f\, g\, h$ satisfies the first necessary condition for sequential weak continuity; for example $f_n(x, y) = \sin(n\, y), g_n(x, y) = \cos(n\, x), h_n(x, y) = \sin(n\, x - n\, y)$ define sequences converging to 0 in $L^\infty(\mathbb{R}^2)$ weak $\star$, satisfying $(f_n)_x = (g_n)_y =$

$$(U_1)_t + (U_1)_x + U_1V_1 - W_2 = 0 \text{ in } \mathbb{R} \times (0, \infty); \ U_1(\cdot, 0) = A_1 \text{ in } \mathbb{R}$$
$$(U_2)_t + (U_2)_x + 2U_2V_1 - 2U_1W_2 = 0 \text{ in } \mathbb{R} \times (0, \infty); \ U_2(\cdot, 0) = A_2 \text{ in } \mathbb{R}.$$
$$(21.3)$$

Multiplying the first equation by $2U_1$ and subtracting from the second, one deduces that $\sigma_u = \sqrt{U_2 - (U_1)^2}$ satisfies the equation $((\sigma_u)^2)_t + ((\sigma_u)^2)_x + 2V_1(\sigma_u)^2 = 0$, or

$$(\sigma_u)_t + (\sigma_u)_x + V_1\sigma_u = 0 \text{ in } \mathbb{R} \times (0, \infty), \tag{21.4}$$

Similarly, taking the limit of the equations for v_n and for $(v_n)^2$ gives

$$(V_1)_t - (V_1)_x + U_1V_1 - W_2 = 0 \text{ in } \mathbb{R} \times (0, \infty); \ V_1(\cdot, 0) = B_1 \text{ in } \mathbb{R}$$
$$(V_2)_t - (V_2)_x + 2U_1V_2 - 2V_1W_2 = 0 \text{ in } \mathbb{R} \times (0, \infty); \ V_2(\cdot, 0) = B_2 \text{ in } \mathbb{R}.$$
$$(21.5)$$

Multiplying the first equation by $2V_1$ and subtracting from the second, one deduces that $\sigma_v = \sqrt{V_2 - (V_1)^2}$ satisfies the equation

$$(\sigma_v)_t + (\sigma_v)_x + U_1\sigma_v = 0 \text{ in } \mathbb{R} \times (0, \infty). \tag{21.6}$$

The equations for σ_u and σ_v show that the oscillations in the sequences u_n or v_n cannot be created, and that the strength of these oscillations decreases in terms of the sole local average of v_n for σ_u, and the sole local average of u_n for σ_v; this is in accordance with the fact that particles from the first or second families disappear by collisions with particles from the opposite family; contrary to what happens with the Carleman model, the equations for σ_u and for σ_v for the Broadwell model are exact. The situation is different for studying the oscillations of w_n, and taking the limit of the equation for w_n gives

$$(W_1)_t - U_1V_1 + W_2 = 0 \text{ in } \mathbb{R} \times (0, \infty); \ W_1(\cdot, 0) = C_1 \text{ in } \mathbb{R}, \tag{21.7}$$

while taking the limit of the equation for $(w_n)^2$ gives

$$(W_2)_t - 2X_{111} + 2W_3 = 0 \text{ in } \mathbb{R} \times (0, \infty); \ W_2(\cdot, 0) = C_2 \text{ in } \mathbb{R}, \tag{21.8}$$

and one should find more about X_{111}. When I was doing this analysis in the early 1980s, I already knew that one cannot expect $X_{111} = U_1V_1W_1$, but I did not understand how to describe the evolution of oscillations, until George PAPANICOLAOU proposed to restrict the class of initial data to periodically modulated functions, a question which I shall describe next. In the general case, I estimated the difference $X_{111} - U_1V_1W_1$ in order to find information on $\sigma_w = \sqrt{W_2 - (W_1)^2}$.

$(h_n)_x + (h_n)_y = 0$, but $f_n g_n h_n = \sin^2(n\,y)\cos^2(n\,x) - \frac{1}{4}\sin(2n\,x)\sin(2n\,y)$ converges to $\frac{1}{4}$ in $L^\infty(\mathbb{R}^2)$ weak $\star$.

Lemma 21.1. *One has the inequality*[2]

$$|X_{111} - U_1 V_1 W_1| \leq \sigma_u \sigma_v \sigma_w. \tag{21.9}$$

Proof: One first notices that $(u_n - U_1)(v_n - V_1)(w_n - W_1)$ converges to $X_{111} - U_1 V_1 W_1$ in $L^\infty(\mathbb{R}^2)$ weak $\star$, because by developing one finds one term $u_n v_n w_n$ which converges to X_{111}, three terms of the form $-u_n v_n W_1$, each of which converges to $-U_1 V_1 W_1$, three terms of the form $u_n V_1 W_1$, each of which converges to $U_1 V_1 W_1$, and one term $-U_1 V_1 W_1$. Then one observes that for every $\alpha > 0$ one has $\pm (u_n - U_1)(v_n - V_1)(w_n - W_1) \leq \frac{\alpha}{2}(u_n - U_1)^2 + \frac{1}{2\alpha}(v_n - V_1)^2 (w_n - W_1)^2$, which at the limit gives $\pm (X_{111} - U_1 V_1 W_1) \leq \frac{\alpha}{2}(\sigma_u)^2 + \frac{1}{2\alpha}(\sigma_v)^2 (\sigma_w)^2$; outside a subset of measure 0 the inequality is true for all positive rationals α and therefore for all real positive α, and then for a point x where all these inequalities are true, one minimizes in $\alpha > 0$ and the minimum is $\sigma_u \sigma_v \sigma_w$. $\qquad\square$

I then deduced a differential inequality for σ_w. Multiplying the equation for W_1 by $2W_1$ and subtracting from the equation for W_2, one obtains $((\sigma_w)^2)_t + 2(X_{111} - U_1 V_1 W_1) + 2(W_3 - W_1 W_2) = 0$; as seen before, the fact that $w_n \geq 0$ implies $(W_2)^2 \leq W_1 W_3$ and therefore $2(W_3 - W_1 W_2) \geq \frac{2W_2}{W_1}(\sigma_w)^2 \geq 2W_1(\sigma_w)^2$, and with Lemma 21.1 one deduces that $((\sigma_w)^2)_t + 2W_1(\sigma_w)^2 \leq 2\sigma_u \sigma_v \sigma_w$, or

$$(\sigma_w)_t + W_1 \sigma_w \leq \sigma_u \sigma_v \text{ in } \mathbb{R} \times (0, \infty). \tag{21.10}$$

This inequality shows that a factor for decreasing the strength of oscillations in w_n is the local average of w_n, in accordance with the fact that particles of the third family disappear by interaction between themselves (as it is really an interaction between the third and fourth family for the four velocities model), but there is a new effect, related to the right side $\sigma_u \sigma_v$: oscillations in w_n could be amplified, and even created if they are not present, but one needs both oscillations in u_n and oscillations in v_n for that, because both σ_u and σ_v must be positive to make an increase in σ_w possible. However, because there is an inequality, one cannot be sure that $\sigma_u > 0$ and $\sigma_v > 0$ is enough to create oscillations, and as I shall show next, it is not always the case, and creation takes place or not according to a resonance effect.

My analysis failed to describe the evolution of the Young measure for a subsequence (u_n, v_n, w_n); it is not always a tensor product, as this would imply $X_{111} = U_1 V_1 W_1$, but its three projections in $(u, v), (v, w), (u, w)$ are tensor products. As no equation is known for $u_n v_n w_n$, this approach does not say if the Young measure can be characterized in terms of the Young measure for (a_n, b_n, c_n), and a further computation done with George PAPANICOLAOU in the early 1980s shows that it is not always so, and it shows that some nonlocal correlations play a role, but our analysis was only done for the case of periodically oscillating initial data, and I could not understand the general

---

[2] In the mid 1990s, Alexander MIELKE, who might not have seen my computations, told me that he could prove the inequality with a better constant in front.

case; however, Guy MÉTIVIER has told me that he has solved that question.[3]

It is useful to understand why I use Young measures, and not be mistaken about what they say and what they cannot say. In the early 1970s, when I was working on homogenization with François MURAT, before I had heard the word itself (that Ivo BABUŠKA had borrowed from nuclear engineers), but after realizing that we had rediscovered and generalized the idea of G-convergence that Sergio SPAGNOLO had developed with Ennio DE GIORGI, we were led to try to find optimal bounds for what physicists call effective coefficients, a term which I learnt much later from George PAPANICOLAOU. I did not know the term Young measures then, and in my 1978 Heriot–Watt lectures I used the term *parametrized measures* which I had heard about in "control theory", in the seminar of Robert PALLU DE LA BARRIÈRE,[4] but the main difficulty was that except in dimension 1, the effective properties of a mixture are not described by proportions alone. I had been quite puzzled then to find that some theoretical physicists, LANDAU and LIFSHITZ,[5,6] pretended to compute a formula for the conductivity of a mixture in terms of the proportions alone, but had I known a little more about the way physicists think, I would have deduced that they were only talking about an approximation. It was clear then for mathematicians in the early 1970s, at least those who paid attention to what I and others had proven in homogenization, that *Young measures are not the right tool for describing microstructures, when there is no underlying one-dimensional pattern*, although they may be useful as a tool for obtaining a partial understanding; in the late 1970s, I had used this tool for expressing the content of the compensated compactness theory, and it was probably the first application of this idea to partial differential equations, outside the restricted geometrical context which Laurence YOUNG had thought about. I had first shown that there are no possible oscillations for some scalar quasi-linear equations in one space variable, but oscillations cannot be killed as fast for semi-linear systems in one space variable, and it was a little surprising then that the compensated compactness theory could help characterize the oscillations in systems of two equations like the Carleman model. It is important to notice that in the compensated compactness theory, Young measures are just used as a passive tool, because they cannot by themselves see the

---

[3] Guy MÉTIVIER, French mathematician, born in 1950. He worked at Purdue University, West Lafayette, IN, at Université de Rennes I, Rennes, France, and at Université de Bordeaux I, Talence, France.

[4] Robert PALLU DE LA BARRIÈRE, French mathematician, born in 1922. He worked in Caen and at Université Paris VI (Pierre et Marie Curie), Paris, France.

[5] Lev Davidovich LANDAU, Azerbaijan-born physicist, 1908–1968. He received the Nobel Prize in Physics in 1962, for his pioneering theories for condensed matter, especially liquid helium. He had worked in Leningrad, in Kharkov, and in Moscow, Russia.

[6] Evgenii Mikhailovich LIFSCHITZ, Russian physicist, 1915–1985. He had worked in Moscow, Russia.

differential structure used to express the equations; Young measures are only a language for expressing what the compensated compactness theory says, and as the compensated compactness uses micro-local objects (and I made this point more precise by the introduction of *H-measures* [18]), the Young measures can only express some of the consequences which do not make use of the differential structure. For a more interesting situation like the Broadwell model, the compensated compactness theory is not powerful enough for describing what is happening, and if one had found a better mathematical tool, some of the consequences could probably be expressed in terms of Young measures, but Young measures cannot be the important part of the argument, and one should not use the term Young measures (or come back to the old term of parametrized measures) as if it had a magical power. More and more, one hears people who replace knowledge by incantation, believing that by using technical words their message will be thought to be deep, a question that FEYNMAN had considered in [15].[7]

In the early 1980s, George PAPANICOLAOU mentioned that when the initial data are periodically modulated, i.e. of the form

$$a_n(x) = a\left(x, \frac{x}{\varepsilon_n}\right),$$
(21.11)

and for a quantity propagating at speed c, he guessed that the solution would have the form

$$A\left(x, \frac{x - ct}{\varepsilon_n}, t\right),$$
(21.12)

and he expected a simple equation for the function $A(x, y, t)$. We checked easily the case of the Carleman model.

Proposition 21.2. *The solutions* (u_n, v_n) *of the Carleman model with initial data*

$$u_n(x, 0) = a\left(x, \frac{x}{\varepsilon_n}\right) \text{ in } \mathbb{R}$$
$$v_n(x, 0) = b\left(x, \frac{x}{\varepsilon_n}\right) \text{ in } \mathbb{R},$$
(21.13)

with $0 \leq a, b \leq M$ *in* $\mathbb{R} \times \mathbb{R}$, *periodic of period 1 in* y *(and smooth enough to present no difficulties with measurability, and for weak limits to be obtained by averaging in* y), *are such that*

---

[7] FEYNMAN described the teaching of his father on that question, saying that when his father was taking him for a walk and observed a bird, he would tell him the name of the bird and give him his imagined version of what people call that bird in various parts of the world, and his father concluded by telling him that if he knew the name of the bird in all these languages, he would still know nothing about the bird. He also described the behaviour of some graduate students in physics, who learned physics as if it was a foreign language, and did not understand the relation with the real world.

$$u_n(x,t) - A\left(x, \tfrac{x-t}{\varepsilon_n}, t\right) \to 0 \ \text{strongly in } \mathbb{R} \times (0, \infty)$$

$$v_n(x,t) - B\left(x, \tfrac{x+t}{\varepsilon_n}, t\right) \to 0 \ \text{strongly in } \mathbb{R} \times (0, \infty), \tag{21.14}$$

where the convergence holds in L^p_{loc} strong for $1 \leq p < \infty$ and L^∞ weak $\star$, and A, B are periodic with period 1 in y and are the solutions of

$$A_t(x,y,t) + A_x(x,y,t) + A^2(x,y,t) - \int_0^1 B^2(x,z,t)\,dz = 0$$
$$\text{in } \mathbb{R} \times (0,1) \times (0,\infty)$$
$$B_t(x,y,t) - B_x(x,y,t) - \int_0^1 A^2(x,z,t)\,dz + B^2(x,y,t) = 0 \tag{21.15}$$
$$\text{in } \mathbb{R} \times (0,1) \times (0,\infty),$$

with initial data

$$A(x,y,0) = a(x,y), B(x,y,0) = b(x,y) \ \text{in } \mathbb{R} \times (0,1). \tag{21.16}$$

Proof: One extracts a subsequence such that $u_n \rightharpoonup U_1$, $v_n \rightharpoonup V_1$, $(u_n)^2 \rightharpoonup U_2$ and $(v_n)^2 \rightharpoonup V_2$ in L^∞ weak $\star$, and one solves

$$A_t(x,y,t) + A_x(x,y,t) + A^2(x,y,t) - V_2(x,t) = 0 \text{ in } \mathbb{R} \times (0,1) \times (0,\infty);$$
$$A(x,y,0) = a(x,y) \text{ in } \mathbb{R} \times (0,1), \tag{21.17}$$

and one wants to show that $u_n(x,t) - A\left(x, \tfrac{x-t}{\varepsilon_n}, t\right)$ tends to 0 strongly in $\mathbb{R} \times (0,\infty)$; one observes that $0 \leq a(x,y) \leq M$ and $0 \leq V_2 \leq M^2$ imply $0 \leq A(x,y,t) \leq M$ for $t > 0$. One defines $\widetilde{u}_n$ by $\widetilde{u}_n(x,t) = A\left(x, \tfrac{x-t}{\varepsilon_n}, t\right)$, and one observes that

$$(\widetilde{u}_n)_t + (\widetilde{u}_n)_x + (\widetilde{u}_n)^2 - V_2 = 0 \text{ in } \mathbb{R} \times (0,\infty); \ \widetilde{u}_n(\cdot, 0) = u_n(\cdot, 0). \tag{21.18}$$

One wants to show that $\widetilde{u}_n - u_n$ converges to 0 strongly, and one writes an equation for $(\widetilde{u}_n - u_n)^2$, namely

$$\left((\widetilde{u}_n - u_n)^2\right)_t + \left((\widetilde{u}_n - u_n)^2\right)_x + 2(\widetilde{u}_n + u_n)(\widetilde{u}_n - u_n)^2 = 2(\widetilde{u}_n - u_n)\left((v_n)^2 - V_2\right), \tag{21.19}$$

and besides using $2(\widetilde{u}_n + u_n)(\widetilde{u}_n - u_n)^2 \geq 0$, one notices that $2(\widetilde{u}_n - u_n)\left((v_n)^2 - V_2\right) \rightharpoonup 0$ in L^∞ weak $\star$ by an application of the div-curl lemma, because $(\widetilde{u}_n - u_n)_t + (\widetilde{u}_n - u_n)_x$ is bounded in L^∞ and $\left((v_n)^2 - V_2\right)_t - \left((v_n)^2 - V_2\right)_x$ is bounded in L^∞; if a subsequence of $(\widetilde{u}_n - u_n)^2$ converges weakly to ℓ, then one finds $\ell_t + \ell_x \leq 0$ and $\ell \mid_{t=0} = 0$ and therefore $\ell = 0$, as one cannot have $\ell < 0$. As a consequence, one deduces that U_2, the weak $\star$ limit of $(u_n)^2$, is the weak $\star$ limit of $A^2\left(x, \tfrac{x-t}{\varepsilon_n}, t\right)$, which is given by averaging with respect to the fast variable, i.e. $U_2(x,t) = \int_0^1 A^2(x,z,t)\,dz$. Similarly, one solves

$$B_t(x,y,t) - B_x(x,y,t) - U_2(x,t) + B^2(x,y,t) = 0 \text{ in } \mathbb{R} \times (0,1) \times (0,\infty);$$
$$B(x,y,0) = b(x,y) \text{ in } \mathbb{R} \times (0,1), \tag{21.20}$$

and one shows that $v_n(x,t) - B\left(x, \frac{x+t}{\varepsilon_n}, t\right)$ tends to 0 strongly in $\mathbb{R} \times (0, \infty)$, and one deduces that $V_2(x,t) = \int_0^1 B^2(x, z, t)\, dz$. This shows that A, B satisfies the desired equations, and because one has uniqueness for that system (which is a locally Lipschitz perturbation of something explicit), one deduces that it is true for the whole sequence. □

The case of the Broadwell model is a little more technical.

Proposition 21.3. *The solutions (u_n, v_n, w_n) of the Broadwell model with initial data*

$$u_n(x, 0) = a\left(x, \frac{x}{\varepsilon_n}\right) \text{ in } \mathbb{R}$$
$$v_n(x, 0) = b\left(x, \frac{x}{\varepsilon_n}\right) \text{ in } \mathbb{R} \qquad (21.21)$$
$$w_n(x, 0) = c\left(x, \frac{x}{\varepsilon_n}\right) \text{ in } \mathbb{R},$$

with $0 \leq a, b, c \leq M$ in $\mathbb{R} \times \mathbb{R}$, periodic of period 1 in y (and smooth enough), are such that

$$u_n(x,t) - A\left(x, \frac{x-t}{\varepsilon_n}, t\right) \to 0 \text{ strongly in } \mathbb{R} \times (0, \infty)$$
$$v_n(x,t) - B\left(x, \frac{x+t}{\varepsilon_n}, t\right) \to 0 \text{ strongly in } \mathbb{R} \times (0, \infty) \qquad (21.22)$$
$$w_n(x,t) - C\left(x, \frac{x}{\varepsilon_n}, t\right) \to 0 \text{ strongly in } \mathbb{R} \times (0, \infty),$$

where the convergence holds in L^p_{loc} strong for $1 \leq p < \infty$ and $L^\infty(\mathbb{R} \times (0, T))$ weak $\star$ for every $0 < T < \infty$, and A, B, C are periodic with period 1 in y and are the solutions of

$$A_t(x, y, t) + A_x(x, y, t) + A(x, y, t) \int_0^1 B(x, z, t)\, dz - \int_0^1 C^2(x, z, t)\, dz = 0$$
$$\text{in } \mathbb{R} \times (0, 1) \times (0, \infty)$$
$$B_t(x, y, t) - B_x(x, y, t) + B(x, y, t) \int_0^1 A(x, z, t)\, dz - \int_0^1 C^2(x, z, t)\, dz = 0$$
$$\text{in } \mathbb{R} \times (0, 1) \times (0, \infty)$$
$$C_t(x, y, t) - \int_0^1 A(x, y - z, t)B(x, y + z, t)\, dz + C^2(x, y, t) = 0$$
$$\text{in } \mathbb{R} \times (0, 1) \times (0, \infty),$$

$$(21.23)$$

with initial data

$$A(x, y, 0) = a(x, y), \; B(x, y, 0) = b(x, y), \; C(x, y, 0) = c(x, y) \text{ in } \mathbb{R} \times (0, 1).$$
$$(21.24)$$

Proof: One extracts a subsequence such that $u_n \rightharpoonup U_1$, $v_n \rightharpoonup V_1$, $w_n \rightharpoonup W_1$, $(u_n)^2 \rightharpoonup U_2$, $(v_n)^2 \rightharpoonup V_2$ and $(w_n)^2 \rightharpoonup W_2$ in L^∞ weak $\star$, and because one has $0 \leq u_n, v_n, w_n \leq F(M, t)$, one deduces that $0 \leq U_1, V_1, W_1 \leq F(M, t)$ and $0 \leq U_2, V_2, W_2 \leq F^2(M, t)$. One solves

$$A_t(x, y, t) + A_x(x, y, t) + A(x, y, t)V_1(x, t) - W_2(x, t) = 0$$
$$\text{in } \mathbb{R} \times (0, 1) \times (0, \infty); \; A(x, y, 0) = a(x, y) \text{ in } \mathbb{R} \times (0, 1), \qquad (21.25)$$

and one wants to show that $u_n(x,t) - A\left(x, \frac{x-t}{\varepsilon_n}, t\right)$ tends to 0 strongly in $\mathbb{R} \times (0, \infty)$; one defines $\tilde{u}_n$ by $\tilde{u}_n(x,t) = A\left(x, \frac{x-t}{\varepsilon_n}, t\right)$, and one observes that

$$(\tilde{u}_n)_t + (\tilde{u}_n)_x + \tilde{u}_n V_1 - W_2 = 0 \text{ in } \mathbb{R} \times (0, \infty); \quad \tilde{u}_n(\cdot, 0) = u_n(\cdot, 0). \quad (21.26)$$

One wants to show that $\tilde{u}_n - u_n$ converges to 0 strongly, and one writes an equation for $(\tilde{u}_n - u_n)^2$, namely

$$\begin{aligned}
&\left((\tilde{u}_n - u_n)^2\right)_t + \left((\tilde{u}_n - u_n)^2\right)_x + 2V_1(\tilde{u}_n - u_n)^2 \\
&= 2u_n(\tilde{u}_n - u_n)(v_n - V_1) + 2(\tilde{u}_n - u_n)\left((w_n)^2 - W_2\right),
\end{aligned} \quad (21.27)$$

and besides using $2V_1(\tilde{u}_n - u_n)^2 \geq 0$, one notices that $2u_n(\tilde{u}_n - u_n)(v_n - V_1) \rightharpoonup 0$ and $2(\tilde{u}_n - u_n)\left((w_n)^2 - W_2\right) \rightharpoonup 0$ in L^∞ weak $\star$ by an application of the div-curl lemma, because $\left(u_n(\tilde{u}_n - u_n)\right)_t + \left(u_n(\tilde{u}_n - u_n)\right)_x$ and $(v_n - V_1)_t - (v_n - V_1)_x$ are bounded in L^∞, and because $(\tilde{u}_n - u_n)_t + (\tilde{u}_n - u_n)_x$ and $\left((w_n)^2 - W_2\right)_t$ are bounded in L^∞; if a subsequence of $(\tilde{u}_n - u_n)^2$ converges weakly to ℓ, then one finds $\ell_t + \ell_x \leq 0$ and $\ell\,|_{t=0} = 0$ and therefore $\ell = 0$; as a consequence, one deduces that $U_1(x,t) = \int_0^1 A(x, z, t)\, dz$. Similarly, one solves

$$\begin{aligned}
&B_t(x, y, t) - B_x(x, y, t) + B(x, y, t)U_1(x, t) - W_2(x, t) = 0 \\
&\text{in } \mathbb{R} \times (0, 1) \times (0, \infty); \quad B(x, y, 0) = b(x, y) \text{ in } \mathbb{R} \times (0, 1),
\end{aligned} \quad (21.28)$$

and one shows that $v_n(x,t) - B\left(x, \frac{x+t}{\varepsilon_n}, t\right)$ tends to 0 strongly in $\mathbb{R} \times (0, \infty)$, and one deduces that $V_1(x,t) = \int_0^1 B(x, z, t)\, dz$.

The next step is more technical, and consists in replacing the term $u_n v_n$ by a simpler term, and considering the solution z_n of the equation

$$(z_n)_t - h_n + (z_n)^2 = 0 \text{ in } \mathbb{R} \times (0, \infty); \quad z_n(\cdot, 0) = w_n(\cdot, 0), \quad (21.29)$$

where

$$\begin{aligned}
&h_n(x, t) = H\left(x, \frac{x}{\varepsilon_n}, t\right) \text{ in } \mathbb{R} \times (0, \infty) \\
&H(x, y, t) = \int_0^1 A(x, y - z, t)B(x, y + z, t)\, dz \text{ in } \mathbb{R} \times (0, 1) \times (0, \infty),
\end{aligned} \quad (21.30)$$

so that one has

$$\begin{aligned}
&z_n(x, t) = C\left(x, \frac{x}{\varepsilon_n}, t\right) \text{ in } \mathbb{R} \times (0, \infty) \\
&C_t(x, y, t) - \int_0^1 A(x, y - z, t)B(x, y + z, t)\, dz + C^2(x, y, t) = 0 \\
&\text{in } \mathbb{R} \times (0, 1) \times (0, \infty); \quad C(x, y, 0) = c(x, y) \text{ in } \mathbb{R} \times (0, 1).
\end{aligned} \quad (21.31)$$

The estimates are technical, because it is not that $u_n v_n - h_n$ is small, but that after integrating in t the difference is small, and one must use bounds on A, B but also their moduli of continuity in t helps. $\square$

For a system of two equations, the assumption that the initial data are periodically modulated is not a big restriction, because in that case the Young measure of the solution is determined by the Young measure of the initial data; apart from measurability questions (which I am not so fond of), for every sequence a_n creating a Young measure, one can create a periodically modulated function a such that a_n and $a\left(x, \frac{x}{\varepsilon_n}\right)$ define the same Young measure; one can also perform rearrangements in the y variable without changing the Young measure. However, for the Broadwell model, the term $\int_0^1 A(y-z)B(y+z)\,dz$ changes if one rearranges A or B, and therefore the oscillations in the solutions depend upon something more precise than Young measures, as there are resonance effects which play a role. One should also notice that if one prepares periodically modulated oscillations with different periods for a, b, c, then the resonance effects cannot occur if some ratios are irrational.

Using Fourier series, i.e. $A(x, y, t) = \sum_{m \in \mathbb{Z}} A_m(x, t) e^{2i\pi m y}$, $B(x, y, t) = \sum_{m \in \mathbb{Z}} B_m(x, t) e^{2i\pi m y}$, $C(x, y, t) = \sum_{m \in \mathbb{Z}} C_m(x, t) e^{2i\pi m y}$, then $\int_0^1 A(x, y - z, t)B(x, y + z, t)\,dz = \sum_{m \in \mathbb{Z}} A_m(x, t)B_m(x, t) e^{2i\pi m y}$, and the system for A, B, C can be written as an infinite system

$$
\begin{aligned}
&(A_0)_t + (A_0)_x + A_0 B_0 - \sum_{k \in \mathbb{Z}} C_k C_{-k} = 0 \\
&(A_m)_t + (A_m)_x + A_m B_0 = 0 \text{ for } m \neq 0 \\
&(B_0)_t - (B_0)_x + A_0 B_0 - \sum_{k \in \mathbb{Z}} C_k C_{-k} = 0 \\
&(B_m)_t - (B_m)_x + A_0 B_m = 0 \text{ for } m \neq 0 \\
&(C_0)_t - A_0 B_0 + \sum_{k \in \mathbb{Z}} C_k C_{-k} = 0 \\
&(C_{2m})_t - A_m B_m = 0 \text{ for } m \neq 0 \\
&(C_{2m+1})_t = 0 \text{ for all } m,
\end{aligned}
\tag{21.32}
$$

with the corresponding Fourier coefficients of a, b, c as initial data. The coefficients A_0, B_0, C_0 are nonnegative, but the other coefficients may be complex, with $A_{-m} = \overline{A_m}$ for example. It is important to observe that such a system is a natural consequence of the Broadwell model, once one follows my philosophy of checking stability with respect to an adapted weak convergence; physicists often derive similar systems for what they call particles, and they invent some games for explaining the equations that they use, but there is no need to invent a game for solving the preceding infinite system, or to use a language of particles for talking about the solution of the system, as any mathematician who has learnt functional analysis knows. Actually, the term particle itself is just a remnant of an 18th century point of view on mechanics (called classical mechanics), which deals with rigid bodies and ordinary differential equations, by opposition to continuum mechanics, which is an 18/19th century point of view on mechanics, and deals with partial differential equations. It is important to understand that there are no particles, but just waves, i.e. partial differential equations with a hyperbolic character.

[Taught on Friday October 26, 2001.]

Notes on names cited in footnotes for Chapter 21, HERIOT,[8] WATT,[9] MIELKE.[10]

---

[8] George HERIOT, Scottish goldsmith, 1563–1624. Heriot–Watt University in Edinburgh, Scotland, is partly named after him.

[9] James WATT, Scottish engineer, 1736–1819. He had worked in Glasgow, Scotland. Heriot–Watt University in Edinburgh, Scotland, is partly named after him.

[10] Alexander MIELKE, German mathematician, born in 1958. He works in Stuttgart, Germany.

22

Generalized Invariant Regions; the Varadhan Estimate

Around 1984, I learnt of a computation by Thomas BEALE, who had shown that for bounded nonnegative data with finite total mass, the solution of the Broadwell model is globally bounded in $L^\infty(\mathbb{R})$ (but his global bound was not expressed in an explicit way). I then simplified a part of his analysis, and developed a method which I called the *generalized invariant region* method.[1] In his analysis, Thomas BEALE introduced two functions, which are potential functions related to the conservation of mass and the conservation of momentum, expressed in the form

$$(u + w)_t + u_x = 0 \text{ in } \mathbb{R} \times (0, \infty)$$
$$(v + w)_t - v_x = 0 \text{ in } \mathbb{R} \times (0, \infty). \tag{22.1}$$

In view of these, it is natural to introduce the functions U and V by

$$U(x,t) = \int_{-\infty}^{x} \big(u(z,t) + w(z,t)\big)\,dz$$
$$V(x,t) = \int_{x}^{+\infty} \big(v(z,t) + w(z,t)\big)\,dz, \tag{22.2}$$

and the important properties of U and V are that they are nonnegative and that both their derivatives in x and in t are expressed in terms of u, v, w, namely

$$\lim_{x \to -\infty} U(x,t) = 0; \ \lim_{x \to +\infty} U(x,t) = \int_{\mathbb{R}} (u_0 + w_0)\,dx$$
$$\lim_{x \to -\infty} V(x,t) = \int_{\mathbb{R}} (v_0 + w_0)\,dx; \ \lim_{x \to +\infty} V(x,t) = 0 \tag{22.3}$$
$$U_x = u + w \geq 0; \ U_t = -u \leq 0; \ V_x = -(v + w) \leq 0; \ V_t = -v \leq 0.$$

I had shown that for bounded nonnegative data with small norm in L^1 the asymptotic behaviour as t tends to ∞ is that u looks like $u_*(x - t)$, v looks

---

[1] In 1985, Takaaki NISHIDA mentioned that the method is the same as one that Tai-Ping LIU had used, for regularization by artificial viscosity of systems of conservation laws, I believe. Henri CABANNES also mentioned that he had used a similar idea in the 1950s.

like $v_*(x + t)$ and w looks like 0, and the integral of u_* is $\int_{\mathbb{R}} (u_0 + w_0)\, dx$, while the integral of v_* is $\int_{\mathbb{R}} (v_0 + w_0)\, dx$. Conservation of mass expresses that $\int_{\mathbb{R}} \big(u(\cdot,t) + v(\cdot,t) + 2w(\cdot,t)\big)\, dx$ is independent of t, and conservation of momentum expresses that $\int_{\mathbb{R}} \big(u(\cdot,t) - v(\cdot,t)\big)\, dx$ is independent of t, and it is equivalent to say that $\int_{\mathbb{R}} \big(u(\cdot,t) + w(\cdot,t)\big)\, dx$ and $\int_{\mathbb{R}} \big(v(\cdot,t) + w(\cdot,t)\big)\, dx$ are independent of t, and the physical interpretation of these quantities is that the first one is the mass which eventually finds its way to $+\infty$ and that the second one is the mass which eventually finds its way to $-\infty$. Actually, although I had only shown that for small initial mass, it is true for any finite initial mass; I am not sure if Thomas BEALE had shown that, but it does follow from an improvement by Raghu VARADHAN, which was shown to me by Kamel HAMDACHE.

The introduction of U and V is then quite natural. The function U increases from 0 to $\int_{\mathbb{R}} (u_0 + w_0)\, dx$ (the mass ending up to $+\infty$), and $U(x,t)$ measures how much of the mass going to $+\infty$ has already gone to the right of the point x at time t, and because $U_t = -u \leq 0$ the flow to the right is irreversible. Similarly, the function V decreases from $\int_{\mathbb{R}} (v_0 + w_0)\, dx$ (the mass ending up to $-\infty$) to 0, and $V(x,t)$ measures how much of the mass going to $-\infty$ has already gone to the left of the point x at time t, and because $V_t = -v \leq 0$ the flow to the left is irreversible. As I shall show in more detail, $U(\cdot,t)$ and $V(\cdot,t)$ permit one to give a measure of the amount of interaction between the particles which will take place after time t.[2]

The method of invariant regions, which does not give any interesting result for nonnegative solutions of the Broadwell model, consists in looking for a set $C \subset \mathbb{R}^3$, necessarily of the form $[0,\alpha] \times [0,\beta] \times [0,\gamma]$, such that if the initial data take their values in C then the solution has values in C for all $t > 0$. A natural improvement is to have α, β, γ functions of t and to ask the stronger requirement that if at time s the values taken belong to $C(s)$ then at any later time t the values taken belong to $C(t)$, and this implies some differential inequalities for α, β, γ which have no globally bounded solution (the requirement is much stronger than the physical one, that initial data taking their values in $C(0)$ give rise to a solution with values in $C(t)$ at time t, a problem that one does not know how to analyse well).

What I call the method of generalized invariant regions consists, in the example of the Broadwell model, in looking for inequalities of the form

---

[2] Before these results, I had already pointed out an analogy with the method that James GLIMM had introduced for quasi-linear systems of conservation laws, where he used an hypothesis of small variation. The relation between his problem and mine is that his estimates were for equations of the form $U_t + \big(F(U)\big)_x = 0$, and that it is $V = U_x$ which satisfies a semi-linear equation $V_t + \nabla F(U).V_x + \nabla^2 F(U) : (V,V) = 0$; however, even around a constant U my condition (S) is not satisfied, because of genuine nonlinearity hypotheses. The analogy between these two questions became much clearer after the estimate of Raghu VARADHAN.

$$0 \leq u(x,t) \leq \alpha\big(t, U(x,t), V(x,t)\big)$$
$$0 \leq v(x,t) \leq \beta\big(t, U(x,t), V(x,t)\big)$$
$$0 \leq w(x,t) \leq \gamma\big(t, U(x,t), V(x,t)\big),$$

(22.4)

and this takes advantage of the fact that one can express the derivatives of U and V in terms of u, v, w.

Traditionally, proving L^∞ estimates consists in comparing the solution to a constant function, but I observed that the solution does not look like a constant function, as for large t the solution u looks like $u_*(x-t)$, for example; however, I noticed that for large t the function U also looks like $U_*(x-t)$, and therefore it seems much more natural to compare u to U in order to obtain an L^∞ bound. Similarly it seems natural to compare v to V. I wrote the inequalities that the general functions α, β, γ must satisfy, but I soon restricted my attention to particular inequalities

$$0 \leq u \leq \lambda(\varepsilon + U), 0 \leq v \leq \mu(\varepsilon + V), 0 \leq w \leq \nu, \qquad (22.5)$$

where $\varepsilon, \lambda, \mu, \nu$ are positive constants. One uses $U_t + U_x = V_t - V_x = w$, and one wants that if $u = \lambda(\varepsilon + U)$ then $u_t + u_x \leq \lambda(U_t + U_x)$ so that the inequality cannot change in the evolution; this gives $w^2 - uv \leq \lambda w$, and considering the worst case $v = 0$, one is led to impose $\nu \leq \lambda$, so that $w \leq \lambda$. Similarly, one wants that if $v = \mu(\varepsilon + V)$ then $v_t - v_x \leq \mu(V_t - V_x)$, which gives $w^2 - uv \leq \mu w$, and considering the worst case $u = 0$, one is led to impose $\nu \leq \mu$; finally, one wants that if $w = \nu$ then $w_t \leq 0$, which gives $uv - w^2 \leq 0$, and one is led to impose $\lambda \mu(\varepsilon + U)(\varepsilon + V) \leq \nu^2$, and because $U_t \leq 0$ and $V_t \leq 0$, it is enough to impose that $\lambda \mu(\varepsilon + U_0)(\varepsilon + V_0) \leq \nu^2$ in $\mathbb{R}$. In the case where the initial data (nonnegative with finite total mass) satisfy

$$U_0 V_0 \leq \theta < 1 \text{ in } \mathbb{R}, \qquad (22.6)$$

then one chooses $\varepsilon > 0$ such that

$$(\varepsilon + U_0)(\varepsilon + V_0) \leq \theta' < 1 \text{ in } \mathbb{R}, \qquad (22.7)$$

and one computes

$$\lambda_0 = \left\| \frac{u_0}{\varepsilon + U_0} \right\|_{L^\infty(\mathbb{R})}, \mu_0 = \left\| \frac{v_0}{\varepsilon + V_0} \right\|_{L^\infty(\mathbb{R})}, \nu_0 = \|w_0\|_{L^\infty(\mathbb{R})}, \qquad (22.8)$$

and one must satisfy the inequalities $\lambda_0 \leq \lambda, \mu_0 \leq \mu, \nu_0 \leq \nu$ and $\nu \leq \lambda, \nu \leq \mu, \lambda \mu \theta' \leq \nu^2$; one may take $\lambda = \mu = \nu = \max\{\lambda_0, \mu_0, \nu_0\}$, for example, and this shows that if $(\varepsilon + U_0)(\varepsilon + V_0) \leq 1$ in $\mathbb{R}$, then

$$\xi(t) = \max\left\{ \left\| \frac{u(\cdot,t)}{\varepsilon + U(\cdot,t)} \right\|_{L^\infty(\mathbb{R})}, \left\| \frac{v(\cdot,t)}{\varepsilon + V(\cdot,t)} \right\|_{L^\infty(\mathbb{R})}, \|w(\cdot,t)\|_{L^\infty(\mathbb{R})} \right\}$$
is nonincreasing in $t \in (0, \infty)$,

(22.9)

and one deduces global L^∞ bounds,

$$0 \le u(x,t) \le \max\{\lambda_0, \mu_0, \nu_0\}\left(\varepsilon + \int_{\mathbb{R}}(u_0 + w_0)\, dx\right) \text{ in } \mathbb{R} \times (0, \infty)$$
$$0 \le v(x,t) \le \max\{\lambda_0, \mu_0, \nu_0\}\left(\varepsilon + \int_{\mathbb{R}}(v_0 + w_0)\, dx\right) \text{ in } \mathbb{R} \times (0, \infty) \qquad (22.10)$$
$$0 \le w(x,t) \le \max\{\lambda_0, \mu_0, \nu_0\} \text{ in } \mathbb{R} \times (0, \infty),$$

recalling that the hypothesis $(\varepsilon + U_0)(\varepsilon + V_0) \le 1$ in $\mathbb{R}$ has been used.

Of course, if the total mass $m = \int_{\mathbb{R}}(u_0 + v_0 + 2w_0)\, dx$ is small enough one has $U_0 V_0 \le \theta < 1$, and more precisely if $m < 2$ one can take $\theta = \frac{m^2}{4}$, because one has $U_0 V_0 \le \left(\int_{\mathbb{R}}(u_0 + w_0)\, dx\right)\left(\int_{\mathbb{R}}(v_0 + w_0)\, dx\right) \le \frac{1}{4}\left(\int_{\mathbb{R}}(u_0 + w_0)\, dx + \int_{\mathbb{R}}(v_0 + w_0)\, dx\right)^2$. However, the condition $U_0 V_0 \le \theta < 1$ can be valid for data with large mass if the initial distribution of mass is adequate, and actually one may have $U_0 V_0 = 0$ everywhere if $w_0 = 0$ and the support of u_0 is entirely to the right of the support of v_0, and in that case the solution is $u(x,t) = u_0(x - t), v(x,t) = v_0(x - t), w(x,t) = 0$. This kind of hypothesis is therefore much better that an hypothesis of small mass, and it has also another interesting feature, that it is not conserved by rearrangement. This type of condition reminds one more of the idea used by James GLIMM for quasi-linear systems of conservation laws, and the analogy became even clearer after an idea of Raghu VARADHAN,[3] who considered the quantity

$$I(t) = \iint_{x<y} \big(u(x,t) + w(x,t)\big)\big(v(y,t) + w(y,t)\big)\, dx\, dy, \qquad (22.11)$$

which measures a potential of interaction left at time t.

Lemma 22.1. *(S.R.S. Varadhan) For initial data which are nonnegative and with finite total mass, $I(t)$ is nonincreasing and*

$$\frac{dI}{dt} = -\int_{\mathbb{R}} (2u\,v + u\,w + v\,w)(x,t)\, dx. \qquad (22.12)$$

Proof: I had noticed that if one applies the div-curl lemma to a sequence satisfying
$$(u + w)_t + u_x = 0 \text{ in } \mathbb{R} \times (0, \infty)$$
$$(v + w)_t - v_x = 0 \text{ in } \mathbb{R} \times (0, \infty), \qquad (22.13)$$
one can pass to the limit in $v(u + w) + u(v + w)$, but I had not found how to use that information; actually, it is exactly the same computation which gives the result of Raghu VARADHAN, but I had not thought of attaching any importance to the functions U or V, and multiplying the first equation by V or the second equation by U gives the desired result:

$$\big(V(u + w)\big)_t + \big(V\,u\big)_x + v(u + w) + (v + w)u = 0$$
$$\big(U(v + w)\big)_t - \big(U\,v\big)_x + u(v + w) + (u + w)v = 0, \qquad (22.14)$$

---

[3] The result was mentioned to me by Kamel HAMDACHE, and I do not know if it had been motivated by simplifying the computations of Thomas BEALE, or had been obtained independently.

if one observes that

$$\int_{\mathbb{R}} (V(u+w)) \, dx = \int_{\mathbb{R}} \left((u+w)(x,t) \int_{x}^{+\infty} (v+w)(y,t) \, dy \right) dx = I(t), \quad (22.15)$$

or

$$\int_{\mathbb{R}} (U(v+w)) \, dx = \int_{\mathbb{R}} \left((v+w)(y,t) \int_{-\infty}^{y} (u+w)(x,t) \, dx \right) dy = I(t), \quad (22.16)$$

and $I(t)$ is easily understood as a measure of the interaction that can take place after time t. □

The estimate of Raghu VARADHAN has at least two interesting consequences.

The first application is that what I had proven for small mass is true for any finite mass; the difficulty that I had was to find a bound for the integral of $u\,v$, and now, by integrating $(V(u+w))_t + (V\,u)_x + v(u+w) + (v+w)u = 0$, one has

$$\int_{\mathbb{R}} \int_{0}^{\infty} uv \, dx \, dt \le I(0) \le \left(\int_{\mathbb{R}} (u_0 + w_0) \, dx \right) \left(\int_{\mathbb{R}} (v_0 + w_0) \, dx \right). \quad (22.17)$$

Then a bound for the integral of w^2 follows, and the solutions belong to the functional spaces that I had introduced, $u \in V_1, v \in V_{-1}, w \in V_0$, which implies the asymptotic behaviour for large t, i.e. u looks like $u_*(x-t)$, v looks like $v_*(x+t)$, and w tends to 0 (as $w_*(x) = 0$ because $w \in L^2$).

The second application is that in the problem with $\varepsilon > 0$, supposed to represent a mean free path between collisions, one had previously found that $\sqrt{u_\varepsilon v_\varepsilon} - w_\varepsilon$ converges strongly to 0 in $L^2(\mathbb{R} \times (0,\infty))$, but one did not know if each term belonged to $L^2(\mathbb{R} \times (0,\infty))$; now the estimate gives $\int_{\mathbb{R}} \int_{0}^{\infty} u_\varepsilon v_\varepsilon \, dx \, dt \le I(0)$, because only the conservation laws have been used in proving Lemma 22.1, and the results are then valid for all $\varepsilon > 0$.

The problem of letting ε tend to 0, which is more a mathematical question than a physical one, is still open in general. What Russell CAFLISCH and George PAPANICOLAOU have proven, is that when the formal limiting equation, which is a quasi-linear hyperbolic system, has a smooth solution for $0 \le t \le T$, then on that interval of time $u_\varepsilon, v_\varepsilon, w_\varepsilon$ converge to u_*, v_*, w_* satisfying $w_* = \sqrt{u_* v_*}$, and $\varrho = u_* + v_* + 2w_*$ and $q = u_* - v_*$ is the smooth solution of the quasi-linear system; it is not known if this is valid after the appearance of a shock for the (ϱ, q) system. Russell CAFLISCH has considered the case of Riemann data for the (ϱ, q) system, in the case where the solution is a single shock, but he has not succeeded in proving that the formal expansion is valid. I have conjectured that it does not always converge to the formal limit, and it was one particular reason why I had studied oscillating sequences of the Broadwell model, but I have also thought that the equation for self-similar solutions (used only locally as they do not have finite total mass) could be the key to some of the missing estimates. Although the Broadwell model is far removed from physics, it is an important training ground for developing better mathematical tools for more interesting models, so that one must consider all these questions as interesting challenges.

The method of generalized invariant regions also gives interesting L^∞ bounds for the Carleman model, and I proved in this way the global Illner–Reed estimate with a bound in $O(m^2 + 1)$ for $C(m)$, and the order cannot be improved because it appears for the self-similar solutions. The study of self-similar solutions, i.e.

$$
\begin{aligned}
u(x,t) &= \tfrac{1}{t} U\!\left(\tfrac{x}{t}\right) \\
v(x,t) &= \tfrac{1}{t} V\!\left(\tfrac{x}{t}\right),
\end{aligned}
\tag{22.18}
$$

which after using the variable $\sigma = \tfrac{x}{t}$ and $\dot{} = \tfrac{d}{d\sigma}$ leads to the system

$$
\begin{aligned}
-U - \sigma \dot{U} + \dot{U} + U^2 - V^2 &= 0 \\
-V - \sigma \dot{V} - \dot{V} - U^2 + V^2 &= 0,
\end{aligned}
\tag{22.19}
$$

was solved by

$$
\begin{aligned}
U &= (1 - \sigma)Z \\
V &= (1 + \sigma)Z \\
(1 - \sigma^2)\dot{Z} - 2\sigma Z + 4\sigma Z^2 &= 0,
\end{aligned}
\tag{22.20}
$$

and $\tfrac{1}{Z}$ satisfies a linear equation, giving

$$
Z = \frac{1}{2 + \gamma(\sigma^2 - 1)},
\tag{22.21}
$$

and the parameter γ must be < 2. For γ near 2, Z behaves like $\frac{1}{\sigma^2 + \varepsilon^2}$ with $\varepsilon > 0$ small, and the L^∞ norm behaves like $\tfrac{1}{\varepsilon^2}$, while the mass $m = \int_{-1}^{+1} Z \, d\sigma$ behaves like $\tfrac{\pi}{\varepsilon}$; for self-similar solutions, the L^∞ norm is then $O(m^2)$ for large m.

For applying the method of generalized invariant regions, one uses

$$
(u + v)_t + (u - v)_x = 0,
\tag{22.22}
$$

and one introduces

$$
W(x,t) = \int_{-\infty}^{x} \big(u(z,t) + v(z,t)\big)\, dz,
\tag{22.23}
$$

so that

$$
W_x = u + v; \quad W_t = v - u,
\tag{22.24}
$$

and in particular $W_t + W_x = 2v$ and $W_t - W_x = -2u$. One looks for bounds of the form

$$
\begin{aligned}
u &\le \frac{A(W)}{t} \\
v &\le \frac{B(W)}{t},
\end{aligned}
\tag{22.25}
$$

which one can easily replace by $u \leq \frac{A(W)}{t+\varepsilon}, v \leq \frac{B(W)}{t+\varepsilon}$ with $\varepsilon > 0$ small in order to avoid the singularity at $t = 0$. When $u = \frac{A(W)}{t}$, one wants to have $u_t + u_x \leq$ the corresponding derivative $-\frac{A(W)}{t^2} + \frac{A'(W)}{t}(W_t + W_x)$, i.e. $v^2 - u^2 \leq -\frac{A(W)}{t^2} + \frac{A'(W)}{t}2v$; checking $v = 0$ gives $A^2 \geq A$, or $A \geq 1$ and checking $v = \frac{B(W)}{t}$ gives

$$2B\, A' \geq A + B^2 - A^2 \text{ and } A \geq 1, \tag{22.26}$$

and similarly, when $v = \frac{B(W)}{t}$, one wants to have $v_t - v_x \leq$ the corresponding derivative $-\frac{A(W)}{t^2} + \frac{A'(W)}{t}(W_t - W_x)$, i.e. $u^2 - v^2 \leq -\frac{B(W)}{t^2} - \frac{B'(W)}{t}2u$; checking $u = 0$ gives $B^2 \geq B$, or $B \geq 1$ and checking $u = \frac{A(W)}{t}$ gives

$$-2A\, B' \geq B + A^2 - B^2, \text{ and } B \geq 1. \tag{22.27}$$

Using the analogy with the computation for self-similar solutions, one chooses

$$\begin{aligned} A(W) &= (1+\sigma)Z(\sigma) \\ B(W) &= (1-\sigma)Z(\sigma), \end{aligned} \tag{22.28}$$

and

$$\frac{dW}{d\sigma} = 2Z(\sigma) \tag{22.29}$$

shows that equality is obtained instead of inequalities; due to the constraints $A \geq 1, B \geq 1$, one cannot use the entire interval $-1 < \sigma < 1$, and as W must be allowed to vary between 0 and m, one must use the self-similar solution with a value of γ corresponding to a larger mass, and $m' = m + 2$ ensures that the integral for the interval where both A and B are ≥ 1 has an integral at least m. This proof provides an L^∞ bound $O(m^2)$ for large m.

[Taught on Monday October 29, 2001.]

23

Questioning Physics; from Classical Particles to Balance Laws

I have been discussing discrete velocity models for a few reasons. One of them is that they are simpler than the Boltzmann equation, which I shall investigate now; although this type of model was introduced by MAXWELL around the same period (around 1860), they seem to have been neglected for a long time; maybe the work of Renée GATIGNOL [16] was one of the first attempts (around 1970) to go beyond a few classical examples and study this type of model in a general way. From the physical point of view, models with all the velocities of the same length lack the possibility of showing temperature effects, but even if this is not the case, there is another defect which was pointed out to me by Clifford TRUESDELL in 1975, that they lack the important property of invariance by rotation; however, I only understood why invariance by rotation could be important after 1990, after I had thought that one way to avoid the angular cut-off hypothesis for the kernel in the Boltzmann equation is to use techniques like those known to specialists of harmonic analysis, like Charles FEFFERMAN and STEIN, for proving the restriction theorem on spheres, and I have mentioned that at the end of Chapter 14.

The defect of having little physical relevance is not so important if one mentions it,[1] and discrete velocity models are still an interesting mathematical arena because there are a few questions which have not been answered yet, suggesting that better mathematical tools must be created.[2]

---

[1] It becomes quite important if someone pretends that these models have any physical relevance, as it either shows some limited understanding of what continuum mechanics or physics are about, or much worse, an intention to mislead. I had started in 1984 to point out a few defects of the Boltzmann equation, but I have made a curious observation, in that case and in other situations: as many mathematicians only want to pretend that the equations that they study are related to continuum mechanics or physics, they close their ears to any information about the defects of the models that they use, and the result is that knowledge spreads at a much slower pace than misinformation does.

[2] Many people mistake development for research, but in research it is difficult to ascertain in advance what the important features for tackling an unsolved problem

The Boltzmann equation was introduced after analysing the behaviour of particles submitted to forces at a distance, and then postulating some probabilistic outputs of collisions (or nearby collisions).

As has been mentioned already, one must know at what step one has postulated a probabilistic game, or made any other assertion which one has not proven, because if one wants to understand more about the gigantic puzzle of the real world, one must first backtrack to a point where one had not yet postulated something about the answer, in order to look for a better way to solve the problem.[3] From this point of view, the discrete velocity models are postulated at too early a stage, and they are lacking the beginning of the derivation of the Boltzmann equation, where one invokes a computation involving two particles and forces at a distance before postulating the form of a kernel; however, although the defect of postulating probabilities occurs later, some defects appear already in the first stage, and they give more reasons why the Boltzmann equation is not really suitable for describing gases which are not rarefied.

Classical mechanics is an 18th century point of view of mechanics, where ordinary differential equations are used as the basic mathematical tool, and it deals with rigid bodies, often assimilated to points. The particles invoked for deriving the Boltzmann equation are assimilated to points and their kinetic energy only has a translation part, unlike in a game of billiards, where balls have a rotational kinetic energy and spin is an important effect to be taken into account for predicting the result of a collision. I shall show later a particular

---

will be. The simplified versions of a problem that are invented or the new ones which are proposed may have lost some important feature of the physical problem, and new obstacles may be created in the "simpler" versions, which are not present in the initial problem. It is part of the reseach work to decide if one should pursue in one direction or investigate in another one, and at the end it might appear that the mathematical problems concerning discrete velocity models, although interesting mathematically, are not really relevant to realistic questions, but one should not forget that the Boltzmann equation itself is not so realistic.

[3] When I learnt about ionic solutions in chemistry, I was puzzled by the type of argument which the teacher used: after applying the law of action of masses he obtained a polynomial equation, and then he assumed that the unknown x was small so that he would neglect all powers of x compared to x and solve a linear equation, and obtaining a value like 10^{-2} he would observe that indeed x was small. Of course, if one considers an equation $x^3 - 3x + \varepsilon = 0$, where ε is a small positive quantity, this argument says that one looks for the simplified equation $-3x + \varepsilon = 0$, and one accepts $x = \frac{\varepsilon}{3}$; however, the equation does have a solution $x = \frac{\varepsilon}{3} + O(\varepsilon^3)$, but also two other solutions $x = \pm\sqrt{3} - \frac{\varepsilon}{6} + O(\varepsilon^2)$, and I find it better to mention that there are theorems about the way roots of polynomials depend upon the coefficients of the polynomial, and that one should check for other solutions even if they are not small, and that one should consider an evolution equation so that a study of stability could be performed around each of the solutions in order to ascertain which ones have some chance to be observed.

case of the Boltzmann equation, called the hard-sphere case, where particles are spheres of radius a which only interact when two spheres collide, i.e. when the distance of two centres is $2a$, but for all other cases one assumes that particles feel a force which depends upon the distance between the particles.

Forces at a distance is the first defect of this approach, but this point of view which goes back to NEWTON was only challenged by POINCARÉ (and maybe EINSTEIN) in the theory of relativity, and this defect was not known then to BOLTZMANN or to MAXWELL in the 1860s. If I have understood correctly,[4] one problem created by the notion of instantaneous forces acting at a distance is the question of instantaneity; it is not difficult for a mathematician to imagine that each particle is paired with an angel who computes the total force that his particle feels by adding the forces created by all the other particles of the universe, because he is in telepathic connection with all the corresponding angels,[5] and this is what a mathematician means by writing $v(x) = \int K(x,y)u(y)\,dy$, independently of the way one will evaluate $u(y)$, the kernel $K(x,y)$, or the integral itself. If POINCARÉ and EINSTEIN understood that there is a problem for putting all the clocks (of the particles) at the same time, physicists do not seem to be as bothered by the notion of distance, and they talk about a universe in expansion while hiding the strange methods used for computing "distance".[6]

---

[4] Physicists like to make fun of mathematicians for not understanding some of the games that they invent, but one reason is that mathematicians are not so good at guessing, and physicists rarely express clearly all the rules of the games that they play; they often discard some old rule and replace it by a new one, and sometimes they even discard completely a game that they have been playing for many years. Mathematicians' duty is to be precise and they are trained to understand implications, but although a theorem proven today will remain true forever, mathematicians should be careful when claiming that what they do is important because it is related to applications; often, they have not learnt enough about the practical applications that they mention, or they do not care much if a model that they use may be soon discarded as obsolete, and it may have already been obsolete before they started working on it.

[5] I follow the French, where the word for angel is masculine (un ange), and gender is a grammatical notion, and it is not related to an early debate, when people had argued if angels were male or female, and they should have thought that they could be both, or after all that they might be neither. Of course, it does not matter at all for my argument if angels exist or not.

[6] The nearby stars move slightly with respect to the background, so their distance is measured by parallax, up to a few light years or about one parsec, I suppose, which is the distance at which the diameter of the earth orbits around the sun, which is about 280 million kilometers, is seen under an angle of one second of arc. After that the distance of the stars is too great to measure, but one has observed some relation between luminosity and distance for those stars which are near enough, and so one switches to measuring luminosity, and one pretends that one is measuring "distance", and far away one switches to something else by way of another observed relation that one postulates to be always true, so that when

As I mentioned earlier, what I call the Maxwell–Heaviside equation is what others call the Maxwell equation, because it was HEAVISIDE who wrote the equation that one uses now, a huge simplification of what MAXWELL had derived, because MAXWELL was thinking in purely mechanical terms for transmitting the electric field and the magnetic field, probably because they correspond to transversal waves which were supposed to propagate only through solids, so I gather that it was related to what physicists called ether for a while, which might be the same as what they call the vacuum nowadays. Although some mathematical theories were first developed because of questions in physics or in continuum mechanics, the results proven for ordinary differential equations or for partial differential equations are not linked to what one thought were good equations for describing the physical world, and mathematicians do not really need to know what were all the philosophical problems that physicists had in changing their intuitive description of the world, but the mathematicians who have doubts about the validity of some models, and who start enquiring about how the equations that one proposes for them to solve had been derived, certainly face quite challenging situations for their talents of detective.

The experience of MICHELSON and MORLEY seemed to show that the velocity of light c does not change in a frame which moves at a constant velocity with respect to a first one,[7,8] and it might have been a reason why POINCARÉ (and maybe EINSTEIN) was led to replace NEWTON's point of view of forces acting instantaneously at a distance, and develop the new point of view where particles feel a field and interact with it, the field being a solution of a hyperbolic system having only the velocity of light c as the characteristic speed, a mathematical consequence being that to relate the measurements in the two frames, one must use the Lorentz group of transformations, instead of the Galilean group of transformations. From a mathematical point of view, the result is that instead of ordinary differential equations one must work with partial differential equations (of hyperbolic type), and this should not have been a surprise to anyone who had understood the passage from classical mechanics, which is an 18th century point of view of mechanics based on ordinary differential equations, to continuum mechanics, which is a 19th century point of view of mechanics, based on partial differential equations. However, the attitude of using a classical mechanics point of view and talking about particles is still prevalent in physics, and one of the reasons why physicists still interpret quantum mechanics in terms of probabilities is that they want

---

astronomers say that the redshift is proportional to distance, one has to wonder if they have not postulated it and use the redshift as a measure of their "distance".

[7] Albert Abraham MICHELSON, Polish-born physicist, 1852–1931. He received the Nobel Prize in Physics in 1907, for his optical precision instruments and the spectroscopic and metrological investigations carried out with their aid. He had worked in Worcester, MA, and in Chicago, IL.

[8] Edward Williams MORLEY, American physicist, 1838–1923. He had worked in Cleveland, OH.

to describe the behaviour of nonexistent particles, while they are actually looking at waves. The mathematical way to understand waves deals with partial differential equations of hyperbolic type, and certainly not with ordinary differential equations, even Hamiltonians, but it might be because of some limiting situations, like *geometrical optics* derived from the scalar wave equation, that physicists may have thought that there was nothing wrong about keeping an 18th century point of view, instead of learning the consequences of the 19th century point of view and going forward.

A similar situation exists when one starts from a problem in *linearized elasticity*, and one derives the Saint-Venant approximation for elongated bodies, and the set of formulas obtained form the basic rules of resistance of materials, which engineers use for computing the behaviour of systems of bars and beams in buildings. There are two attitudes if one needs to deal with a structure like the hyperbolic paraboloids used as cooling towers for power plants;[9] the first one is to go back to the theory of linearized elasticity and to derive equations valid for thin shells, and then to discretize the equations obtained in order to perform numerical simulations; the second one is to imagine the structure as an assemblage of a huge number of bars and beams. Obviously, the second solution resembles the first after one has performed a discretization, but one learns in numerical analysis that not all discretizations are good,[10] but although good engineers have often invented interesting numerical schemes, it seems that the proofs that a numerical scheme converges always rely on the first approach and the identification of an adapted variational framework.

The particles that the physicists use are like the bars and beams that the engineers use for computing a thin shell structure; contrary to the appearance, the structure is not full of holes but resists the wind, and these strange bars and beams which oppose the wind are a little similar to the strange particles which manage to be in many places at the same time. Obviously, this type of difficulty disappears if one understands that continuum mechanics recreates classical mechanics in some limiting cases, and there is no doubt that chronologically, continuum mechanics was partly obtained as a limit of classi-

---

[9] The cooling towers are about one hundred metres high and thin enough so that one uses shell theory for studying their elastic behaviour, or better their visco-elastic behaviour (because concrete is a visco-elastic material), in particular for the way they react to the wind. From afar, one does not always see that they do not touch the ground, and they are built on pylons, because their purpose is to create an upward draught of air, and I had thought that this shape had been found very efficient for creating a strong draught, but I was told that the shape has been used since the 19th century for a much simpler reason, because it is very easy to build, and that is because of the two families of straight lines which generate these hyperboloids.

[10] There are questions of consistency to check, or one may approach the solution of a different equation, and there are conditions of stability to check, or the numerical scheme may diverge.

cal mechanics,[11] like in the work of D. BERNOULLI and of CAUCHY, that will help us understand a little more about forces; what a force is will not be found in this way, and the physicists' description of what happens at the atomic level is not so clear. In the early 1980s, I was already wondering about what a force is, and I asked the question to a few people; Robin KNOPS pointed out that some definitions can be circular,[12] because a force is something which is measured with a dynamometer, and a dynamometer is based on the theory of linearized elasticity, and what one has measured is a displacement, so at the end one has not really defined what a force is, but some terms are called forces in the equations of linearized elasticity, or the equations of finite elasticity. I heard later that experiments cannot be independent of a theory for interpreting the result of the experiment, and this shows why physics is necessarily quite different from mathematics. I have understood some questions about "particles" because of some mathematical results for H-measures [18], which I had developed for another purpose, and I hope to derive a better mathematical tool, which will explain more questions about "particles" and about the "forces that bind them".

In some simple linear partial differential equations, the relation between forces acting at a distance and the equivalent effect of a field can be seen easily. If one considers a repartition of fixed electric charges ϱ, and one uses the Maxwell–Heaviside equation for the vacuum and without a magnetic field, so that $curl(E) = 0$, $div(D) = \varrho$ and $D = \varepsilon_0 E$, then using $E = -grad(V)$ for defining the electrostatic potential V chosen to be 0 at infinity, one has the *Laplace/Poisson equation* $-\varepsilon_0 \Delta V = \varrho$. Using the elementary solution $\frac{1}{4\pi r}$ of $-\Delta$ in $\mathbb{R}^3$, one has $V = \frac{1}{4\pi \varepsilon_0 r} \star \varrho$, i.e.

$$V(x) = \int_{\mathbb{R}^3} \frac{1}{4\pi \varepsilon_0 |x - y|} \varrho(y)\, dy, \tag{23.1}$$

showing that a charge q' at y creates a potential $\frac{q'}{4\pi \varepsilon_0 |x-y|}$ at x; in that case the force on a charge q at x is $q E$, and it looks as if the charge q' at y is creating a force of magnitude $\frac{|q q'|}{4\pi \varepsilon_0 |x-y|^2}$ on the particle at x, the force being repulsive if the two charges have the same sign, and attractive if they have opposite sign.

Forces inversely proportional to the square of the distance suggest then the presence of a Laplacian in an equation, and when one knows that an elementary solution of $-\Delta + \alpha^2$ is $\frac{e^{-\alpha r}}{4\pi r}$, which had been introduced by YUKAWA,[13] for

---

[11] The Euler equation for ideal fluids was guessed directly, and not obtained after a limiting process, I believe.

[12] Robin John KNOPS, English mathematician, born in 1932. He worked at Heriot–Watt University, Edinburgh, Scotland.

[13] Hideki YUKAWA, Japanese physicist, 1907–1981. He received the Nobel Prize in Physics in 1949, for his prediction of the existence of mesons on the basis of theoretical work on nuclear forces. He had worked in Kyoto, Japan.

describing the short range of nuclear forces, one understands that this type of truncated potential may appear because of the presence of a term of order zero in an equation with a Laplacian. François MURAT and Doina CIORANESCU have studied the apparition of a zero-order term by homogenization,[14] which they call a strange term coming from nowhere, but George PAPANICOLAOU, who has studied a probabilistic version with Raghu VARADHAN, mentioned that such examples of screening effects are common in physics, for example in plasmas, where one uses $\alpha = \frac{1}{R_D}$, where R_D is called the Debye radius.[15] However, there are other potentials which physicists use, like the Lennard-Jones potential,[16] which is attractive with a force in r^{-6} when particles are far apart and repulsive with a force in r^{-12} when particles are close together, for which I do not know any relation with a system of partial differential equations.

Physicists also use forces which do not depend only upon the position of a particle but also upon its velocity v, and the electrostatic force $q\,E$ already mentioned is actually a truncated form of the Lorentz force $q(E + v \times B)$. Of course, one needs to use the Maxwell–Heaviside equation

$$div(B) = 0; \; B_t + curl(E) = 0$$
$$div(D) = \varrho; \; -D_t + curl(H) = j, \tag{23.2}$$

and as the density of charge ϱ is interpreted as an average of point charges q_i and the current density j is interpreted as an average of $q_i v_i$, there is a density of Lorentz force

$$\varrho\,E + j \times B, \tag{23.3}$$

whose power density is $(j.E)$.[17]

---

[14] Doina POP-CIORANESCU, Romanian-born mathematician. She works at CNRS (Centre National de la Recherche Scientifique) and Université Paris VI (Pierre et Marie Curie), Paris, France.

[15] Petrus (Peter) Josephus Wilhelmus DEBYE, Dutch-born physicist, 1884–1966. He received the Nobel Prize in Chemistry in 1936, for his contributions to our knowledge of molecular structure through his investigations on dipole moments and on the diffraction of X-rays and electrons in gases. He had worked in Zürich, Switzerland, in Utrecht, The Netherlands, in Göttingen, in Leipzig and in Berlin, Germany, and then at Cornell University, Ithaca, NY.

[16] Sir John Edward LENNARD-JONES, British chemist, 1894–1954. He had worked in Bristol and in Cambridge, England.

[17] The Maxwell–Heaviside equation can be expressed in terms of differential forms, one 2-form ω_2 having coefficients E and B whose exterior derivative is 0 (and therefore the exterior derivative of a 1-form ω_1 having coefficients V and A, the scalar and vector potentials), and another 2-form ω_2' having coefficients D and H, whose exterior derivative is a 3-form ω_3 having coefficients ϱ and j. Using the Euclidean structure of $\mathbb{R}^3$ one can associate a 1-form ω_1' to ω_3, and the exterior product of ω_2 and ω_1' (or an interior product of ω_2 and ω_3) is a 3 form ω_3' whose coefficients are $\varrho\,E + j \times B$ and $(j.E)$. What puzzled me in the mid 1970s, after I had learnt about this formulation from Joel ROBBIN, was that the

I shall show that if one denotes by $f(\mathbf{x}, \mathbf{v}, t)$ a density of particles at point $\mathbf{x}$ and time t and having velocity $\mathbf{v}$, then in the case where there are no forces acting on the particles, f satisfies the equation $\frac{\partial f}{\partial t} + \mathbf{v}.\frac{\partial f}{\partial \mathbf{x}} = 0$, so that $f(\mathbf{x}, \mathbf{v}, t) = f(\mathbf{x} - t\,\mathbf{v}, v, 0)$.

Let us consider what happens when there are forces acting on the particles. If one tries to understand what forces are, one is bound to stumble upon other concepts which have not been defined in a clear way, like mass. In trying to understand what forces between particles are, and what the mass of a particle is, one difficulty is that there are no particles in the real world and they are only idealizations; only waves exist and any explanation of origin must start at a very small scale (where physicists talk of quantum effects, which I have proposed to look at in a different way), and then one must explain what important quantities are needed at mesoscopic levels and at our macroscopic level. Unfortunately, there is much left to be understood in this direction, and one is bound to use the intermediate description of continuum mechanics (where questions appearing at microscopic levels and mesoscopic levels are often mentioned), and its simplification of classical mechanics (where one always forgets to mention the restrictive assumptions which are made); rigid particles are in the realm of classical mechanics and I shall start in that way, but the limiting behaviour for letting a number of particles tend to infinity will take us into the realm of continuum mechanics, if not further.

Let $M(t)$ denote the position of a particle, which then has velocity $v = \frac{dM(t)}{dt}$ and acceleration $a = \frac{d^2 M(t)}{dt^2}$; if there are forces acting on the particle, Newton's law is then $force = mass \times acceleration$, i.e. $m\frac{d^2 M(t)}{dt^2} = F(t)$, and mass is just a positive parameter. A force is actually known by its work,[18] or its power,[19] and multiplying the equation by $\frac{dM(t)}{dt}$ one finds that $\frac{d}{dt}\left(\frac{m}{2}\left|\frac{dM(t)}{dt}\right|^2\right) = \left(F(t).\frac{dM(t)}{dt}\right)$, and $\frac{m}{2}\left|\frac{dM(t)}{dt}\right|^2$ is the kinetic energy of the particle. Relativistic effects are not taken into account here, and in that case a particle with rest mass m_0 is said to have a mass depending upon its velocity by the formula $m = \frac{m_0}{\sqrt{1-v^2/c^2}}$, and its energy is given by the formula $e = m\,c^2$, usually attributed to EINSTEIN, but which had been in print before,[20] so that for v small compared to the velocity of light c one has $e - e_0 \approx \frac{m_0}{2}v^2$, giving the classical formula for the kinetic energy, but I think that it does not really make much sense using a classical mechanics framework for discussing "relativistic particles".[21]

---

weak topology is natural for $\omega_1, \omega_2, \omega_2'$ and ω_3, but $\varrho E + j \times B$ and $(j.E)$ are not among the sequentially weakly continuous functionals, and it suggested that the weak topology is not adapted to forces.

[18] $work = force \times displacement$.

[19] $power = force \times velocity$.

[20] It seems that POINCARÉ had used it in 1900, and DE PRETTO in 1903.

[21] FEYNMAN wrote that, because of the Lorentz compression of length in the direction of the movement, he thought of electrons moving at a velocity near the

The force usually depends upon the position of the particle, and some-times upon its velocity, like for the *Coriolis force*,[22] which was actually first introduced by LAGRANGE, or the Lorentz force in electromagnetism,[23] and one often talks of a force field defined everywhere and not only at places where there are particles, and one may think that the force field at a point could be measured if one could add a new particle at that point.

If one considers many particles with the same mass m and the same charge q, feeling the Lorentz force created by an electric field E and a magnetic induction field B (depending upon (x, t)), then each particle position satisfies an equation

$$m \frac{d^2 M}{dt^2} = q\left(E + \frac{dM}{dt} \times B\right), \tag{23.4}$$

and if there are many particles and one takes a limit for an infinite number of particles while keeping the ratio $\frac{q}{m} = \gamma$, the limit density $f(\mathbf{x}, \mathbf{v}, t)$ satisfies the equation

$$\frac{\partial f}{\partial t} + \sum_{i=1}^{3} v_i \frac{\partial f}{\partial x_i} + \gamma \sum_{i=1}^{3} \left(E_i + \sum_{j,k=1}^{3} \varepsilon_{i,j,k} v_j B_k\right) \frac{\partial f}{\partial v_i} = 0. \tag{23.5}$$

Let us deduce the equations for *fluid quantities*, like the density ϱ and the momentum P related to the (macroscopic) velocity u by $P = \varrho u$, defined by

$$\varrho(\mathbf{x}, t) = \int_{\mathbb{R}^3} f(\mathbf{x}, \mathbf{v}, t) \, d\mathbf{v}$$
$$P_i(\mathbf{x}, t) = \int_{\mathbb{R}^3} v_i f(\mathbf{x}, \mathbf{v}, t) \, d\mathbf{v} \text{ for } i = 1, 2, 3. \tag{23.6}$$

Integrating the equation in $\mathbf{v}$ over $\mathbb{R}^3$ gives the conservation of mass (or con-servation of charge)

$$\frac{\partial \varrho}{\partial t} + \sum_{i=1}^{3} \frac{\partial P_i}{\partial x_i} = 0, \tag{23.7}$$

at least if f tends to 0 fast enough as $\mathbf{v}$ tends to ∞, so that integrals in $\mathbf{v}$ of derivatives in $\mathbf{v}$ are 0, because one finds that $\gamma\left(\sum_i E_i \int_{\mathbb{R}^3} \frac{\partial f}{\partial v_i} \, d\mathbf{v} + \sum_{i,j,k} \varepsilon_{i,j,k} B_k \int_{\mathbb{R}^3} v_j \frac{\partial f}{\partial v_i} \, d\mathbf{v}\right) = 0$, because $\varepsilon_{i,j,k} v_j \frac{\partial f}{\partial v_i} = \varepsilon_{i,j,k} \frac{\partial (v_j f)}{\partial v_i}$, as the com-pletely antisymmetric tensor $\varepsilon_{i,j,k}$ is such that $\varepsilon_{i,j,k} = 0$ if two of the indices i, j, k are equal. If one multiplies by v_k before integrating in $\mathbf{v}$, one obtains the equation expressing the *balance of momentum*,

---

velocity of light c as flat pancakes, but using an idea of BOSTICK, I think they may look more like flat doughnuts.

[22] Gaspard Gustave DE CORIOLIS, French mathematician, 1792–1843. He had worked in Paris, France.

[23] Using the formula $(u.\nabla)u = -u \times curl\, u + grad\, \frac{|u|^2}{2}$, Euler equation for an incom-pressible ideal (inviscid) fluid, $\varrho_0(\partial_t + u.\nabla)u + grad\, p = f$ (and $div\, u = 0$) takes the form of the Lorentz force $f = \varrho_0(E + u \times B)$ with $E = \partial_t u + grad\left(\frac{p}{\varrho_0} + \frac{|u|^2}{2}\right)$ and $B = -curl\, u$, which satisfy the corresponding part of Maxwell–Heaviside equation, $div\, B = 0$ and $\partial_t B + curl\, E = 0$.

$$\frac{\partial P_k}{\partial t} + \sum_{i=1}^{3} \frac{\partial R_{i,k}}{\partial x_i} = \gamma(\varrho E + P \times B)_k$$
$$R_{i,k} = \int_{\mathbb{R}^3} v_i v_k f \, dv \text{ for } i, k = 1, 2, 3,$$

(23.8)

and if one defines the symmetric *Cauchy stress tensor* σ by

$$\sigma_{i,k} = - \int_{\mathbb{R}^3} (v - u_i)(v - u_k) f \, d\mathbf{v} \text{ for } i, k = 1, 2, 3, \qquad (23.9)$$

then

$$\sum_{i=1}^{3} \frac{\partial R_{i,k}}{\partial x_i} = \sum_{i=1}^{3} \frac{\partial(\varrho u_i u_k)}{\partial x_i} - \sum_{i=1}^{3} \frac{\partial \sigma_{i,k}}{\partial x_i} \text{ for } k = 1, 2, 3. \qquad (23.10)$$

Similar computations were done in the 1860s by BOLTZMANN, but in the Boltzmann equation a force different from the Lorentz force appears, which is supposedly computed from the interaction of pairs of particles (and that implicitly assumes that one is dealing with a rarefied gas).

[Taught on Wednesday October 31, 2001.]

Notes on names cited in footnotes for Chapter 23, BOSTICK,[24] DE PRETTO,[25] and for the preceding footnotes, STEVENS.[26]

---

[24] Winston Harper BOSTICK, American physicist, 1916–1991. He had worked at Stevens Institute of Technology, Hoboken, NJ.

[25] Olinto DE PRETTO, Italian industrialist, 1857–1921.

[26] Edwin Augustus STEVENS, American engineer and philanthropist, 1795–1868. The Stevens Institute of Technology, Hoboken, NJ, is named after him.

24

Balance Laws; What Are Forces?

When one considers a finite number of particles, with particle i having mass m^i, and position $M^i(t)$, and feeling a force $F^i(t)$, conservation of mass is just the fact that the masses m^i are independent of t.

The condition $\frac{dm^i}{dt} = 0$ for all i is equivalent to the equation

$$\frac{\partial \varrho}{\partial t} + \sum_{k=1}^{3} \frac{\partial P_k}{\partial x_k} = 0 \text{ in the sense of distributions,} \tag{24.1}$$

where one writes formally

$$\varrho = \sum_i m^i \delta_{(M^i(t),t)}$$
$$P = \sum_i m^i \frac{dM^i}{dt} \delta_{(M^i(t),t)}, \tag{24.2}$$

but ϱ and P should be considered as Radon measures (or distributions) in (x,t), acting as

$$\langle \varrho, \varphi \rangle = \sum_i \int_{\mathbb{R}} m^i \varphi(M^i(t), t)\, dt$$
$$\langle P, \varphi \rangle = \sum_i \int_{\mathbb{R}} m^i \frac{dM^i(t)}{dt} \varphi(M^i(t), t)\, dt, \tag{24.3}$$

for all φ which are continuous with compact support (or C^∞ with compact support).

In order to check (24.1), one takes φ to be C^1 with compact support, and one has

$$\left\langle \frac{\partial \varrho}{\partial t} + \sum_{k=1}^{3} \frac{\partial P_k}{\partial x_k}, \varphi \right\rangle = -\left\langle \varrho, \frac{\partial \varphi}{\partial t} \right\rangle - \sum_{k=1}^{3} \left\langle P_k, \frac{\partial \varphi}{\partial x_k} \right\rangle$$
$$= -\sum_i \int_{\mathbb{R}} m^i \frac{d[\varphi(M^i(t),t)]}{dt}\, dt = \sum_i \int_{\mathbb{R}} \frac{dm^i}{dt} \varphi(M^i(t), t)\, dt, \tag{24.4}$$

so that, assuming the position of the particles to be distinct, one sees that (24.1) is equivalent to $\frac{dm^i}{dt} = 0$ for all i.

When one lets the number of particles tend to infinity, one rescales the masses of the particles in order to have the corresponding ϱ (density of mass) and the corresponding P (density of linear momentum) converge to Radon measures (or distributions), and equation (24.1) stays valid at the limit, as it is written in the sense of distributions.

For a subsequence of ϱ to converge to a Radon measure it is sufficient that for every compact K and every $T > 0$ there exists a constant $C(K;T)$ such that $\sum_{\{i|M^i(t)\in K\}} m^i \leq C(K;T)$ for $0 < t < T$ (assuming that the initial time of interest is 0). For a subsequence of P to converge to a Radon measure it is sufficient that for every compact K and every $T > 0$ there exists a constant $C_1(K;T)$ such that $\sum_{\{i|M^i(t)\in K\}} m^i \left|\frac{dM^i}{dt}\right| \leq C_1(K;T)$ for $0 < t < T$.

Because the masses m^i are positive, the condition for a subsequence of ϱ to converge to a distribution is the same, from the remark of Laurent SCHWARTZ that nonnegative distributions coincide with nonnegative Radon measures; however, it is different for the density of linear momentum, and it may happen that a subsequence of P converges to a distribution without converging to a Radon measure, or even that it converges to a Radon measure in the sense of distributions but not in the sense of Radon measures; the same considerations will arise when dealing with forces.

To study the balance of linear momentum, one uses the equations of motion

$$m^i \frac{d^2 M^i}{dt^2} = F^i \text{ for all,} \qquad (24.5)$$

but it is not yet important to know what the forces F^i are, i.e. I shall not use the information

$F^i = \sum_{j \neq i} F^{i,j}$, with $F^{i,j}$ the force exerted on particle i by particle j
$F^{i,j}$ is in the direction of the particle j, with $F^{i,j} + F^{j,i} = 0$ for all $i \neq j$
$F^{i,j}$ depending only upon the distance between particle i and particle j.
$$(24.6)$$

Besides ϱ and P already used, the equation of balance of momentum uses the tensor R and the resultant force F defined by

$$R = \sum_i m^i \left(\frac{dM^i}{dt} \otimes \frac{dM^i}{dt}\right) \delta_{(M^i(t),t)} \qquad (24.7)$$
$$F = \sum_i F^i \delta_{(M^i(t),t)},$$

which should be considered as Radon measures (or distributions) in (x,t), and the equation of balance of linear momentum takes the form

$$\frac{\partial P_\ell}{\partial t} + \sum_{k=1}^{3} \frac{\partial R_{k,\ell}}{\partial x_k} = F_\ell \text{ for } \ell = 1, 2, 3, \qquad (24.8)$$

because

$$\left\langle \frac{\partial P_\ell}{\partial t} + \sum_{k=1}^{3} \frac{\partial R_{k,\ell}}{\partial x_k}, \varphi \right\rangle = -\sum_i \int_{\mathbb{R}} m^i \frac{dM_\ell^i}{dt} \frac{\partial \varphi}{\partial t}(M^i(t),t)\, dt$$
$$-\sum_{i,k} \int_{\mathbb{R}} m^i \frac{dM_k^i}{dt} \frac{dM_\ell^i}{dt} \frac{\partial \varphi}{\partial x_k}(M^i(t),t)\, dt \qquad (24.9)$$
$$= \sum_i \int_{\mathbb{R}} m^i \frac{d^2 M_\ell^i}{dt^2} \varphi(M^i(t),t)\, dt = \langle F_\ell, \varphi \rangle,$$

as a consequence of

$$\frac{d}{dt}\left(\frac{dM_\ell^i}{dt}\varphi(M^i(t),t)\right) = \frac{dM_\ell^i}{dt}\left(\frac{\partial\varphi}{\partial t}(M^i(t),t) + \sum_{k=1}^{3}\frac{dM_k^i}{dt}\frac{\partial\varphi}{\partial x_k}(M^i(t),t)\right)$$
$$+\frac{d^2M_\ell^i}{dt^2}\varphi(M^i(t),t).$$

(24.10)

A limit in the sense of Radon measures requires that $\sum_{\{i|M^i(t)\in K\}}|F^i(t)| \le C_2(K,T)$ for $0 < t < T$, and this is questionable, and it will be discussed later.

The macroscopic velocity u of transport of mass is defined by writing $P_k = \varrho\, u_k$ for $k = 1,2,3$, and then one writes

$$R_{k,\ell} = \varrho\, u_k u_\ell - \sigma_{k,\ell}, \text{ for } k,\ell = 1,2,3,$$

(24.11)

and σ is the Cauchy stress tensor, which is symmetric.

The basic idea in kinetic theory is to introduce a density of particles $f(\mathbf{x},\mathbf{v},t)$ which sees the position and the velocity of the particles, and then one writes

$$\varrho(\mathbf{x},t) = \int_{\mathbb{R}^3} f(\mathbf{x},\mathbf{v},t)\,d\mathbf{v}$$
$$P(\mathbf{x},t) = \int_{\mathbb{R}^3} \mathbf{v}\, f(\mathbf{x},\mathbf{v},t)\,d\mathbf{v}$$
$$R(\mathbf{x},t) = \int_{\mathbb{R}^3} (\mathbf{v}\otimes\mathbf{v})f(\mathbf{x},\mathbf{v},t)\,d\mathbf{v}.$$

(24.12)

From these formulas, one deduces that the Cauchy stress tensor is given by

$$\sigma = -\int_{\mathbb{R}^3}\left((\mathbf{v}-\mathbf{u})\otimes(\mathbf{v}-\mathbf{u})\right)f(\mathbf{x},\mathbf{v},t)\,d\mathbf{v},$$

(24.13)

and in the case of a gas at equilibrium, the density is a Gaussian and the Cauchy stress tensor reduces to a hydrostatic pressure, i.e. $\sigma_{i,j} = -p\,\delta_{i,j}$.

In this new point of view, the particle i will be a Dirac mass of weight m^i at the point $\left(M^i(t), \frac{dM^i(t)}{dt}, t\right)$ in the $(\mathbf{x},\mathbf{v},t)$ space, and in order to derive an equation for f one needs to understand more about the forces.

I have mentioned that forces may correspond to objects which are not necessarily Radon measures, but are distributions in the sense of Laurent SCHWARTZ. One classical notion in physics is that of a dipole, and it is the limit of a sequence $k(\delta_a - \delta_b)$ when the points a and b get very near and the coefficient k tends to ∞ in such a way that $|k|\,|a - b|$ converges to a (nonzero) constant; Laurent SCHWARTZ had noticed that these objects are just derivatives (in the sense of distributions, of course) of Dirac masses.[1] Recalling that $\langle\delta_a,\varphi\rangle = \varphi(a)$, one has $\langle\frac{\partial\delta_a}{\partial x_j},\varphi\rangle = -\langle\delta_a,\frac{\partial\varphi}{\partial x_j}\rangle = -\frac{\partial\varphi}{\partial x_j}(a)$. For example, in one dimension, the sequence $\mu_n = n(\delta_{1/n} - \delta_0)$ is not bounded in the space of Radon measures $\mathcal{M}(\mathbb{R})$, and if one uses the Banach space of Radon measures with finite total mass (dual of $C_0(\mathbb{R})$, the space of continous functions tending to 0 at ∞, with the sup norm), the norm of μ_n is $2n$.

---

[1] DIRAC might have used this intuition for using derivatives of his "function".

However, μ_n is bounded in the space of distributions of order ≤ 1, because for φ a function of class C^1 with compact support, one has $|\langle \mu_n, \varphi \rangle| \leq \max |\varphi'|$, and actually μ_n converges to $-(\delta_0)'$, because $\langle \mu_n, \varphi \rangle = n\left(\varphi\left(\frac{1}{n}\right) - \varphi(0)\right) = \varphi'(0) + o(1) \rightarrow \varphi'(0) = \langle -(\delta_0)', \varphi \rangle$; this is consistent with the fact that if $w_n(x) = n$ for $0 < x < \frac{1}{n}$ and 0 elsewhere, then w_n converges to δ_0 in the sense of Radon measures, and in the sense of distributions one has $(w_n)' = -\mu_n$.

One is in a similar situation, where limits may not exist in the sense of Radon measures but may exist in the sense of distributions, when particles get very near, and forces between them become large, for example when one deals with forces depending upon the distance as negative powers of the distance.

It is important then to have some idea about what are reasonable hypotheses concerning forces, and for this I shall show a model, which has been used by D. BERNOULLI for approximating the movement of a string by that of small masses linked by springs; of course, this is a model for a solid, and one should be careful about the way one uses it for questions about liquids, or about gases. BERNOULLI was interested in the frequencies of vibrations, as for a violin string, and he did not derive the string equation (i.e. the wave equation in one dimension), and D'ALEMBERT is credited for writing the one-dimensional wave equation,[2] but I do not think that he derived it by going further than BERNOULLI's analysis, and he may just have written an equation that would have solutions of the form $u(x,t) = f(x - ct)$ for an arbitrary (smooth) function f, as well as solutions of the form $v(x,t) = g(x + ct)$ for an arbitrary (smooth) function g, and checking that the equation that he had written, $u_{tt} - c^2 u_{xx} = 0$, had the general solution $f(x-ct) + g(x+ct)$. POISSON may have been the first to work on the three-dimensional wave equation.[3]

[Taught on Friday November 2, 2001.]

---

[2] Jean LE ROND, known as D'ALEMBERT, French mathematician, 1717–1783. He had worked in Paris, France.

[3] His motivation may have been the study of pressure waves in gases.

25

D. Bernoulli: from Masslets and Springs to the 1-D Wave Equation

In a gas, the forces between particles can become quite large when two particles are near, but these forces are of the same magnitude but opposite in direction and there is then some cancellation effect; in a limiting process, when the number of particles gets large and one rescales their masses, one will find that some sequence may converge in the sense of distributions but not in the sense of Radon measures.

In order to study this phenomenon, I shall use a discrete model of a vibrating string, by considering small masses connected with springs, an idea going back to D. BERNOULLI, but in order to do the analysis completely I shall consider longitudinal waves,[1] while the vibration of a violin string is a transversal wave,[2] for which the analysis uses a linearization, and at the limit one obtains the string equation, i.e. the one-dimensional wave equation. I shall show later the corresponding analysis for two- or three-dimensional bodies, which uses a linearization too; it was first used by CAUCHY, and it generates the equation for linearized elasticity.

I consider small masses $m_1, \ldots, m_{N-1}$ moving on a line between 0 and L, and occupying positions $0 < x_1(t) < \ldots < x_{N-1}(t) < L$, and I use $x_0(t) = 0$ and $x_N(t) = L$. For $j = 1, \ldots, N$, there is a spring with constant $\kappa_j > 0$ and equilibrium length $\ell_j > 0$ between the masses at x_{j-1} and at x_j (but x_0 and x_N are actually walls, which do not move). The increase in length of spring j is $x_j - x_{j-1} - \ell_j$, so that the force exerted at x_j is $-\kappa_j(x_j - x_{j-1} - \ell_j)$ and the force exerted at x_{j-1} is $\kappa_j(x_j - x_{j-1} - \ell_j)$; one deduces that the equation for the movement of the mass j is

---

[1] One generates a longitudinal wave in a metallic bar by hitting it with a hammer, in the direction of the length, so that the motion of the points and the direction of propagation of the wave are along the length of the bar.

[2] Because the movement of a point is perpendicular to the string, along which the wave propagates.

$$m_j \frac{d^2 x_j}{dt^2} + \kappa_j(x_j - x_{j-1} - \ell_j) - \kappa_{j+1}(x_{j+1} - x_j - \ell_{j+1}) = 0, \text{ for } j = 1, \ldots, N-1,$$
(25.1)

and one must pay attention to the fact that, once these equations are written, it is not clear if the evolution will enforce $x_{j-1}(t) < x_j(t)$ for $j = 1, \ldots, N$ and all $t > 0$.[3] If one denotes by y_j the equilibrium position of mass j, then one must have

$$\kappa_j(y_j - y_{j-1} - \ell_j) - \kappa_{j+1}(y_{j+1} - y_j - \ell_{j+1}) = 0,$$
$$\text{for } j = 1, \ldots, N-1, \text{ with } y_0 = 0, y_N = L,$$
(25.2)

and the existence of a solution of (25.2) is equivalent to its uniqueness, as it is a linear system with $N - 1$ equations for $N - 1$ unknowns, but one must check if the solution satisfies $y_{j-1} < y_j$ for $j = 1, \ldots, N$. To prove uniqueness, one considers the homogeneous version of (25.2),

$$\kappa_j(z_j - z_{j-1}) - \kappa_{j+1}(z_{j+1} - z_j) = 0, \text{ for } j = 1, \ldots, N-1, \text{ with } z_0 = z_N = 0,$$
(25.3)

then multiplying by z_j and summing in j gives $\sum_{j=1}^{N} \kappa_j(z_j - z_{j-1})^2 = 0$, and therefore all the z_j are equal and must be 0 as $z_0 = z_N = 0$.

Existence being proven, let F_L be defined by

$$F_L = \kappa_j(z_j - z_{j-1}) = \kappa_j(y_j - y_{j-1} - \ell_j) \text{ for all } j = 1, \ldots, N, \quad (25.4)$$

which uses (25.3), so that F_L is the force that one should apply at the point L in order to maintain equilibrium with the last point at position L; one finds easily from (25.4) that

$$\left(\frac{1}{\kappa_1} + \ldots + \frac{1}{\kappa_N} \right) F_L = L - (\ell_1 + \ldots + \ell_N), \quad (25.5)$$

so that

$$\text{if } L \geq \ell_1 + \ldots + \ell_N, \text{ then } F_L \geq 0 \text{ and } y_{j-1} < y_j \text{ for } j = 1, \ldots, N, \quad (25.6)$$

and this is the case when the springs under no tension have lengths that do not add up to L and one must stretch them. In the case when the springs under no tension have total length larger than L, and one must compress the springs (remembering that the model has ruled out buckling), one wants to have $y_{j-1} < y_j$ for $j = 1, \ldots, N$, which is $\ell_j + \frac{F_L}{\kappa_j} > 0$ for $j = 1, \ldots, N$, or

$$F_L > - \min_{j=1,\ldots,N} \kappa_j \ell_j, \quad (25.7)$$

and using (25.5) one deduces that

---

[3] The reason is that the law of force applied to a spring to compress it has been linearized, and as a consequence only a finite force is necessary to squeeze it to zero length. Of course, buckling is not taken into account in the model.

$$\ell_1 + \ldots + \ell_N < L + \left(\frac{1}{\kappa_1} + \ldots + \frac{1}{\kappa_N}\right) \min_{j=1,\ldots,N} \kappa_j \ell_j \tag{25.8}$$
$$\text{implies } y_{j-1} < y_j \text{ for } j = 1,\ldots,N,$$

and the condition (25.8) is automatically true if all the ℓ_j are equal and all the κ_j are equal.

Writing

$$x_j(t) = y_j + z_j(t) \text{ for } j = 0,\ldots,N, \tag{25.9}$$

one obtains the equations

$$m_j \frac{d^2 z_j}{dt^2} + k_j(z_j - z_{j-1}) - k_{j+1}(z_{j+1} - z_j) = 0, \text{ for } j = 1,\ldots,N-1. \tag{25.10}$$

Multiplying by $\frac{dz_j}{dt}$ and summing in j gives

$$\sum_{j=1}^{N-1} \frac{m_j}{2}\left(\frac{dz_j}{dt}\right)^2 + \sum_{j=1}^{N} \frac{k_j}{2}(z_j - z_{j-1})^2 = constant, \tag{25.11}$$

but this constant is not always the *total energy*, sum of the *kinetic energy*, which is here

$$K(t) = \sum_{j=1}^{N-1} \frac{m_j}{2}\left(\frac{dx_j}{dt}\right)^2 = \sum_{j=1}^{N-1} \frac{m_j}{2}\left(\frac{dz_j}{dt}\right)^2, \tag{25.12}$$

and of the *potential energy* (i.e. the elastic energy stored inside the springs), which is here

$$\mathcal{P}(t) = \sum_{j=1}^{N} \frac{\kappa_j}{2}(x_j - x_{j-1} - \ell_j)^2 = \sum_{j=1}^{N} \frac{\kappa_j}{2}(z_j - z_{j-1} + y_j - y_{j-1} - \ell_j)^2$$
$$= \sum_{j=1}^{N} \frac{\kappa_j}{2}\left(z_j - z_{j-1} + \frac{F_L}{\kappa_j}\right)^2$$
$$= \sum_{j=1}^{N} \frac{\kappa_j}{2}(z_j - z_{j-1})^2 + \frac{F_L}{2}(L - (\ell_1 + \ldots + \ell_N)), \tag{25.13}$$

because $\sum_{j=1}^{N}(z_j - z_{j-1}) = z_N - z_0 = 0$, and (25.5) gives $F_L\left(\frac{1}{\kappa_1} + \ldots + \frac{1}{\kappa_N}\right) = L - (\ell_1 + \ldots + \ell_N)$, so the constant is the total energy only in the case when $L = \ell_1 + \ldots + \ell_N$, corresponding to $F_L = 0$.

One may naively think that if one starts with the springs under no tension, occupying the length $\ell_1 + \ldots + \ell_N$ and one applies the force F_L until the end point is at position L, then the work of the force is $F_L(L - (\ell_1 + \ldots + \ell_N))$ and this is not the value $\frac{F_L}{2}(L - (\ell_1 + \ldots + \ell_N))$ which appears in the preceding formula, but if one behaves in such a naive way, one will not end up with mass j at position y_j and with velocity 0, of course. If at time 0 one starts with $x_1(0) = \ell_1, x_2(0) = \ell_1 + \ell_2, \ldots, x_N(0) = \ell_1 + \ldots + \ell_N$, with $\frac{dx_j}{dt}(0) = 0$ for $j = 1,\ldots,N$, and one applies a force $F(t)$, then the system is changed and one no longer has $x_N(t) = L$ but $m_N \frac{d^2 x_N}{dt^2} + \kappa_N(x_N - x_{N-1} - \ell_N) = F(t)$, while before there was no mass m_N involved; between time 0 and T the work of the force is going to be $\int_0^T F(t)\frac{dx_N}{dt}\,dt$; multiplying equation j by $\frac{dx_j}{dt}$ and summing in j one obtains that $\frac{d}{dt}\left[\mathcal{K}(t) + \frac{m_N}{2}\left(\frac{dx_N}{dt}\right)^2 + \mathcal{P}(t)\right] = F(t)\frac{dx_N}{dt}$, and therefore the work done between time 0 and T is $\mathcal{K}(T) + \frac{m_N}{2}\left(\frac{dx_N}{dt}(T)\right)^2 + \mathcal{P}(T)$; if at time T one has succeeded in having $x_N(T) = L$ and $\frac{dx_N}{dt}(T) = 0$, then the work done by the force is exactly the total energy of the system that one had considered from the start. The possibility of finding a force $F(t)$ such that at time T the $x_j(T)$ and $\frac{dx_j}{dt}(T)$ take given values for $j = 1, \ldots, N$ is a question of controllability; there is an algebraic characterization for that and it can be checked that the system is indeed controllable if one has $\kappa_j > 0$ for all j.[4]

For simplicity, let us assume now that all the masses are equal to m and all the strengths of the springs are equal to κ, and one looks at periodic solutions of the form $x(t) = y + e^{i\omega t}a$, where y is the equilibrium solution, and one finds that one must have $M_0 a = \omega^2 a$, where M_0 is the symmetric matrix defined by

$$
\begin{aligned}
&(M_0)_{i,j} = 0 \text{ for } j \neq i \text{ or } j \neq i \pm 1 \\
&(M_0)_{i,i} = \tfrac{2\kappa}{m} \text{ for } i = 1, \ldots, N-1 \\
&(M_0)_{i,i-1} = (M_0)_{i,i+1} = \tfrac{-\kappa}{m} \text{ for } i = 1, \ldots, N-1,
\end{aligned}
\tag{25.14}
$$

and one either discards the condition for $(M_0)_{1,0}$ and $(M_0)_{N-1,N}$ in this list, or one considers that one must have $a_0 = a_N = 0$. Recalling trigonometric formulas, one can find explicitly the eigenvectors and eigenvalues of M_0 by choosing $p = 1, \ldots, N-1$, and then a defined by

$$
a_i = \sin\left(\frac{i\,p\,\pi}{N}\right) \text{ for } i = 1, \ldots, N-1,
\tag{25.15}
$$

which gives the eigenvalue

$$
\omega_p^2 = 2\frac{\kappa}{m}\left[1 - \cos\left(\frac{p\,\pi}{N}\right)\right] = \frac{4\kappa}{m}\sin^2\left(\frac{p\,\pi}{2N}\right),
\tag{25.16}
$$

so the corresponding frequencies of vibration of the system are

$$
\nu_p = \frac{\omega_p}{2\pi} = \frac{\sqrt{\kappa}}{\pi\sqrt{m}}\sin\left(\frac{p\,\pi}{2N}\right),
\tag{25.17}
$$

---

[4] In the case of a system $\frac{dX}{dt} = A\,X + B\,u$, with X of size M, the necessary and sufficient condition for controllability is that the rank of the matrix with block columns $Y = (B \quad A\,B \quad \ldots \quad A^{M-1}B)$ is M. Here $M = 2N$ with $X = \begin{pmatrix} x \\ \frac{dx}{dt} \end{pmatrix}$ and $A = \begin{pmatrix} 0 & I \\ -M_0 & 0 \end{pmatrix}$, where M_0 is a tridiagonal matrix, and $B = e_{2N}$; $A\,B$ puts e_N in the span of the columns of Y, and then $A^2 B$ puts $M_0 e_N$, which is a combination of e_{2N-1} and e_{2N} if $\kappa_N > 0$, so e_{2N-1} is in the span and $A^3 B$ puts e_{N-1} and $A^4 B$ puts $M_0 e_{N-1}$, which adds e_{2N-2} in the span if $\kappa_{N-1} > 0$, and so on.

and therefore if N is large the lowest frequency is $\nu_1 \approx \frac{\sqrt{k}}{2N\sqrt{m}}$.

Then, the problem is to let N tend to infinity, while rescaling correctly m and κ. To rescale mass, it is natural to take $m = \frac{m_*}{N}$, where m_* is the mass of the vibrating string that one is trying to model, which one thinks as divided into N equal parts. The rescaling for κ is less intuitive; as *force* is *mass* $\times$ *acceleration*, the unit for κ is *mass* $\times$ *time*$^{-2}$ and if mass is $\frac{m_*}{N}$ one may choose *time* $= \frac{time_*}{N}$, which corresponds to $\kappa = N\kappa_*$, and this corresponds to keeping the lowest frequency almost fixed, and it might be the way BERNOULLI thought. One may also choose to have a constant (nonzero) speed of propagation so *length* and *time* should be rescaled in the same way, but BERNOULLI could not have thought in terms of wave speed, as he had not derived the one-dimensional wave equation. One may also ask that forces stay bounded and do not tend to 0, because from the experimental evidence of tuning a violin (which was probably the intuition of BERNOULLI) some high tension must be put on the strings but certainly not an infinite one; a piece of the string of length $\frac{L}{N}$ might show an increase in length of the same order $O(\frac{1}{N})$ but for a force $O(1)$ and therefore κ must be $O(N)$. Another way of looking at the problem would be to have a bounded kinetic energy and a bounded potential energy; for the kinetic energy one has N terms which are $O(\frac{1}{N})$ as m is $O(\frac{1}{N})$ and $\frac{dx_j}{dt}$ is of order $O(1)$, and for the potential energy one expects $x_j - x_{j-1} - \ell_j$ to be $O(\frac{1}{N})$ and in order to have $\kappa(x_j - x_{j-1} - \ell_j)^2$ also of order $O(\frac{1}{N})$, one needs to have κ of order $O(N)$; I do not know if thinking in terms of potential energy was a natural approach for BERNOULLI.

All the preceding considerations consist in arguing about physical intuition, but one should be aware that one's physical intuition might be wrong, and in facing a physical problem that one is not sure about, one should turn to the mathematical side and prove various theorems, under different hypotheses. Suppose then that one scales $m = \frac{m_*}{N}$, as this part is clear, and that one uses a sequence $\kappa(N)$ which one is ready to let behave in various ways as $N \to \infty$; suppose that one starts from an initial datum where $x_j(0) = y_j = \frac{jL}{N}$, with $\frac{dx_j}{dt} = O(1)$ for $j = 1, \ldots, N-1$, so that $\mathcal{K}(0) = O(1)$ and $\mathcal{P}(0) = 0$; from the solution of the differential system one may construct a function u_N which at any time t is continuous and piecewise affine with $u_N(\frac{jL}{N}, t) = x_j(t) - y_j$, and one wonders what happens to u_N as $N \to \infty$.

If $\frac{\kappa(N)}{N} \to 0$, one finds that every weak limit u_∞ of a subsequence extracted from the sequence u_N satisfies $\frac{\partial^2 u}{\partial t^2} = 0$; this is the case where there is only kinetic energy and no elastic behaviour giving rise to a nonzero potential energy. If $\frac{\kappa(N)}{N} \to \infty$, one finds that every weak limit u_∞ satisfies $\frac{\partial^2 u}{\partial x^2} = 0$; this is the case where the string behaves as a rigid body. If $\frac{\kappa(N)}{N} \to \kappa \neq 0$, one finds that every weak limit u_∞ satisfies $\frac{\partial^2 u}{\partial t^2} - c^2 \frac{\partial^2 u}{\partial x^2} = 0$, with $c^2 = \frac{\kappa L^2}{m_*}$; this is the case corresponding to a vibrating string (without dissipation, while the real one slowly loses energy). These results can be proven with standard variational techniques, commonly used in the abstract part of numerical analysis, where the problem considered is usually the opposite one, starting from the wave equation and wanting to approach its solution, but the ideas are the same and use the bound on the total energy in a crucial way; notice that the choice $x_j(0) = y_j$ for all j was chosen as a simplification, but if one wants to start with a nonzero potential energy, then it is better to take it bounded if one wants a subsequence of u_N to converge in a reasonable functional space.

One important result in the preceding computation is that forces were $O(1)$, and this suggests that one must consider the convergence in the sense of distributions for the sequence of Radon measures denoted before $\sum_i F_i \delta_{M_i(t)}$, as it may not stay bounded in the space of Radon measures. However, one should pay attention that the preceding model is a model for a one-dimensional solid, and one should think about the differences between fluids and gases. If one considers water (H_2O), then a mole weighs around 18 grams and occupies around 18 cm^3 if it is liquid,[5] and slightly more if it is solid,[6] while the corresponding amount of water vapour occupies 22.4 dm^3, so there is a factor 1,244 for increase in volume, corresponding to a factor around 10.8 for increase in distance between molecules; the number of molecules is huge, about the Avogadro number, 6.022×10^{23} in 22.4 dm^3, but I am not able to determine the size of the forces between molecules, for which one certainly needs more experimental information. For example, one transforms one gram of ice into water at 0°C with 80 calories, then one needs about 100 calories to heat it up to the boiling temperature of 100°C under the usual atmospheric pressure of 1 bar ($= 10^5$ pascals), and then 537 calories to transform it into vapour, and one calorie is 4.18 joules. One difficulty is that a part of this energy that one has supplied has been used for breaking bonds and a part used for giving internal energy to the gas, whose origin is supposed to be the kinetic energy of molecules. In understanding bonds, physicists play with Lennard-Jones potentials, which have the form $\frac{A}{r^6}\left(\frac{\ell^6}{r^6} - 2\right)$, where ℓ is a characteristic length, where the potential attains it minimum, and $\frac{A}{\ell^6}$ is the energy in the bond; however, it seems to me to be one possible model among millions, so that for real forces between molecules I cannot really explain anything for sure.

Anyway, one important thing to observe is that *internal forces* require going beyond Radon measures and this is how the Cauchy stress tensor σ appears, corresponding to a force g given by $g_i = \sum_j \frac{\partial \sigma_{i,j}}{\partial x_j}$, and distributions of order 1 have therefore appeared in a natural way.

---

[5] I suppose that it was one basic idea of the French scientists who developed the metric system at the end of the 18th century, that one just had to carry a graded stick to know the unit of length, and then the unit of mass was derived, with a kilogram being the mass of a litre of water, occupying a cube of side 10 centimetres.

[6] The case of water is special, as for almost all other liquids there is a loss of volume during solidification, and I have only heard of bismuth as an other exception.

In the point of view of looking at particles in the $(\mathbf{x}, \mathbf{v}, t)$ space, one considers the Radon measure denoted formally by

$$f = \sum_i m^i \delta_{(M^i(t), V^i(t))} \text{ with } V^i = \frac{dM^i}{dt}, \tag{25.18}$$

but it is actually a Radon measure in $(\mathbf{x}, \mathbf{v}, t)$ acting on a function φ by

$$\langle f, \varphi \rangle = \int_0^T \sum_i m^i \varphi\left(M^i(t), \frac{dM^i(t)}{dt}, t\right) dt. \tag{25.19}$$

In this framework one is led to denote

$$\begin{aligned}
&\Phi(M, V, t) \text{ the force acting on a particle} \\
&\text{of unit mass at } M \text{ having velocity } V \text{ at time } t,
\end{aligned} \tag{25.20}$$

and under suitable hypotheses one can derive an equation expressing both conservation of mass and the balance of linear momentum, and this equation is

$$\frac{\partial f}{\partial t} + \sum_{j=1}^3 v_j \frac{\partial f}{\partial x_j} + \sum_{j=1}^3 \Phi_j \frac{\partial f}{\partial v_j} = 0, \text{ under the hypothesis } div_v \Phi = 0. \tag{25.21}$$

Indeed, for a finite sum and for a function φ of class C^1 with compact support, one has

$$\left\langle \frac{\partial f}{\partial t}, \varphi \right\rangle = -\left\langle f, \frac{\partial \varphi}{\partial t} \right\rangle = -\sum_i \int_0^T m^i \frac{\partial \varphi}{\partial t}\left(M^i(t), \frac{dM^i(t)}{dt}, t\right) dt, \tag{25.22}$$

and if one notices that

$$\begin{aligned}
\frac{d}{dt}\left[\varphi\left(M^i(t), \frac{dM^i(t)}{dt}, t\right)\right] &= \frac{\partial \varphi}{\partial x}\left(M^i(t), \frac{dM^i(t)}{dt}, t\right) \cdot \frac{dM^i}{dt} \\
&+ \frac{\partial \varphi}{\partial v}\left(M^i(t), \frac{dM^i(t)}{dt}, t\right) \cdot \frac{d^2 M^i}{dt^2} + \frac{\partial \varphi}{\partial t}\left(M^i(t), \frac{dM^i(t)}{dt}, t\right),
\end{aligned} \tag{25.23}$$

one deduces that

$$\left\langle \frac{\partial f}{\partial t}, \varphi \right\rangle = \sum_{k=1}^3 \left\langle v_k f, \frac{\partial \varphi}{\partial x_k} \right\rangle + \sum_{k=1}^3 \left\langle \Phi_k f, \frac{\partial \varphi}{\partial v_k} \right\rangle, \tag{25.24}$$

from which one deduces that

$$\frac{\partial f}{\partial t} + \sum_{j=1}^3 v_j \frac{\partial f}{\partial x_j} + \sum_{j=1}^3 \Phi_j \frac{\partial f}{\partial v_j} + f \, div_v \Phi = 0. \tag{25.25}$$

The Lorentz force in electromagnetism, or the Coriolis force in moving frames, satisfy $div_v \Phi = 0$.

Another aspect that should be kept in mind is that one should actually be interested not only in the density of particles in some regions of $\mathbf{x}$ space or $(\mathbf{x}, \mathbf{v})$ space, but also in correlations of distances between particles; for example if particles are all spheres of radius a, one looks at the probability of finding particles whose centres are at a distance d, for $d \geq 2a$. Using a *molecular dynamics* approach, i.e. computing the evolution of a large number of particles with interactions following a given force law, one can compute a few averages and correlations of positions, and compare to experimental measurements (done for example by using neutron scattering), and one can fit the best values of the parameters of the force law used (two for Lennard-Jones potentials). This approach has the advantages and defects of all numerical methods in the absence of a well developed theory, that it can only provide conjectures. There is a *Percus–Yevick equation* for correlations,[7,8] but I have not studied the subject enough to judge its validity.

[Taught on Monday November 5, 2001.]

---

[7] Jerome Kenneth PERCUS, American mathematician. He works at NYU (New York University), New York NY.
[8] George J. YEVICK, American physicist. He works at Stevens Institute of Technology, Hoboken NJ.

26

Cauchy: from Masslets and Springs to 2-D Linearized Elasticity

We have seen that in the case of equal small masses m and springs of equal strength κ, the equations are

$$m \frac{d^2 x_i}{dt^2} + \kappa(2x_i - x_{i-1} - x_{i+1}) = 0$$
$$\text{for } i = 1, \ldots, N-1, \text{ with } x_0(t) = 0, \ x_N(t) = L. \tag{26.1}$$

By rescaling,

$$m = \frac{m_*}{N}, \ \kappa = N\kappa_*, \text{ give } m_* \frac{d^2 x_i}{dt^2} + N^2\kappa_*(2x_i - x_{i-1} - x_{i+1}) = 0$$
$$\text{for } i = 1, \ldots, N-1. \tag{26.2}$$

Letting N tend to $+\infty$, gives an approximation of the wave equation

$$\frac{\partial^2 u}{\partial t^2} - c^2 \frac{\partial^2 u}{\partial x^2} = 0, \text{ with } c^2 = \frac{\kappa_* L^2}{m_*}. \tag{26.3}$$

Choosing $\ell = \frac{L}{N}$ as mesh size (often denoted Δx or h), and defining $x_j(t) = u(j\,\ell, t)$, one checks the consistency of the scheme by considering a smooth solution of (25.3) and using the Taylor expansion of u at the point $(j\,\ell, t)$, one obtains $x_{j+1}(t) = u((j+1)\ell, t) \approx u + \ell \frac{\partial u}{\partial x} + \frac{\ell^2}{2} \frac{\partial^2 u}{\partial x^2} + o(\ell^2)$ and $x_{j-1}(t) = u((j-1)\ell, t) \approx u - \ell \frac{\partial u}{\partial x} + \frac{\ell^2}{2} \frac{\partial^2 u}{\partial x^2} + o(\ell^2)$, where u, $\frac{\partial u}{\partial x}$ and $\frac{\partial^2 u}{\partial x^2}$ are evaluated at the point $(j\,\ell, t)$, and therefore as $\frac{N^2\kappa_*}{m_*}\ell^2 = c^2$, one deduces that $\frac{N^2\kappa_*}{m_*}(2x_i - x_{i-1} - x_{i+1}) = -c^2 \frac{\partial^2 u}{\partial x^2} + o(1)$.

This computation is the consistency of the difference scheme with respect to the wave equation, and it helps in proving that, if the numerical scheme converged, the limit would satisfy the wave equation; that the numerical scheme does converge is related to a different property, the stability of the numerical scheme, and this property can be deduced from the bound on the total energy that was derived for the discrete approximation. It is actually a general remark, due to Peter LAX, that for linear partial differential equations, the consistency and the stability of a numerical scheme usually imply its convergence; of course, the linearity is a crucial assumption in this remark.[1]

The wave equation $\frac{\partial^2 u}{\partial t^2} - c^2 \frac{\partial^2 u}{\partial x^2} = 0$ is conservative, but real materials are slightly dissipative, and for this reason one often considers the dissipative model

$$\frac{\partial^2 u}{\partial t^2} - c^2 \frac{\partial^2 u}{\partial x^2} + \frac{1}{\tau} \frac{\partial u}{\partial t} = 0, \qquad (26.4)$$

where τ is a characteristic time. Multiplying by $\frac{\partial u}{\partial t}$, one deduces that[2]

$$\frac{dE(t)}{dt} + \frac{1}{\tau} \int_0^L \left(\frac{\partial u}{\partial t} \right)^2 dx = 0, \text{ with } E(t) = \int_0^L \left[\frac{1}{2} \left(\frac{\partial u}{\partial t} \right)^2 + \frac{c^2}{2} \left(\frac{\partial u}{\partial x} \right)^2 \right] dx,$$
the total energy at time t,

$$(26.5)$$

so that $E(t)$ is nonincreasing. Actually, the total energy tends to 0 exponentially, and one way to see this is to multiply the equation by $\frac{\partial u}{\partial t} + \frac{\varepsilon}{\tau} u$ and one obtains

$$\frac{dF}{dt} + G = 0, \text{ with } F(t) = \int_0^L \left[\frac{1}{2} \left(\frac{\partial u}{\partial t} \right)^2 + \frac{c^2}{2} \left(\frac{\partial u}{\partial x} \right)^2 + \frac{\varepsilon}{\tau} u \frac{\partial u}{\partial t} + \frac{\varepsilon}{2\tau^2} u^2 \right] dx$$
$$\text{and } G(t) = \int_0^L \left[\frac{1-\varepsilon}{\tau} \left(\frac{\partial u}{\partial t} \right)^2 + \frac{\varepsilon c^2}{\tau} \left(\frac{\partial u}{\partial x} \right)^2 \right] dx,$$

$$(26.6)$$

and one then notices that for $0 < \varepsilon < 1$ both F and G are equivalent to the energy, so that the differential inequality implies their exponential decay.

Another way to see the exponential decay of total energy is to decompose functions on the basis of eigenvectors of $-\frac{\partial^2 u}{\partial x^2}$ with Dirichlet conditions, i.e. $e_n = \sqrt{2} \sin \frac{n\pi x}{L}$ for $n = 1, \dots$. Looking for solutions of the form $e^{\lambda t} e_n(x)$ gives the equation $\lambda^2 + \frac{\lambda}{\tau} + \frac{c^2 n^2 \pi^2}{L^2} = 0$; for $n \geq \frac{L}{2c\pi\tau}$ the real part of λ is $\frac{-1}{2\tau}$ and for $1 \leq n < \frac{L}{2c\pi\tau}$ the values of λ are real and negative. This shows that the exponential decay is uniform above some threshold. In elastic bars (which are modelled in a different way), the decay of modes has been studied by David RUSSELL,[3] and as the experimental evidence is not compatible with a dissipative term in $\frac{\partial u}{\partial t}$, he has proposed heuristic convolution terms.

Actually, the model studied with small masses and springs is not really a good model for the motion of a violin string, because I have studied

---

[1] The stability permits one to extract a subsequence which converges weakly, or weakly $\star$ in some adapted functional space; using linearity and transposition for making the translations act on the test functions, the consistency makes the corresponding translation operator converge to the transposed partial differential operator.

[2] This computation requires enough smoothness for the solution, and the proof that the result is indeed true with initial data of finite total energy requires a little more care.

[3] David L. RUSSELL, American mathematician. He worked at University of Wisconsin, Madison, WI, and at Virginia Tech (Virginia Polytechnic Institute and State University), Blacksburg VA.

longitudinal waves, where the displacements of material points are in the direction of the propagating wave, while the waves in a violin string are transversal waves, where the displacements of material points are in a direction perpendicular to the propagating wave. It would have been better to consider the masses moving in a two-dimensional (or three-dimensional) space, and ask that the mass initially at $i\ell$ (with $\ell = \frac{L}{N}$) is at the point (x_i, y_i), but as the increase in length is $\sqrt{(x_i - x_{i-1})^2 + (y_i - y_{i-1})^2} - \ell$, the formulas become more difficult to study (and one loses the important linearity hypothesis in the argument of Peter LAX about consistency and stability).

A limitation of many models used for solids is that they only consider nearest neighbour interactions, and this is not compatible with the belief that particles attract or repel each other depending upon their distances, except for the case of very short-range potentials, which is not what Lennard-Jones potentials are about, for example. If one was studying vibrations around an equilibrium position, then it would be like having point i and point j linked by a spring of weight $\kappa_{i,j}$, and one would have to consider the potential energy $\sum_{i\neq j} \frac{\kappa_{i,j}}{2}(x_j - x_i - (j - i)\ell)^2$, and if u was a smooth function and $x_i(t) = u(i\ell, t)$, then besides quantities proportional to $\int_0^L \left|\frac{\partial u}{\partial x}\right|^2 dx$ one could well see quantities of the type $\iint_{(0,L)\times(0,L)} |u(x) - u(y)|^2 w(x, y)\, dx\, dy$ for some weight function w, and it is worth mentioning that the norms of fractional Sobolev spaces show similar quantities.[4]

I consider now the two-dimensional linearized elasticity, as studied by CAUCHY. He considered a square lattice with small masslets of size m at the points $(i\ell, j\ell)$ for integers i, j, with springs of strength κ along the horizontal and vertical lines, but also springs of strength κ' along the two diagonal directions.[5] The scaling is now $\ell = \frac{L}{N}$ and $m = \frac{m_*}{N^2}$, corresponding to a finite density of mass at the limit $N \to \infty$, but $\kappa = \kappa_*$ independent of N, because $\frac{\kappa}{m}$ must have the dimension $\frac{1}{time^2}$, and time scales as length in order to have a fixed velocity of propagation of waves; another way to interpret the scaling for κ is that one does not impose forces $O(1)$ at each point of the boundary, but a force per unit of length, i.e. a (two-dimensional) pressure, so that forces

---

[4] For $0 < s < 1$ and $1 \leq p < \infty$, the Sobolev space $W^{s,p}(\mathbb{R}^N)$ is the space of function $u \in L^p(\mathbb{R}^N)$ such that $\iint_{\mathbb{R}^N \times \mathbb{R}^N} \frac{|u(x) - u(y)|^p}{|x-y|^{N+sp}}\, dx\, dy < \infty$. For a bounded open set $\Omega \subset \mathbb{R}^N$ with smooth boundary, one must be careful with boundary conditions, and for $0 < s \leq \frac{1}{p}$ one has $W_0^{s,p}(\Omega) = W^{s,p}(\Omega)$, while for $\frac{1}{p} < s < 1$ one has $W_0^{s,p}(\Omega) \neq W^{s,p}(\Omega)$, but in the case $s = \frac{1}{p}$ there is another natural space, the Lions–Magenes space $W_{00}^{1/p,p}(\Omega)$, for which the functions extended by 0 outside Ω belong to $W^{1/p,p}(\mathbb{R}^N)$, also equal to the space of $u \in W^{1/p,p}(I)$ satisfying $\frac{u}{d^{1/p}} \in L^p(\Omega)$ where d is the distance to the boundary $\partial\Omega$.

[5] Without the diagonal springs, the lattice is quite weak, and the infinite lattice has a family of equilibria, where all the squares become lozenges, without increasing the length of any of the springs; with the diagonal springs these equilibria disappear.

are $O\left(\frac{1}{N}\right)$; for a three-dimensional problem, mass scales as $m = \frac{m_*}{N^3}$ and κ scales as $\kappa = \frac{\kappa_*}{N}$, and forces are $O\left(\frac{1}{N^2}\right)$, so that pressure is $O(1)$.

The displacement has two components, denoted u^1 and u^2, and one uses the notation $u_{i,j}^k$ for $u^k(i\,\ell, j\,\ell)$; one assumes that u^1 and u^2 are smooth functions in order to use Taylor expansion for identifying the partial differential equation governing the motions of the masslets in the limit $N \to \infty$. One makes an assumption of linearized elasticity, corresponding to the approximation that the directions of the springs are almost fixed and only the displacements in the directions of the springs are felt. Considering the forces at the point $(i\,\ell, j\,\ell)$, there are two horizontal forces $\kappa(u_{i+1,j}^1 - u_{i,j}^1)$ and $\kappa(u_{i-1,j}^1 - u_{i,j}^1)$, two vertical forces $\kappa(u_{i,j+1}^2 - u_{i,j}^2)$ and $\kappa(u_{i,j-1}^2 - u_{i,j}^2)$, and forces along the diagonals, but it is only for a particular value of $\frac{\kappa'}{\kappa}$ that one finds the behaviour of an isotropic (linearized) elastic material. CAUCHY's approach only gave the case $\lambda = \mu$ in the more general family of isotropic (linearized) elastic materials proposed by LAMÉ,[6] where the Cauchy stress tensor has the form

$$\sigma_{i,j} = \mu\left(\frac{\partial u_i}{\partial x_j} + \frac{\partial u_j}{\partial x_i}\right) + \lambda\,\delta_{i,j}\sum_k \frac{\partial u_k}{\partial x_k}, \tag{26.7}$$

and the general equations of linearized elasticity are

$$\varrho\,\frac{\partial^2 u_i}{\partial t^2} - \sum_j \frac{\partial \sigma_{i,j}}{\partial x_j} = 0 \text{ for all } i, \tag{26.8}$$

which in the isotropic case (26.7) give the *Lamé equation*

$$\varrho\,\frac{\partial u_i}{\partial t^2} - \mu\,\Delta\,u_i - (\lambda + \mu)\frac{\partial [div(u)]}{\partial x_i} = 0 \text{ for all } i, \tag{26.9}$$

which imply that both $div(u)$ and $curl(u)$ satisfy wave equations, but with different speeds of propagation; the P-waves (or pressure waves) correspond to the wave equation for $div(u)$ and have velocity $\sqrt{\frac{2\mu+\lambda}{\varrho}}$ and the S-waves (or shear waves) correspond to the wave equation for $curl(u)$ and have the velocity $\sqrt{\frac{\mu}{\varrho}}$, which in practice is smaller.[7]

[Taught on Wednesday November 7, 2001.]

Notes on names cited in footnotes for Chapter 26, MAGENES.[8]

---

[6] Gabriel LAMÉ, French mathematician, 1795–1870. He had worked in St. Petersburg, Russia and in Paris, France.

[7] Because most real materials have $\lambda > 0$. In seismology, one makes the approximation that the ground is linearly elastic (and even isotropic!), and it is useful that P-waves travel faster than S-waves, because in earthquakes the P-waves are not dangerous and they signal the danger coming, as it is the S-waves which destroy the buildings which have not been designed carefully.

[8] Enrico MAGENES, Italian mathematician, born in 1923. He worked at Università di Pavia, Pavia, Italy.

27

The Two-Body Problem

The preceding computations, motivated by understanding the magnitude of forces of interaction, have dealt with quite simplified models where only linearized elastic effects were involved, so that no large displacement could be taken into account, where there was no temperature, and where a much too simple crystalline framework was involved,[1] but they have shown the difference of order of magnitude in one, two or three dimensions.

The lack of temperature is an important restriction for a model which is supposed to be realistic, but temperature is actually an equilibrium concept, and not much is understood about nonequilibrium situations, but if one increases temperature slowly it is reasonable to assume that one will move along equilibria without noticeable dynamical effects. What is observed for real materials is that after increasing the temperature of a solid one eventually reaches a critical temperature where there is a change of phase, either a transformation into a solid phase with a different organization of atoms, or a transformation into a liquid phase, and at a higher critical temperature the liquid transforms into a gas.[2] An important effect is the latent heat, which seems to be the energy necessary for breaking bonds, which are either interpreted in terms of classical mechanics, in which case one talks about the minimum energy that one needs for escaping from the attraction of a stable

---

[1] Crystals are not good for elasticity, and polycrystals, which show different grains with various crystalline orientations are observed, but not much is understood about how grain boundaries move (as it is certainly not a local question!).

[2] The experimental physicists who have studied phase transitions have considered the various crystalline orientations that a solid may prefer under various conditions (of temperature and pressure), the temperature of fusion (or sublimation) of a solid, the temperature of boiling of a liquid, with the *latent heat* involved, and the question of triple points, where one goes continuously from one phase to another without latent heat involved (for water, it happens at a temperature of 374° Celsius and a pressure of 220 bars).

equilibrium, or in terms of quantum mechanics, where physicists' ideas always look a little strange.

According to the classical point of view, there are not many bonds left between atoms in a gas, and particles may move quite freely, but one should remember that only rare gases are made of individual atoms, and oxygen or nitrogen prefer binary molecules O_2 or N_2, for example. One basic assumption in kinetic theory is to avoid molecules, which besides kinetic energy show rotational energy and internal energy of vibration (when the distance of the two atoms forming the molecule varies). Another basic assumption of kinetic theory is to consider only binary interactions between particles, and one talks of collisions or nearby collisions, and one estimates the probability of such collisions, and such a description could only be reasonable for a rarefied gas.[3] For a rarefied plasma, where particles are electrically charged (lighter electrons and heavier ions), one usually considers that there are no collisions at all, and one works with the *Vlasov equation*,[4] coupled with the Maxwell–Heaviside equation, of course, or a simplified version of it, as the Laplace/Poisson equation. Physicists often mention particles which cannot be discerned one from the other, and most of the mathematical work in kinetic theory deals with a gas made of identical atomic particles, despite the fact that in applications most gases are mixtures of different molecules, but that particles cannot be discerned is not a bad hypothesis at all, because talking about particles is just an approximation for describing localized waves, and these particles do not really exist as classical ones.

One considers then only two classical particles rushing towards each other without noticing the crowd of other particles around them, but most of the time they do not really collide. In the case where particles are rigid spheres of radius a, the two particles collide only if the distance of their centres becomes $\leq 2a$ at some time; in other words, if a particle is fixed and one wants to know if another moving particle will collide it, one considers that the moving particle sweeps a circular cylinder of section $4\pi a^2$ and will hit the fixed particle only if its centre belongs to the cylinder. If particles attract each other with a law depending upon their distances, then particles which are too far apart are essentially undisturbed, i.e. they do not acquire much kinetic energy because of a close encounter, and one talks about a scattering cross-section (which would be πa^2 in the case of rigid spheres) by considering the particles which would change their direction by more than $\frac{\pi}{2}$, i.e. are reflected backwards.

---

[3] Because one basic assumption in the kinetic theory of gases is that the gas is rarefied, I strongly disagree with the physical interpretation of letting the mean free path between collisions tend to 0, which one calls the *fluid dynamical limit*, and I suggest considering that as a strictly mathematical problem, because it is bad physics (which probably explains the interest of some political group for that type of questions). Actually, I conjecture that the *Hilbert expansion* is false in general, because of the appearance of oscillations.

[4] Anatoliĭ Aleksandrovich VLASOV, Russian physicist, 1908–1975.

The motion of two particles in a field of central forces was solved long ago in classical mechanics (while the n-body problem is still not so well understood), and the equations are

$$m_1 \frac{d^2 \mathbf{M}_1}{dt^2} = \mathbf{F}_1, \ m_2 \frac{d^2 \mathbf{M}_2}{dt^2} = \mathbf{F}_2, \tag{27.1}$$

and with the only information that

$$\mathbf{F}_1 + \mathbf{F}_2 = \mathbf{0}, \tag{27.2}$$

one already finds that the centre of gravity $\mathbf{G}$, defined by

$$(m_1 + m_2)\mathbf{G} = m_1\mathbf{M}_1 + m_2\mathbf{M}_2, \tag{27.3}$$

moves with constant speed, because adding the equations gives

$$(m_1 + m_2)\frac{d^2 \mathbf{G}}{dt^2} = 0, \text{ so that } \frac{d\mathbf{G}}{dt} = constant. \tag{27.4}$$

If one moves with the centre of gravity, which one then takes as the origin, i.e. $\mathbf{G} = \mathbf{0}$, then the information that the forces are along the line joining the two particles leads to the first two Kepler laws; the angular momentum computed at $\mathbf{G} = \mathbf{0}$ is

$$\Omega = m_1\mathbf{M}_1 \wedge \frac{d\mathbf{M}_1}{dt} + m_2\mathbf{M}_2 \wedge \frac{d\mathbf{M}_2}{dt}, \tag{27.5}$$

which is $\frac{m_1}{m_2}(m_1 + m_2)\mathbf{M}_1 \wedge \frac{d\mathbf{M}_1}{dt}$, so that

$$\frac{d\Omega}{dt} = m_1\mathbf{M}_1 \wedge \mathbf{F}_1 + m_2\mathbf{M}_2 \wedge \mathbf{F}_2 = 0, \tag{27.6}$$

because both $\mathbf{F}_1$ and $\mathbf{F}_2$ are along M_1M_2, which are parallel to $0M_1$ and $0M_2$ by the choice $\mathbf{G} = \mathbf{0}$. If $\Omega \neq 0$, i.e. particles are not both moving on a line, then $\mathbf{M}_1$ and $\frac{d\mathbf{M}_1}{dt}$ are in the plane perpendicular to Ω, giving the first Kepler law, that the two particles move in a plane, while the second Kepler law that the area swept by GM_1 is proportional to time comes precisely from the fact that $\mathbf{M}_1 \wedge \frac{d\mathbf{M}_1}{dt}$ is a constant vector. The third Kepler law only holds for forces in $\frac{1}{distance^2}$, that particles follow ellipses (or more generally conic sections) with $\mathbf{G}$ a focus with a precise relation between size and period.

I was told that KEPLER had postulated that the planets follow ellipses with the sun at a focus, and he had needed precise astronomical measurements for discovering how the planets moved on these ellipses, like those made by BRAHE.[5] When I visited Klaus KIRCHGÄSSNER in Stuttgart in 1987,[6] he had

---

[5] Tyge BRAGE (Tycho BRAHE), Danish-born astronomer, 1546–1601. He had worked in Prague, now capital of the Czech republic.

[6] Klaus KIRCHGÄSSNER, German mathematician, born in 1931. He worked in Stuttgart, Germany.

told me that BRAHE did not want to give his measurements to KEPLER, who had then managed to steal them after BRAHE had died. Of course, KEPLER had wrongly guessed that the sun is at a focus, because if there was only one planet it would be the centre of gravity of the sun/planet pair which would be at a focus, but the centre of gravity of the solar system falls inside the sun anyway. The orbits are not exactly ellipses, because there are many planets, and LAGRANGE had been the first to develop a theory of perturbations for studying that question, which became useful after Uranus had been found in 1781 by HERSCHEL in a systematic survey of the sky,[7] because its irregular motion led to the hypothesis that it was perturbed by another planet.[8]

Actually, the situation considered in kinetic theory is not to think about trajectories as ellipses but as hyperbolas (as in the trajectories of some comets), and consider the limiting velocities before and after a "collision" (which should only be thought of as a near collision). This creates a picture of a gas which does not allow for the possibility of having particles moving with their cohort of satellites, like small solar systems or better like double stars (or multiple ones) if all particles are considered identical. This restriction is related to the postulate that the only type of energy considered for a gas is translational kinetic energy.

In the study of the possible outputs of a "collision", the timing is usually not considered, and it is assumed that two particles with velocities $\mathbf{v}$ and $\mathbf{w}$ at a point $\mathbf{x}$ and at a time t may transform into two particles with velocities $\mathbf{v}'$ and $\mathbf{w}'$ in an instantaneous way, at the same point $\mathbf{x}$ and at the same time t; moreover one postulates some probability distribution among the outputs.

Because one assumes that all particles are identical, conservation of mass is just the fact that two particles colliding give two particles as the output; conservation of linear momentum is equivalent to

$$\mathbf{v} + \mathbf{w} = \mathbf{v}' + \mathbf{w}', \tag{27.7}$$

conservation of angular momentum is automatically verified because the two particles are at the same point, before and after the collision, and conservation of kinetic energy is equivalent to

$$|\mathbf{v}|^2 + |\mathbf{w}|^2 = |\mathbf{v}'|^2 + |\mathbf{w}'|^2, \tag{27.8}$$

and one deduces that

---

[7] William HERSCHEL, German-born astronomer, 1738–1822. He had worked in England.

[8] Both J.C. ADAMS and LE VERRIER successfully applied the theoretical work of LAGRANGE and found the correct position of that planet, but CHALLIS failed to see it; LE VERRIER was better served by GALLE, and got full credit for the discovery, and the right to call it Neptune, although it had actually been observed before, by LALANDE in 1795, and even by Galileo in 1613.

$$|\mathbf{v}' - \mathbf{w}'|^2 = |\mathbf{v} - \mathbf{w}|^2. \tag{27.9}$$

Writing $\mathbf{v}' = \mathbf{v} + a\,\boldsymbol{\alpha}$ with $a \in \mathbb{R}$ and $\boldsymbol{\alpha} \in \mathbb{R}^3$ with $|\boldsymbol{\alpha}| = 1$, one must have $\mathbf{w}' = \mathbf{w} - a\,\boldsymbol{\alpha}$, and then $|\mathbf{w}' - \mathbf{v}'|^2 = |\mathbf{v} - \mathbf{w} + 2a\,\boldsymbol{\alpha}|^2 = |\mathbf{v} - \mathbf{w}|^2$, so that $4a(\mathbf{v} - \mathbf{w}, \boldsymbol{\alpha}) + 4a^2 = 0$, and apart from the trivial solution $a = 0$, one must have $a = (\mathbf{w} - \mathbf{v}, \boldsymbol{\alpha})$, and as the case $a = 0$ is then obtained by choosing $\boldsymbol{\alpha}$ perpendicular to $\mathbf{w} - \mathbf{v}$, the general solution is

$$\text{for } \boldsymbol{\alpha} \in \mathbb{S}^2, \begin{cases} \mathbf{v}' = \mathbf{v} + (\mathbf{w} - \mathbf{v}, \boldsymbol{\alpha})\boldsymbol{\alpha} \\ \mathbf{w}' = \mathbf{w} - (\mathbf{w} - \mathbf{v}, \boldsymbol{\alpha})\boldsymbol{\alpha} \end{cases}, \tag{27.10}$$

and in particular one has

$$\mathbf{w}' - \mathbf{v}' = (I - 2\boldsymbol{\alpha} \otimes \boldsymbol{\alpha})(\mathbf{w} - \mathbf{v}). \tag{27.11}$$

If one defines the angle $\theta \in [0, \pi]$ by

$$(\mathbf{w} - \mathbf{v}, \boldsymbol{\alpha}) = |\mathbf{w} - \mathbf{v}| \cos\theta, \tag{27.12}$$

then $\theta = \frac{\pi}{2}$ corresponds to $\mathbf{v}' = \mathbf{v}$ and $\mathbf{w}' = \mathbf{w}$, which happens if particles miss each other in the collision, while the case $\theta = 0$ corresponds to $\mathbf{v}' = \mathbf{w}$ and $\mathbf{w}' = \mathbf{v}$. In the frame of the centre of gravity $\mathbf{G}$, where $\mathbf{v} + \mathbf{w} = \mathbf{0}$, one has $\mathbf{v}' = (I - 2\boldsymbol{\alpha} \otimes \boldsymbol{\alpha})v$, so one sees two particles arriving with the velocity of approach $\frac{|\mathbf{w} - \mathbf{v}|}{2}$ and leaving with the same velocity but in a direction making an angle 2θ. In the frame linked with the centre of gravity there is a symmetry around the line of approach of the particles, and therefore one postulates that the various angles θ are obtained as outputs with a probability which only depends upon $|\mathbf{w} - \mathbf{v}|$ and θ.

[Taught on Friday November 9, 2001.]

Notes on names cited in footnotes for Chapter 27, J.C. ADAMS,[9] LE VER-RIER,[10] GALLE,[11] LALANDE.[12]

---

[9] John Couch ADAMS, English astronomer, 1819–1892. He had worked in Cambridge, England.

[10] Urbain Jean Joseph LE VERRIER, French astronomer, 1811–1877. He had worked in Paris, France.

[11] Johann Gottfried GALLE, German astronomer, 1812–1910. He had worked in Berlin, Germany.

[12] Joseph-Jérôme LE FRANÇOIS DE LA LANDE, French astronomer, 1732–1807. He had worked at Collège de France, Paris, France.

28

The Boltzmann Equation

If there were no forces on the particles, the evolution equation for the density of particles $f(\mathbf{x}, \mathbf{v}, t)$ would be the free transport equation

$$\frac{\partial f}{\partial t} + \mathbf{v}.\frac{\partial f}{\partial \mathbf{x}} = 0. \tag{28.1}$$

The presence of collisions transforms this equation into a form

$$\frac{\partial f}{\partial t} + \mathbf{v}.\frac{\partial f}{\partial \mathbf{x}} = Q(f, f), \tag{28.2}$$

where the nonlinearity $Q(f, f)$ takes into account the disappearance of particles with velocity $\mathbf{v}$ by collision against particles with velocity $\mathbf{w}$ (creating particles with velocities $\mathbf{v}'$ and $\mathbf{w}'$), but also the appearance of particles with velocity $\mathbf{v}$ (by collisions of particles with velocities $\mathbf{v}'$ and $\mathbf{w}'$). $Q(f, f)$ is chosen to be quadratic in f, by an argument that the probability of collision of particles with velocity $\mathbf{v}_1$ and particles with velocity $\mathbf{v}_2$ is proportional to $f(\mathbf{x}, \mathbf{v}_1, t) f(\mathbf{x}, \mathbf{v}_2, t)$, the product of the densities of the two types of particles, and this assumes that some independence property holds. I think that this argument only makes sense for a rarefied gas, where the picture is like that of *hyperbolic orbits* of some comets, but if one is not in a rarefied situation, either one thinks in terms of classical mechanics, and the simple description using $f(\mathbf{x}, \mathbf{v}, t)$ seems too naive, and it seems natural to add correlations of position to the description, or one thinks from a modern point of view where there are only waves which in some limiting situation may look as "particles"; one is not in such a limiting situation and one must understand better about the wave nature of these particles that one is dealing with.[1]

---

[1] Using analogies with my H-measures [18], and their variants, which are quadratic micro-local objects, the function $f(\mathbf{x}, \mathbf{v}, t)$ looks like the density of a such a micro-local measure, and if the underlying equation was a linear hyperbolic system in x with a quadratic conservation law, I would expect a linear transport equation

Formally

$$Q(f,f) = \int_{\mathbb{R}^3 \times \mathbb{S}^2} k(\mathbf{v}, \mathbf{w}, \boldsymbol{\alpha}) \big(f(\mathbf{x}, \mathbf{v}', t) f(\mathbf{x}, \mathbf{w}', t) - f(\mathbf{x}, \mathbf{v}, t) f(\mathbf{x}, \mathbf{w}, t) \big) \, d\mathbf{w} \, d\boldsymbol{\alpha},$$
with notation (27.10),

$$(28.3)$$

i.e. $\mathbf{v}' = \mathbf{v} + (\mathbf{w} - \mathbf{v}, \boldsymbol{\alpha})\boldsymbol{\alpha}$ and $\mathbf{w}' = \mathbf{w} - (\mathbf{w} - \mathbf{v}, \boldsymbol{\alpha})\boldsymbol{\alpha}$, and where the kernel k is nonnegative. Due to symmetries, the kernel $k(\mathbf{v}, \mathbf{w}, \boldsymbol{\alpha})$ has the form

$$k(\mathbf{v}, \mathbf{w}, \boldsymbol{\alpha}) = K(|\mathbf{v} - \mathbf{w}|, \theta), \text{ with notation (27.12),} \qquad (28.4)$$

i.e. $((\mathbf{w} - \mathbf{v}), \boldsymbol{\alpha}) = |\mathbf{w} - \mathbf{v}| \cos\theta$, and an analytic expression of K can be deduced from the precise force law used for attraction (or repulsion) of particles, and more precisely, for an attractive force,

> a law in $\frac{1}{distance^s}$ gives K proportional to $|\mathbf{v} - \mathbf{w}|^\gamma, \gamma = \frac{s-5}{s-1}$,
> $s = 5, \gamma = 0$, giving $K(\theta)$, is referred to as *Maxwellian molecules*, (28.5)
> $s = +\infty, \gamma = +1$, is referred to as the *hard-sphere* case.

However, the main problem is that the kernel tends to ∞ for $\theta = \frac{\pi}{2}$, and for an attractive force,

> a law in $\dfrac{1}{distance^s}$ gives a singularity in $\dfrac{1}{|\cos\theta|^\nu}, \nu = \dfrac{s+1}{s-1}.$ (28.6)

Of course, if $\theta = \frac{\pi}{2}$ then one has $f(\mathbf{x}, \mathbf{v}', t) f(\mathbf{x}, \mathbf{w}', t) - f(\mathbf{x}, \mathbf{v}, t) f(\mathbf{x}, \mathbf{w}, t) = 0$ (because $\mathbf{v}' = \mathbf{v}$ and $\mathbf{w}' = \mathbf{w}$), and therefore one has an indeterminate form in the integrand. One way to avoid this problem is to use the angular cut-off assumption made by Harold GRAD, which consists in changing the kernel near $\theta = \frac{\pi}{2}$ so that it becomes integrable in θ.

The Boltzmann equation is postulated, and one should not exaggerate its importance and pretend (as too many seem to believe) that starting from the Boltzmann equation and deducing by purely formal considerations other (postulated) equations used for describing the behaviour of *real fluids*, like the Euler equation or the Navier–Stokes equation, gives more credence to these equations. One may be interested in purely mathematical questions concerning the Boltzmann equation, and one interesting mathematical question is to avoid making the angular cut-off assumption; in doing that it seems that one should be able to estimate

> cancellations in the difference $f(\mathbf{x}, \mathbf{v}', t) f(\mathbf{x}, \mathbf{w}', t) - f(\mathbf{x}, \mathbf{v}, t) f(\mathbf{x}, \mathbf{w}, t),$
> (28.7)

---

in (x, ξ) (and there is no problem about denoting this dual variable v), but the underlying equation should be a semi-linear hyperbolic system in x instead, with the same kind of quadratic conservation laws, and more general objects than H-measures must be developed for the analysis, i.e. I do not think that one should search for a nonlinear equation for the density of an H-measure at all.

but most of the time one estimates $f(\mathbf{x}, \mathbf{v}', t) f(\mathbf{x}, \mathbf{w}', t)$ and $f(\mathbf{x}, \mathbf{v}, t) f(\mathbf{x}, \mathbf{w}, t)$ independently, so that no cancellation can be studied, and one is led to limit the strength of the kernel for the purpose of proving results, and this is not a very scientific point of view.

The problem of θ being near $\frac{\pi}{2}$ is that of grazing collisions, for which particles only change their velocity very slightly in the interaction, and the result of many such small changes in velocity is often described by a diffusion in velocity space, giving rise to the Fokker–Planck equation

$$\frac{\partial f}{\partial t} + \mathbf{v}.\frac{\partial f}{\partial \mathbf{x}} - \kappa \Delta_{\mathbf{v}} f = 0. \tag{28.8}$$

However, some people write a nonlinear Fokker–Planck equation with a diffusion depending upon f, so that Maxwellian distributions satisfy it, or derive such a nonlinear equation from the Boltzmann equation, which is not very logical, as grazing collisions are not well taken care of in the Boltzmann equation.

Because of the invariance of the number of particles in each collision, one deduces that

$$\int_{\mathbb{R}^3} Q(f, f) \, d\mathbf{v} = 0, \tag{28.9}$$

if the Fubini theorem can be applied, of course; similarly, because $\mathbf{v}' + \mathbf{w}' = \mathbf{v} + \mathbf{w}$ for each collision, one deduces that

$$\int_{\mathbb{R}^3} v_j Q(f, f) \, d\mathbf{v} = 0 \text{ for } j = 1, 2, 3, \tag{28.10}$$

and because $|\mathbf{v}'|^2 + |\mathbf{w}'|^2 = |\mathbf{v}|^2 + |\mathbf{w}|^2$ for each collision, one deduces that

$$\int_{\mathbb{R}^3} |\mathbf{v}|^2 Q(f, f) \, d\mathbf{v} = 0. \tag{28.11}$$

From these equalities, one deduces conservation laws for fluid quantities defined by integration in $\mathbf{v}$. One defines the *density of mass* $\varrho(\mathbf{x}, t)$ by

$$\varrho(\mathbf{x}, t) = \int_{\mathbb{R}^3} f(\mathbf{x}, \mathbf{v}, t) \, d\mathbf{v}, \text{ a.e. } \mathbf{x} \in \mathbb{R}^3, \tag{28.12}$$

the (macroscopic) *velocity* $\mathbf{u}(\mathbf{x}, t)$ by

$$\varrho(\mathbf{x}, t) u_j(\mathbf{x}, t) = \int_{\mathbb{R}^3} v_j f(\mathbf{x}, \mathbf{v}, t) \, d\mathbf{v}, \text{ for } j = 1, 2, 3, \text{ a.e. } \mathbf{x} \in \mathbb{R}^3, \tag{28.13}$$

the *Cauchy stress tensor* σ by

$$\sigma_{i,j} = -\int_{\mathbb{R}^3} \left(v_i - u_i(\mathbf{x}, t)\right)\left(v_j - u_j(\mathbf{x}, t)\right) f(\mathbf{x}, \mathbf{v}, t) \, d\mathbf{v}, \text{ a.e. } \mathbf{x} \in \mathbb{R}^3, \tag{28.14}$$

the *internal energy per unit of mass* $e(\mathbf{x}, t)$ by

$$\varrho(\mathbf{x}, t) e(\mathbf{x}, t) = \int_{\mathbb{R}^3} \frac{|\mathbf{v} - \mathbf{u}(\mathbf{x}, t)|^2}{2} f(\mathbf{x}, \mathbf{v}, t) \, dv, \text{ a.e. } \mathbf{x} \in \mathbb{R}^3, \tag{28.15}$$

the *density of total energy* $E(\mathbf{x}, t)$ by

$$E(\mathbf{x}, t) = \frac{\varrho(\mathbf{x}, t) |\mathbf{u}(\mathbf{x}, t)|^2}{2} + \varrho(\mathbf{x}, t) e(\mathbf{x}, t), \text{ a.e. } \mathbf{x} \in \mathbb{R}^3, \tag{28.16}$$

and the *heat flux* $\mathbf{q}(\mathbf{x}, t)$ by

$$q_i(\mathbf{x}, t) = \int_{\mathbb{R}^3} (v_i - u_i(\mathbf{x}, t)) \frac{|\mathbf{v} - \mathbf{u}(\mathbf{x}, t)|^2}{2} f(\mathbf{x}, \mathbf{v}, t) \, dv \tag{28.17}$$
$$\text{for } i = 1, 2, 3, \text{ a.e. } \mathbf{x} \in \mathbb{R}^3.$$

By integrating the Boltzmann equation in $\mathbf{v}$, conservation of mass becomes

$$\frac{\partial \varrho}{\partial t} + \sum_{j=1}^{3} \frac{\partial (\varrho u_j)}{\partial x_j} = 0 \text{ in } \mathbb{R}^3, \tag{28.18}$$

by multiplying the Boltzmann equation by v_i and integrating in $\mathbf{v}$, the balance of linear momentum becomes

$$\frac{\partial (\varrho u_i)}{\partial t} + \sum_{j=1}^{3} \frac{\partial (\varrho u_i u_j)}{\partial x_j} - \sum_{j=1}^{3} \frac{\partial \sigma_{i,j}}{\partial x_j} = 0 \text{ for } i = 1, 2, 3, \text{ in } \mathbb{R}^3, \tag{28.19}$$

and by multiplying the Boltzmann equation by $\frac{|\mathbf{v}|^2}{2}$ and integrating in $\mathbf{v}$, the balance of energy becomes

$$\frac{\partial E}{\partial t} + \sum_{j=1}^{3} \frac{\partial (E u_j)}{\partial x_j} - \sum_{i,j=1}^{3} \frac{\partial (\sigma_{i,j} u_i)}{\partial x_j} + \sum_{j=1}^{3} \frac{\partial q_j}{\partial x_j} = 0 \text{ in } \mathbb{R}^3, \tag{28.20}$$

and conservation of angular momentum then follows from the symmetry of the Cauchy stress tensor.

There is an important identity which is always valid,

$$2\varrho(\mathbf{x}, t) e(\mathbf{x}, t) + \sum_{i=1}^{3} \sigma_{i,i}(\mathbf{x}, t) = 0, \tag{28.21}$$

and in the case of a gas at local equilibrium, one has

$$\sigma_{i,j} = -p \, \delta_{i,j}, \text{ for } i, j = 1, 2, 3, \text{ where } p \text{ is the } \textit{pressure}, \tag{28.22}$$

so that

$$p(\mathbf{x}, t) = \frac{2\varrho(\mathbf{x}, t) e(\mathbf{x}, t)}{3} \quad \text{at equilibrium for such a gas.} \tag{28.23}$$

Property (28.23) is valid for perfect gases, but not for real gases, whose equation of state is not compatible with the preceding relation between ϱ, p and e, so that real gases are not so well described by the Boltzmann equation.

Definition 28.1. *A function φ defined on $\mathbb{R}^3$ is a* collision invariant, *if it satisfies*

$$\varphi(\mathbf{v}') + \varphi(\mathbf{w}') = \varphi(\mathbf{v}) + \varphi(\mathbf{w}), \quad \text{whenever (27.7) and (27.8) are satisfied.} \tag{28.24}$$

Each function φ defined by

$$\varphi(\mathbf{v}) = a\,|\mathbf{v}|^2 + (\mathbf{b}.\mathbf{v}) + c \text{ for all } \mathbf{v} \in \mathbb{R}^3, \tag{28.25}$$

for $a, c \in \mathbb{R}$ and $\mathbf{b} \in \mathbb{R}^3$ is a collision invariant, and one may wonder if there are other collision invariants besides those given by (28.24). BOLTZMANN had shown that if φ is of class C^1 then every collision invariant has this form, and GRONWALL removed the smoothness hypothesis.[2] Below I follow the proof given by Clifford TRUESDELL and Robert MUNCASTER in their book [22],[3] where they mention that Lennart CARLESON and FROSTMAN included a proof of theirs when they edited the posthumous book of CARLEMAN [1].

Proposition 28.2. *Every measurable collision invariant has the form (28.25).*

Proof: One looks for

$$\varphi(v) + \varphi(w) = \psi(v + w, |v|^2 + |w|^2) \text{ for all } v, w \in \mathbb{R}^3, \tag{28.26}$$

and ψ must be measurable on $\{(u, s) \mid s \geq \frac{|u|^2}{2}\}$,[4] and as one may add a constant to φ, one assumes that $\varphi(0) = 0$ so that $\psi(0, 0) = 0$, and using $w = 0$ gives

$$\psi(v, |v|^2) = \varphi(v), \text{ so that } \psi(v, |v|^2) + \psi(w, |w|^2) = \psi(v + w, |v|^2 + |w|^2)$$
$$\text{for all } v, w \in \mathbb{R}^3. \tag{28.27}$$

---

[2] Thomas Hakon GRÖNWALL, Swedish-born mathematician, 1877–1932. He had worked as an engineer, then at Princeton University, Princeton, NJ, and at Columbia University, New York, NY.

[3] Robert Gary MUNCASTER, American mathematician, born in 1948. He works at University of Illinois, Urbana-Champaign, IL.

[4] If $w = u - v$, then $|v|^2 + |u - v|^2$ is minimum for $v = \frac{u}{2}$ so that ψ is only evaluated at points (u, s) with $s \geq \frac{|u|^2}{2}$. That ψ is measurable can be seen from an explicit choice, for example if $u \neq 0$ by taking $v = a\,u$, $w = (1-a)u$, with $2a^2 - 2a + 1 = \frac{s}{|u|^2}$ and choosing the root $a \geq \frac{1}{2}$, and if $u = 0$ by taking $v = -w = \frac{\sqrt{s}}{\sqrt{2}} e$ for a fixed unit vector e.

Using $w = -v$ gives

$$\psi(0, 2|v|^2) = \psi(v, |v|^2) + \psi(-v, |v|^2) \text{ for all } v, w \in \mathbb{R}^3, \tag{28.28}$$

and then in the case where $v.w = 0$, one deduces that

$$
\begin{aligned}
\psi(0, 2|v|^2 + 2|w|^2) &= \psi(0, 2|v + w|^2) \\
&= \psi(v + w, |v + w|^2) + \psi(-v - w, |v + w|^2) \\
&= \psi(v + w, |v|^2 + |w|^2) + \psi(-v - w, |v|^2 + |w|^2) \\
&= \psi(v, |v|^2) + \psi(w, |w|^2) + \psi(-v, |v|^2) + \psi(-w, |w|^2) \\
&= \psi(0, 2|v|^2) + \psi(0, 2|w|^2),
\end{aligned}
\tag{28.29}
$$

so that

$$\psi(0, a) + \psi(0, b) = \psi(0, a + b) \text{ for all } a, b \geq 0, \tag{28.30}$$

and it is classical that (28.30) implies that there exists a constant C such that

$$\psi(0, a) = C\, a \text{ for all } a \geq 0. \tag{28.31}$$

One defines the (measurable) function g by

$$g(v) = \psi(v, |v|^2) - \psi(0, |v|^2), \text{ for all } v \in \mathbb{R}^3, \tag{28.32}$$

so that g is odd by (28.28), and additive on orthogonal pairs by (28.27) and (28.31), and it remains to show that g is additive.

Let m and n be unit vectors which are orthogonal, then

$$
\begin{aligned}
g(\alpha^2 m + \alpha\beta n) &= g(\alpha^2 m) + g(\alpha\beta n) \\
g(\beta^2 m - \alpha\beta n) &= g(\beta^2 m) - g(\alpha\beta n),
\end{aligned}
\tag{28.33}
$$

but as $\alpha^2 m \pm \alpha\beta n$ and $\beta^2 m \mp \alpha\beta n$ are orthogonal one deduces that

$$
\begin{aligned}
g\big((\alpha^2 + \beta^2)m\big) &= g(\alpha^2 m + \alpha\beta n) + g(\beta^2 m - \alpha\beta n) = g(\alpha^2 m) + g(\beta^2 m) \\
g\big((\alpha^2 - \beta^2)m\big) &+ g(2\alpha\beta n) = g(\alpha^2 m + \alpha\beta n) + g(-\beta^2 m + \alpha\beta n) \\
&= g(\alpha^2 m) - g(\beta^2 m) + 2g(\alpha\beta n),
\end{aligned}
\tag{28.34}
$$

and as $g(2x) = 2g(x)$ by the preceding case, one deduces that

$$g\big((\alpha^2 - \beta^2)m\big) = g(\alpha^2 m) - g(\beta^2 m). \tag{28.35}$$

This shows that $g(x + y) = g(x) + g(y)$ if x and y are parallel. Then for two arbitrary vectors v and w one writes $w = \alpha v + z$ with z orthogonal to v and therefore $g(v + w) = g\big((1 + \alpha)v\big) + g(z) = g(v) + g(\alpha v) + g(z) = g(v) + g(w)$, and the classical result then implies that g is linear. $\qquad\square$

Another important observation of BOLTZMANN is that

$$\int_{\mathbb{R}^3 \times \mathbb{R}^3} f(\mathbf{x}, \mathbf{v}, t) \log\big(f(\mathbf{x}, \mathbf{v}, t)\big) \, d\mathbf{x} \, d\mathbf{v} \text{ is nonincreasing with time,} \tag{28.36}$$

and this follows from

$$\int_{\mathbb{R}^3} Q(f,f) \log f \, dv \le 0, \tag{28.37}$$

which is proven by showing that

$$\int_{\mathbb{R}^3 \times \mathbb{R}^3} K(|\mathbf{v} - \mathbf{w}|, \theta)(f(\mathbf{v}')f(\mathbf{w}') - f(\mathbf{v})f(\mathbf{w})) \log f(\mathbf{v}) \, d\mathbf{v} \, d\mathbf{w} \le 0 \\ \text{for } \mathbf{x}, t, \boldsymbol{\alpha} \text{ given.} \tag{28.38}$$

One observes that the kernel is invariant by the exchange of $\mathbf{v}$ and $\mathbf{w}$, and also by the change of variables $(\mathbf{v}, \mathbf{w}) \mapsto (\mathbf{v}', \mathbf{w}')$ and that $d\mathbf{v}' \, d\mathbf{w}' = d\mathbf{v} \, d\mathbf{w}$, so that the integral considered is

$$\frac{-1}{4} \int_{\mathbb{R}^3 \times \mathbb{R}^3} K(|\mathbf{v} - \mathbf{w}|, \theta)(f(\mathbf{v}')f(\mathbf{w}') - f(\mathbf{v})f(\mathbf{w})) \\ \left(\log f(\mathbf{v}') + \log f(\mathbf{w}') - \log f(\mathbf{v}) - \log f(\mathbf{w})\right) d\mathbf{v} \, d\mathbf{w}, \tag{28.39}$$

and then one uses the fact that $K \ge 0$ and that with $a = f(\mathbf{v}')f(\mathbf{w}')$ and $b = f(\mathbf{v})f(\mathbf{w})$ one has $(\log a - \log b)(a - b) \ge 0$ (if $a, b > 0$, and by continuity if a or b is 0).

One also deduces from the preceding computation that

$$\begin{array}{l} \text{if } \int_{\mathbb{R}^3 \times \mathbb{R}^3} f(\mathbf{x}, \mathbf{v}, t) \log(f(\mathbf{x}, \mathbf{v}, t)) \, d\mathbf{x} \, d\mathbf{v} \text{ is constant} \\ \text{then } f(\mathbf{v}')f(\mathbf{w}') - f(\mathbf{v})f(\mathbf{w}) = 0, \end{array} \tag{28.40}$$

because $(\log a - \log b)(a - b) = 0$ implies $a = b$ (and one has assumed that $K > 0$); this means that $\log f$ is a collision invariant, and therefore that

$$\log f(\mathbf{x}, \mathbf{v}, t) = a(\mathbf{x}, t)|\mathbf{v}|^2 + (\mathbf{b}(\mathbf{x}, t).\mathbf{v}) + c(\mathbf{x}, t), \text{ i.e. } f \text{ is a local Maxwellian} \tag{28.41}$$

which implies $Q(f, f) = 0$, and one needs to have $a < 0$ in order to have f integrable in $\mathbf{v}$, so that $\varrho(\mathbf{x}, t)$ is defined. Of course, if

$$Q(f, f) = 0 \text{ implies } \int_{\mathbb{R}^3 \times \mathbb{R}^3} f(\mathbf{x}, \mathbf{v}, t) \log(f(\mathbf{x}, \mathbf{v}, t)) \, d\mathbf{x} \, d\mathbf{v} = constant, \tag{28.42}$$

so that the two conditions are "equivalent" (one should check that integrability properties of the solution f are sufficient for the Fubini theorem to be applicable for proving that this "equivalence" is true).

If $Q(f, f) = 0$ then f satisfies a free transport equation, and therefore $g(\mathbf{x}, \mathbf{v}, t) = \log f(\mathbf{x}, \mathbf{v}, t) = a(\mathbf{x}, t)|\mathbf{v}|^2 + (\mathbf{b}(\mathbf{x}, t).\mathbf{v}) + c(\mathbf{x}, t)$ also satisfies a free transport equation. In the expression of $\frac{\partial g}{\partial t} + \mathbf{v}.\frac{\partial g}{\partial \mathbf{x}} = 0$, the coefficient of $v_i|\mathbf{v}|^2$ is $\frac{\partial a}{\partial x_i} = 0$, the coefficient of $v_i v_j$ is $\frac{\partial b_i}{\partial x_j} + \frac{\partial b_j}{\partial x_i} = 0$ if $i \ne j$ and $\frac{\partial b_i}{\partial x_i} + \frac{\partial a}{\partial t} = 0$ if $i = j$, the coefficient of v_i is $\frac{\partial b_i}{\partial t} + \frac{\partial c}{\partial x_i} = 0$, and the constant coefficient is $\frac{\partial c}{\partial t} = 0$. One deduces that a is a function of t alone and c is a function of $\mathbf{x}$ alone. Using the identity

$$2\frac{\partial^2 b_i}{\partial x_j \partial x_k} = \frac{\partial}{\partial x_j}\left(\frac{\partial b_i}{\partial x_k} + \frac{\partial b_k}{\partial x_i}\right) - \frac{\partial}{\partial x_i}\left(\frac{\partial b_k}{\partial x_j} + \frac{\partial b_j}{\partial x_k}\right) + \frac{\partial}{\partial x_k}\left(\frac{\partial b_j}{\partial x_i} + \frac{\partial b_i}{\partial x_j}\right) \\ \text{for all } i, j, k = 1, 2, 3, \tag{28.43}$$

one finds that $\frac{\partial^2 b_i}{\partial x_j \partial x_k} = 0$ for all $i, j, k = 1, 2, 3$, because a depends only upon t, and therefore one has $\mathbf{b}(\mathbf{x}, t) = M(t)\mathbf{x} + \mathbf{b}^0(t)$ and $M(t) + M^T(t) + 2\frac{da}{dt}I = 0$. Using $\frac{dM}{dt}\mathbf{x} + \frac{d\mathbf{b}^0}{dt} + grad_x c = 0$, one deduces that $\frac{dM}{dt}$ is symmetric and independent of t and that $\frac{d\mathbf{b}^0}{dt}$ is independent of t. One has $\frac{dM}{dt} + \frac{d^2 a}{dt^2}I = 0$, so that $M(t) = -\frac{da}{dt}I + N$ with N independent of t and satisfying $N^T + N = 0$; a must be a quadratic in t and $\mathbf{b}^0$ affine in t, and then $c(\mathbf{x}) = \frac{1}{2}\frac{d^2 a}{dt^2}|\mathbf{x}|^2 + \frac{d\mathbf{b}^0}{dt}.\mathbf{x} + c_0$ for a constant c_0. All this shows that $g(\mathbf{x}, \mathbf{v}, t)$ is a linear combination of $|\mathbf{x} - t\,\mathbf{v}|^2$, $(\mathbf{x} - t\,\mathbf{v}).\mathbf{v}$, $x_i v_j - x_j v_i$ for all $i \neq j$, $|\mathbf{v}|^2$, v_i for all i and 1 (solutions of the free transport equation must have the form $h(\mathbf{x} - t\,\mathbf{v}, \mathbf{v})$, and one should notice that $x_i v_j - x_j v_i$ can be written as $(x_i - v_i t)v_j - (x_j - v_j t)v_i$).

Of course, if f is a stationary solution of the Boltzmann equation, i.e. a solution independent of t, then it must be a global Maxwellian, $f(\mathbf{x}, \mathbf{v}) = e^{a|\mathbf{v}|^2 + (\mathbf{b}.\mathbf{v}) + c}$ with $a < 0$, and it is useful to relate the coefficients $a, \mathbf{b}, c$ to the macroscopic quantities $\varrho, \mathbf{u}, e$ defined by $\varrho = \int_{\mathbb{R}^3} f(\mathbf{v})\,d\mathbf{v}$, $\varrho\,\mathbf{u} = \int_{\mathbb{R}^3} \mathbf{v} f(\mathbf{v})\,d\mathbf{v}$ and $\varrho e = \int_{\mathbb{R}^3} \frac{|\mathbf{v} - \mathbf{u}|^2}{2} f(\mathbf{v})\,d\mathbf{v}$. From $\int_{\mathbb{R}} e^{-\pi x^2}\,dx = 1$ one deduces by a change of variable that $\int_{\mathbb{R}} e^{-a x^2}\,dx = \sqrt{\frac{\pi}{a}}$ for $a > 0$, and by an integration by parts that $\int_{\mathbb{R}} x^2 e^{-a x^2}\,dx = \frac{1}{2a}\sqrt{\frac{\pi}{a}}$ for $a > 0$; one deduces that

$$\int_{\mathbb{R}^3} e^{-a|w|^2}\,dw = \frac{\pi^{3/2}}{a^{3/2}}, \quad \text{and} \quad \int_{\mathbb{R}^3} |w|^2 e^{-a|w|^2}\,dv = \frac{3\pi^{3/2}}{2a^{5/2}} \text{ for } a > 0. \quad (28.44)$$

One may then write the global Maxwellian distribution as

$$f(\mathbf{v}) = \varrho\frac{a^{3/2}}{\pi^{3/2}}e^{-a|\mathbf{v} - \mathbf{u}|^2}, \quad \text{and it gives } e = \frac{3}{2a}. \quad (28.45)$$

In this model, one has $\sigma_{i,j} = -p\,\delta_{i,j}$ and $p = \frac{2\varrho e}{3}$, and if one uses the relation

$$de + p\,d\left(\frac{1}{\varrho}\right) = T\,ds, \quad (28.46)$$

where T is the absolute temperature and s the entropy per unit of mass, this gives $de - \frac{2e\,d\varrho}{3\varrho} = T\,ds$, so that $\frac{1}{T}$ is an integrating factor of $de - \frac{2e\,d\varrho}{3\varrho}$, and one of these integrating factors is $\frac{1}{e}$, giving $s_0 = \log e - \frac{2}{3}\log \varrho + constant$ (and the other multiplying factors are then of the form $\frac{\varphi(s_0)}{e}$). What BOLTZMANN found is that one may define s by

$$\varrho s = \int_{\mathbb{R}^3} f(v)\log\big(f(v)\big)\,dv. \quad (28.47)$$

Indeed, it is $\int_{\mathbb{R}^3} \varrho\frac{a^{3/2}}{\pi^{3/2}}e^{-a|v - u|^2}\big(\log \varrho + \frac{3}{2}\log a - \frac{3}{2}\log \pi - a|v - u|^2\big)\,dv = \varrho\big(\log \varrho + \frac{3}{2}\log a - \frac{3}{2}\log \pi - \frac{1}{2}\big)$, so that $s = \log \varrho + \frac{3}{2}\log a - \frac{3}{2}\log \pi - \frac{1}{2} = \frac{2}{3}s_0 + constant$. As the unit of temperature was already chosen, the term

$a\,|v-u|^2$ in the exponential is written as $\frac{1}{kT}\frac{|v-u|^2}{2}$, where k is the Boltzmann constant.[5]

Based on the knowledge of equilibrium solutions for the Boltzmann equation, with or without exterior force potentials, BOLTZMANN devised the rules now used in statistical physics; one only considers systems in thermal equilibrium in this framework, and one postulates that the state of a system is indexed by the absolute temperature T, and the rule says that there is a "probability" to find the system in a state of energy W, which is proportional to $exp(-\frac{W}{kT})$. Of course, the basic rule of this game makes no sense but for large systems whose parts are connected enough to interact and settle quickly to a unique temperature.[6]

[Taught on Monday November 12, 2001.]

---

[5] $k = 1.3807\ 10^{-23}$ joule kelvin^{-1}; the joule is the unit for energy, newton metre, or kilogram metre^2 second^{-2}.

[6] Specialists of plasma physics have observed that in their experiments lighter electrons tend to settle quickly to some temperature, while heavier ions tend to settle quickly to another temperature, and their experiments do not last long enough for these two temperatures to come together.

29

The Illner–Shinbrot and the Hamdache Existence Theorems

In the early 1980s, I had asked my student Kamel HAMDACHE to try to extend to the Boltzmann equation the method that I had created for some discrete velocity models in one space dimension, namely use functions satisfying $0 \leq f(\mathbf{x}, \mathbf{v}, t) \leq F(\mathbf{x} - t\mathbf{v}, \mathbf{v})$, and the problem was to discover a good class of functional spaces for the functions F so that a fixed point argument could be used for small initial data.

The first to obtain a result in this direction were Reinhard ILLNER and SHINBROT,[1] who in 1983 treated the case $K(|\mathbf{v} - \mathbf{w}|, \theta) = |\mathbf{v} - \mathbf{w}| \kappa(\theta)$ with κ integrable, corresponding to hard spheres; their choice was to take $F(\mathbf{x}, \mathbf{v}) = e^{-\alpha |\mathbf{x}|^2}$, i.e. Maxwellians in $\mathbf{x}$ (instead of the classical Maxwellians in $\mathbf{v}$), and they proved global existence for small (nonnegative) initial data.

Then Kamel HAMDACHE extended their result by considering $F(\mathbf{x}, \mathbf{v}) = e^{-\alpha |\mathbf{x}|^2} h(\mathbf{v})$ for $h \in L^p$ with $p \neq \infty$; in the case of forces in $\frac{1}{distance^s}$ with angular cut-off, he was able to treat the case of small (nonnegative) initial data and prove global existence for $s > \frac{7}{3}$ (the value of p depending upon s).

In the summer of 1984, at a meeting in Santa Fe, NM, I checked with him that, without the hypothesis of smallness for the (nonnegative) initial data, one can prove a local existence theorem, which requires $2 < s < \infty$ (we did not publish this result).

Then Kamel HAMDACHE extended the method to the case with diffusion in $\mathbf{x}$ or diffusion in $\mathbf{v}$ (the Fokker–Planck equation), in such a way that he could let the diffusion coefficient tend to 0 and recover the results without diffusion, but his solution is more technical in that case and it uses a family of explicit solutions which are exponentials of quadratic functions in $(\mathbf{x}, \mathbf{v})$; he remarked that in the case of the Boltzmann equation, the choice of F is such that $f(\mathbf{x}, \mathbf{v}, t) = F(\mathbf{x} - t\mathbf{v}, \mathbf{v})$ satisfies both $Q(f, f) = 0$ and the Boltzmann

---

[1] Marvin SHINBROT, American-born mathematician, 1928–1987. He had worked in Victoria, British Columbia.

equation, and having changed the linear part in order to include diffusion terms, he had to use a class of explicit solutions of the linear equation.

I shall sketch the basic idea behind the computations of Reinhard ILLNER & SHINBROT and of Kamel HAMDACHE.

One considers an iterative method $f^{(n)} \mapsto f^{(n+1)}$ defined by the equation

$$\frac{\partial f^{(n+1)}}{\partial t} + \mathbf{v} \cdot \frac{\partial f^{(n+1)}}{\partial \mathbf{x}} = \int_{\mathbb{R}^3 \times \mathbb{S}^2} K(|\mathbf{v} - \mathbf{w}|, \theta) \big(f^{(n)}(\mathbf{v}') f^{(n)}(\mathbf{w}') - f^{(n+1)}(\mathbf{v}) f^{(n)}(\mathbf{w}) \big) \, d\mathbf{w} \, d\alpha,$$
(29.1)

with $f^{(n+1)}(\mathbf{x}, \mathbf{v}, 0) = g(\mathbf{x}, \mathbf{v})$, and this method is chosen because if $g \geq 0$ and $f^{(n)} \geq 0$ then one has $f^{(n+1)} \geq 0$; indeed, $f^{(n+1)}$ satisfies a linear equation

$$\frac{\partial f^{(n+1)}}{\partial t} + \mathbf{v} \cdot \frac{\partial f^{(n+1)}}{\partial \mathbf{x}} + a^{(n)} f^{(n+1)} = b^{(n)}, \quad \text{with } f^{(n+1)}(\mathbf{x}, \mathbf{v}, 0) = g(\mathbf{x}, \mathbf{v}),$$
(29.2)

and $g \geq 0$ with $b^{(n)} \geq 0$ implies $f^{(n+1)} \geq 0$; that this is the case if $f^{(n)} \geq 0$ follows from

$$b^{(n)}(\mathbf{x}, \mathbf{v}, t) = \int_{\mathbb{R}^3 \times \mathbb{S}^2} K(|\mathbf{v} - \mathbf{w}|, \theta) f^{(n)}(\mathbf{x}, \mathbf{v}', t) f^{(n)}(\mathbf{x}, \mathbf{w}', t) \, d\mathbf{w} \, d\alpha. \quad (29.3)$$

The sign of $a^{(n)}$ is not so important for proving that $f^{(n+1)} \geq 0$, but it is useful for obtaining an upper bound for $f^{(n+1)}$, and indeed $f^{(n)} \geq 0$ implies $a^{(n)} \geq 0$, because

$$a^{(n)} = \int_{\mathbb{R}^3 \times \mathbb{S}^2} K(|\mathbf{v} - \mathbf{w}|, \theta) f^{(n)}(\mathbf{w}) \, d\mathbf{w} \, d\alpha. \quad (29.4)$$

One deduces that $0 \leq f^{(n+1)} \leq \varphi^{(n+1)}$, where $\varphi^{(n+1)}$ is the solution of

$$\frac{\partial \varphi^{(n+1)}}{\partial t} + \mathbf{v} \cdot \frac{\partial \varphi^{(n+1)}}{\partial \mathbf{x}} = b^{(n)} \quad \text{with } \varphi^{(n+1)}(\mathbf{x}, \mathbf{v}, 0) = g(\mathbf{x}, \mathbf{v}), \quad (29.5)$$

and $\varphi^{(n+1)}$ is given explicitly by

$$\begin{aligned}
\varphi^{(n+1)}(\mathbf{x}, \mathbf{v}, t) &= g(\mathbf{x} - t\,\mathbf{v}, \mathbf{v}) + \int_0^t b^{(n)}(\mathbf{x} - (t-s)\mathbf{v}, \mathbf{v}, s) \, ds \\
&= g(\mathbf{x} - t\,\mathbf{v}, \mathbf{v}) + \int_{\mathbb{R}^3 \times \mathbb{S}^2} K(|\mathbf{v} - \mathbf{w}|, \theta) \\
&\quad \left(\int_0^t f^{(n)}(\mathbf{x} - (t-s)\mathbf{v}, \mathbf{v}', s) f^{(n)}(\mathbf{x} - (t-s)\mathbf{v}, \mathbf{w}', s) \, ds \right) d\mathbf{w} \, d\alpha.
\end{aligned}$$
(29.6)

One wants to find a function F such that if $0 \leq f^{(n)}(\mathbf{x}, \mathbf{v}, t) \leq F(\mathbf{x} - t\,\mathbf{v}, \mathbf{v})$ then one has $0 \leq f^{(n+1)}(\mathbf{x}, \mathbf{v}, t) \leq F(\mathbf{x} - t\,\mathbf{v}, \mathbf{v})$; of course, one will also need the mapping $f^{(n)} \mapsto f^{(n+1)}$ to be a strict contraction in an adapted norm, but that is essentially the same type of estimate which is needed. Of course, it is enough to show that $\varphi^{(n+1)}(\mathbf{x}, \mathbf{v}, t) \leq F(\mathbf{x} - t\,\mathbf{v}, \mathbf{v})$, and because

$$\varphi^{(n+1)}(\mathbf{x}, \mathbf{v}, t) \le g(\mathbf{x} - t\,\mathbf{v}, \mathbf{v}) + \int_{\mathbb{R}^3 \times \mathbb{S}^2} K(|\mathbf{v} - \mathbf{w}|, \theta)$$
$$\left(\int_0^t F(\mathbf{x} - (t-s)s\mathbf{v} - s\,\mathbf{v}', \mathbf{v}')F(\mathbf{x} - (t-s)s\mathbf{v} - s\,\mathbf{w}', \mathbf{w}')\, ds \right) d\mathbf{w}\, d\alpha,$$
$$\tag{29.7}$$

it is enough to find F satisfying

$$g(\omega, \mathbf{v}) + \int_{\mathbb{R}^3 \times \mathbb{S}^2} K(|\mathbf{v} - \mathbf{w}|, \theta)$$
$$\left(\int_0^t F(\omega + s\,(\mathbf{v} - \mathbf{v}'), \mathbf{v}')F(\omega + (\mathbf{v} - s\,\mathbf{w}'), \mathbf{w}')\, ds \right) d\mathbf{w}\, d\alpha \le F(\omega, \mathbf{v}).$$
$$\tag{29.8}$$

One now chooses

$$F(\omega, \mathbf{v}) = e^{-\alpha |\omega|^2} h(\mathbf{v}), \text{ with } \alpha > 0, \tag{29.9}$$

and one notices that

$$F(\omega + s\,(\mathbf{v} - \mathbf{v}'), \mathbf{v}')F(\omega + s\,(\mathbf{v} - \mathbf{w}'), \mathbf{w}') = h(\mathbf{v}')h(\mathbf{w}')e^{-\alpha |\omega|^2} e^{-\alpha |\omega - s\,(\mathbf{v} - \mathbf{w})|^2},$$
$$\tag{29.10}$$

because

$$|\omega + s\,(\mathbf{v} - \mathbf{v}')|^2 + |\omega + (v - w')s|^2 = |\omega|^2 + |\omega - (v - w)s|^2. \tag{29.11}$$

One deduces that

if $g(\omega, \mathbf{v}) \le e^{-\alpha |\omega|^2} g_0(\mathbf{v})$, one must find h such that
$$g_0(\mathbf{v}) + \int_{\mathbb{R}^3 \times \mathbb{S}^2} K(|\mathbf{v} - \mathbf{w}|, \theta) h(\mathbf{v}')h(\mathbf{w}') \left(\int_0^t e^{-\alpha |\omega - s\,(\mathbf{v} - \mathbf{w})|^2}\, ds \right) d\mathbf{w}\, d\alpha \le h(\mathbf{v}).$$
$$\tag{29.12}$$

Then, for a unit vector $\mathbf{e}$ parallel to $\mathbf{v} - \mathbf{w}$, one uses[2]

$$\int_0^t e^{-\alpha |\omega - s\,(\mathbf{v} - \mathbf{w})|^2}\, ds \le \int_0^\infty e^{-\alpha |\omega - s\,(\mathbf{v} - \mathbf{w})|^2}\, ds = \frac{1}{|\mathbf{v} - \mathbf{w}|} \int_0^\infty e^{-\alpha |\omega - s\,\mathbf{e}|^2}\, ds,$$
$$\tag{29.13}$$

(and $+\infty$ if $\mathbf{w} = \mathbf{v}$), and the supremum of $\int_0^\infty e^{-\alpha |\omega - s\,\mathbf{e}|^2}\, ds$ is obtained by letting ω tend to $+\infty$ in the direction of $\mathbf{e}$, and the supremum is $\int_{\mathbb{R}} e^{-\alpha x^2}\, dx = \sqrt{\frac{\pi}{\alpha}}$.

In the case considered by Reinhard ILLNER and SHINBROT, where $K(|\mathbf{v} - \mathbf{w}|, \theta) = |\mathbf{v} - \mathbf{w}|\, k(\theta)$ with k integrable, if $g_0(\mathbf{v}) \le \beta_0$ then one may take $h(\mathbf{v}) = \beta$ with $\beta \le \beta_0 + C\,\beta^2$, with $C = \sqrt{\frac{\pi}{\alpha}} \int_{\mathbb{R}^3 \times \mathbb{S}^2} k(\theta)\, d\mathbf{w}\, d\alpha$.

It is a purely mathematical problem to consider a gas filling out the whole space and to wonder about what happens to such a gas which at time 0 has a finite mass, finite momentum and finite kinetic energy, but an important feature for real gases is that they must be contained,[3] and the boundary conditions are important.

---

[2] If one only looks at local existence, then one also uses $\int_0^t e^{-\alpha |\omega - s\,(\mathbf{v} - \mathbf{w})|^2}\, ds \le t$.

[3] One may consider that the atmosphere around the earth is not contained and that indeed a few particles escape the earth's gravitational field.

For a discrete velocity model like the Broadwell model, restricted to having $x \in (0, L)$, one may consider a purely mathematical question like periodic solutions in space, i.e. $u(L, t) = u(0, t)$ and $v(L, t) = v(0, t)$, which is like considering $(0, L)$ as a circle, but more realistic boundary conditions are $u(0, t) = v(0, t)$ and $u(L, t) = v(L, t)$, which express the fact that particles bounce on the boundary of the interval. In a three-dimensional setting, this is the case of specular reflection, whose expression involves the (exterior) normal $\mathbf{n}$; if $(\mathbf{v}, \mathbf{n}) > 0$ the particle is hitting the boundary, but if $(\mathbf{v}, \mathbf{n}) < 0$ it is coming from the boundary, and the particle hitting the boundary with velocity $\mathbf{v}_{in}$ comes out instantaneously with velocity $\mathbf{v}_{out}$, and conservation of energy gives $|\mathbf{v}_{in}| = |\mathbf{v}_{out}|$ and the change in momentum $\mathbf{v}_{in} - \mathbf{v}_{out}$ is considered to be parallel to $\mathbf{n}$,[4] so that the formula is $\mathbf{v}_{out} = (I - 2\mathbf{n} \otimes \mathbf{n})\mathbf{v}_{in}$, equivalent to $\mathbf{v}_{in} = (I - 2\mathbf{n} \otimes \mathbf{n})\mathbf{v}_{out}$. The boundary condition is then $f(\mathbf{x}, \mathbf{v}, t) = f(\mathbf{x}, (I - 2\mathbf{n} \otimes \mathbf{n})\mathbf{v}, t)$ for all $\mathbf{v}$.

MAXWELL had already imagined another type of boundary condition, that the particles hitting the boundary are first absorbed by the boundary and then are (immediately) re-emitted by the boundary in all directions, according to Lambert's law,[5] and with the distribution in velocity of the Maxwellian distribution corresponding to the temperature of the boundary. Reality seems to be between these two extremes.[6]

[Taught on Wednesday November 14, 2001.]

---

[4] The exchanges of momentum by all these particles hitting the boundary are responsible for the pressure.

[5] Johann Heinrich LAMBERT, French-born mathematician, 1728–1777. He had worked in Berlin, Germany.

[6] At a meeting in Grado, Italy, in 1986, I heard about an experiment which had been done on the space shuttle, for which particles had a very high velocity and arrived all with the same incidence on a plate, and were reflected in various directions; the highest probability of reflection was near the specular reflection, but a few particles were reflected in quite odd directions. The explanation seems to be that particles may enter inside the boundary and interact with the atoms there, and this process might be very sensitive to the velocity of the particles, the angle of incidence, and the nature of the material of which the boundary is made; physicists involve questions of quantum mechanics in these calculations, and one is then reminded that particles are just localized waves anyway; of course, one sees that one should not expect the nonspecular reflections to be instantaneous.

30

The Hilbert Expansion

There is a formal procedure, called the Hilbert expansion, which considers the Boltzmann equation with a small parameter ε, often called the mean free path between collisions. Another parameter is used in bounded domains, the Knudsen number,[1] which is a dimensionless number, the ratio of a characteristic length of the container to the mean free path between collisions. One considers

$$\frac{\partial f}{\partial t} + \mathbf{v} \cdot \frac{\partial f}{\partial \mathbf{x}} = \frac{1}{\varepsilon} Q(f, f),$$ (30.1)

and the Hilbert expansion postulates that

$$f(\mathbf{x}, \mathbf{v}, t) = f_0(\mathbf{x}, \mathbf{v}, t) + \varepsilon f_1(\mathbf{x}, \mathbf{v}, t) + \varepsilon^2 f_2(\mathbf{x}, \mathbf{v}, t) + \ldots,$$ (30.2)

and formally one finds that $Q(f, f) = 0$, so that f is a local Maxwellian, and the macroscopic parameters solve the Euler equation, for an ideal fluid. A variant of this formal procedure, the Chapman–Enskog procedure, makes the Navier–Stokes equation appear (with a small viscosity).

Of course, one should always be careful with formal expansions, because there is no good reason to believe that the solution will appear the way that one postulates, and it may happen that the expansion is valid in some cases but not in others; actually, it is known that there are boundary layer effects to

---

[1] Martin Hans Christian KNUDSEN, Danish Physicist, 1871–1949. He had worked in Copenhagen, Denmark.

consider too, either near the boundary or near the initial time.[2,3] I conjecture that there might be oscillation effects in some cases, whose presence would render the expansion wrong.[4] Anyway, letting ε tend to 0 is only a mathematical question, because the assumptions used for deriving the Boltzmann equation were that the gas was rarefied, and that pairs of particles could interact without being bothered by other particles, i.e. that there were only two-body problems to consider and no n-body problems with $n \geq 3$.

One has

$$Q(f, f) = Q(f_0, f_0) + 2\varepsilon \, B(f_0, f_1) + \varepsilon^2 \big(Q(f_1, f_1) + 2B(f_0, f_2)\big) + \dots, \quad (30.3)$$

where B is the symmetric bilinear mapping defining the quadratic mapping Q. The only term in ε^{-1} in the equation, is $Q(f_0, f_0)$, and one is led to impose that

$$Q(f_0, f_0) = 0, \text{ so that } f_0(\mathbf{x}, \mathbf{v}, t) = \varrho(\mathbf{x}, t)\frac{a(\mathbf{x},t)^{3/2}}{\pi^{3/2}}e^{-a(\mathbf{x},t)\,|\mathbf{v}-\mathbf{u}(\mathbf{x},t)|^2}$$
$$\text{with } a(\mathbf{x}, t) = \frac{1}{2k\,T(\mathbf{x},t)}, \text{ or } e(\mathbf{x}, t) = \frac{3}{2a(\mathbf{x},t)}. \quad (30.4)$$

Looking at the terms in ε^0, one deduces that

$$\frac{\partial f_0}{\partial t} + \mathbf{v}.\frac{\partial f_0}{\partial \mathbf{x}} = 2B(f_0, f_1), \quad (30.5)$$

and the problem is to find an equation for f_0 alone. One observes that, whatever f_1 is, one has

$$\begin{aligned}\int_{\mathbb{R}^3} B(f_0, f_1)\, d\mathbf{v} &= 0 \\ \int_{\mathbb{R}^3} v_i B(f_0, f_1)\, d\mathbf{v} &= 0 \text{ for } i = 1, 2, 3, \\ \int_{\mathbb{R}^3} |v|^2 B(f_0, f_1)\, d\mathbf{v} &= 0,\end{aligned} \quad (30.6)$$

---

[2] One may start from initial data which are not local Maxwellians, and in the Broadwell model it means that $u_0 v_0 - w_0^2 \neq 0$ on a set of positive measure. In that case, the intuition is that, because of the factor $\frac{1}{\varepsilon}$, there is a boundary layer in time where the transport does not play any role (at least for the first term). For the Broadwell model, it means that one studies the ordinary differential equation $\frac{du}{dt} = \frac{dv}{dt} = -\frac{dw}{dt} = -u\,v + w^2$, $u(0) = a \geq 0$, $v(0) = b \geq 0$, $w(0) = c \geq 0$, with an accelerated time, and as $u + v + w = a + b + c$ and $u - v = a - b$, one can solve the system explicitly by a quadrature, but because $u\,\log(u) + v\,\log(v) + 2w\,\log(w)$ is a Lyapunov function which stops decreasing only where $u\,v - w^2 = 0$, one can compute the limit as $t \to \infty$ without writing the solution.

[3] Aleksandr Mikhailovich LYAPUNOV, Russian mathematician, 1857–1918. He had worked in Kharkov and in St Petersburg, Russia, and in Odessa (then in Russia, now in Ukraine).

[4] Some people have shown that the expansion is valid under some assumptions, but if there were oscillations their assumptions would not hold, and it would not mean that their proofs are wrong (i.e. they are not proofs), but that their results may not be applicable in some cases. In other words, these statements say that if there are no problems, then everything is OK, and there are different ways to express the hypothesis that there are no problems.

as these terms are the coefficients of ε in the identities

$$\begin{aligned}
&\int_{\mathbb{R}^3} Q(f,f)\,d\mathbf{v} = 0 \\
&\int_{\mathbb{R}^3} v_i Q(f,f)\,d\mathbf{v} = 0 \text{ for } i = 1,2,3, \\
&\int_{\mathbb{R}^3} |v|^2 Q(f,f)\,d\mathbf{v} = 0.
\end{aligned} \tag{30.7}$$

One then multiplies (30.5) by 1, by v_i and by $|\mathbf{v}|^2$ and one integrates in $\mathbf{v}$, and this gives equations satisfied by the macroscopic quantities ϱ, $\mathbf{u}$, and e; in these equations the pressure p appears, defined by $2\varrho\,e = 3p$, because the fact that f_0 is a function of $|\mathbf{v} - \mathbf{u}|^2$ gives $\sigma_{i,j} = -p\,\delta_{i,j}$ for $i,j = 1,2,3$; one should notice that it also gives the heat flux $\mathbf{q} = 0$. This shows that the macroscopic quantities defined by f_0 satisfy the Euler equation with the equation of state $2\varrho\,e = 3p$, i.e.

$$\frac{\partial \varrho}{\partial t} + \sum_{j=1}^{3} \frac{\partial(\varrho\,u_j)}{\partial x_j} = 0, \tag{30.8}$$

for conservation of mass,

$$\frac{\partial(\varrho\,u_i)}{\partial t} + \sum_{j=1}^{3} \frac{\partial(\varrho\,u_i\,u_j)}{\partial x_j} + \frac{\partial p}{\partial x_i} = 0 \text{ for } i = 1,2,3, \tag{30.9}$$

for the balance of linear momentum and

$$\frac{\partial E}{\partial t} + \sum_{j=1}^{3} \frac{\partial((E+p)u_j)}{\partial x_j} = 0, \tag{30.10}$$

for the balance of energy, where $E = \varrho\left(\frac{|\mathbf{u}|^2}{2} + e\right)$.

The limiting problem is a quasi-linear hyperbolic system of conservation laws, and one knows that discontinuities may happen in finite time for this kind of equation, if the data are too large for example; however, in space dimension > 1 there is a dispersion effect which may win over the nonlinear tendency of creating discontinuities (shocks or contact discontinuities), and there are small smooth data for which the solution exists for all time and stays smooth. For some situation of this kind, Takaaki NISHIDA has shown that the solution of the Boltzmann equation exists for all time and converges as ε tends to 0 to the (smooth) solution of the Euler equation.[5]

Russell CAFLISCH and George PAPANICOLAOU have worked out the analogous result for the (one-dimensional) Broadwell model, but for the finite time

---

[5] In order to give a meaning to such comparisons, one associates to a function f defined on $\mathbb{R}^3 \times \mathbb{R}^3 \times (0,\infty)$ its moments $\varrho = \int_{\mathbb{R}^3} f\,d\mathbf{v}$, $\varrho\,\mathbf{u} = \int_{\mathbb{R}^3} f\,\mathbf{v}\,d\mathbf{v}$, $\frac{\varrho|\mathbf{u}|^2}{2} + \varrho\,e = \int_{\mathbb{R}^3} f\,\frac{|v|^2}{2}\,d\mathbf{v}$, which give macroscopic quantities ϱ, $\mathbf{u}$, e which one may compare to those appearing in the Euler equations, and to three functions ϱ, $\mathbf{u}$, e defined on $\mathbb{R}^3 \times (0,\infty)$ one associates the local Maxwellian f having these characteristics, which one may compare to the one appearing in the Boltzmann equation.

where a solution of the quasi-linear hyperbolic system of conservation laws has a smooth solution. For the case of a Riemann problem giving rise to a single shock solution of the quasi-linear hyperbolic system, Russell CAFLISCH has tried without success to perform the same analysis and show that a solution of the Broadwell model does exist and converges as ε tends to 0 to the discontinuous solution of the quasi-linear hyperbolic system of conservation laws. I have suggested that as ε tends to 0 the sequence of solutions might develop oscillations and might converge only in a weak topology to a different function, in which case some effective equation would have to be discovered and studied.

The Chapman–Enskog procedure is slightly different from the Hilbert expansion, and creates the (compressible) Navier–Stokes equation, where the Cauchy stress tensor is given by

$$\sigma_{i,j} = 2\mu\,\varepsilon_{i,j} - p\,\delta_{i,j}, \text{ with } \varepsilon_{i,j} = \frac{1}{2}\left(\frac{\partial u_i}{\partial x_j} + \frac{\partial u_j}{\partial x_i}\right) \text{ for } i,j = 1,2,3, \quad (30.11)$$

and the viscosity μ is > 0; when μ tends to 0 one formally finds the Euler equation again, but flows with small μ (or high Reynolds number) may show turbulent effects,[6] and an effective equation for turbulent flows is not known; although one should always be careful not to exchange the order of limits without first proving that one is allowed to do so, it lends credence to the possibility of oscillations in the sequence of solutions of the Boltzmann equation when ε tends to 0.

Most mathematicians working on the Navier–Stokes equation nowadays use a simplified incompressible model,[7] following the pioneering work of Jean LERAY in the 1930s, followed in the 1950s by Eberhard HOPF and by Olga LADYZHENSKAYA. In three space dimensions, global existence of smooth solutions for the incompressible Navier–Stokes equation is conjectured,[8] and for

---

[6] Osborne REYNOLDS, Irish-born mathematician, 1842–1912. He had worked in Manchester, England.

[7] The simplification comes from the fact that ϱ and μ are independent of the temperature, and therefore one may solve the equation for u independently of the equation for balance of energy. Incompressibility is expressed by $\varrho = constant$, which implies $div(u) = 0$; the condition $div(u) = 0$ is also true for a mixture of fluids if each one is incompressible but the fluids are not miscible, because $\frac{d\varrho}{dt} = 0$ in this case (as usual, $\frac{d}{dt} = \frac{\partial}{\partial t} + \sum_{j=1}^{3} u_j \frac{\partial}{\partial x_j}$).

[8] Jean LERAY seems to have thought that singularities may appear, and that this was related to turbulent flows, but turbulence has not much to do with regularity (or with letting t tend to ∞ as many deluded mathematicians think), but has been related to fluctuations in velocity at least since REYNOLDS. Jindřich NEČAS told me at some time that he thought that singularities do occur, but later that he was not so sure anymore. I believe that solutions stay smooth, but I insist that it is a mathematical problem without much physical relevance, because the mathematical difficulty is that there could be large gradients that one does not

the Euler equation it was usually thought that singularities would appear in a finite time (but it is not so clear now that it is so); in the early 1980s, Shmuel KANIEL had proposed an approach,[9] which he was not able to follow completely, for proving smoothness of solutions, and one interesting feature (which I was hearing for the first time then) was to create a kinetic equation with equilibria described by rectangular curves instead of Gaussian curves, namely

$$f(\mathbf{x}, \mathbf{v}, t) = a(\mathbf{x}, t) \text{ if } |\mathbf{v} - \mathbf{u}(\mathbf{x}, t)| \leq r(\mathbf{x}, t), \text{ and}$$
$$f(\mathbf{x}, \mathbf{v}, t) = 0 \text{ if } |\mathbf{v} - \mathbf{u}(\mathbf{x}, t)| > r(\mathbf{x}, t); \tag{30.12}$$

one deduces that

$$\varrho = \int_{|\mathbf{v}-\mathbf{u}| \leq r} a \, d\mathbf{v} = \frac{4\pi \, a \, r^3}{3}, \text{ so that } \int_{|\mathbf{v}-\mathbf{u}| \leq r} a \, \mathbf{v} \, d\mathbf{v} = \varrho \, \mathbf{u}$$
$$\varrho e = \int_{|\mathbf{v}-\mathbf{u}| \leq r} \frac{|\mathbf{v}-\mathbf{u}|^2}{2} a \, d\mathbf{v} = \frac{2\pi \, a \, r^5}{5}, \text{ so that } e = \frac{3r^2}{10}. \tag{30.13}$$

This idea was later used with more success in one dimension for proving the existence of some quasi-linear hyperbolic systems of conservation laws, by Pierre-Louis LIONS, Benoît PERTHAME & Eitan TADMOR.[10,11]

[Taught on Friday November 16, 2001.]

Notes on names cited in footnotes for Chapter 30, NEČAS.[12]

---

know how to control, but that would mean a lot of energy dissipated by viscosity, and in a real fluid it would make the temperature increase, and therefore the viscosity would decrease and evacuation of the heat would become easier then, and this realistic scenario (for which the flow may look turbulent) cannot occur in the mathematical problem where the equation of balance of energy has been decoupled, because the viscosity has been chosen to be independent of temperature.

[9] Shmuel KANIEL, Israeli mathematician. He works at The Hebrew University, Jerusalem, Israel.

[10] Benoît PERTHAME, French mathematician. He worked in Orléans, and works now at Université Paris VI (Pierre et Marie CURIE), Paris, France.

[11] Eitan TADMOR, Israeli-born mathematician. He has worked at UCLA (University of California at Los Angeles), Los Angeles, CA, and at University of Maryland, College Park, MD.

[12] Jindřich NEČAS, Czech-born mathematician, 1929–2002. He had worked at Northern Illinois University, De Kalb, IL, and at Charles University, Prague, first in Czechoslovakia, then capital of the Czech Republic.

31

Compactness by Integration

In proving existence for some problems of transport, there is an interesting effect of compactness by integration, called the *averaging lemma*, which was first mentioned to me by Benoît PERTHAME as a question, which he solved afterward,[1] that if a sequence f_n converges weakly to f_∞, and is such that $\frac{\partial f_n}{\partial t} + \mathbf{v} \cdot \frac{\partial f_n}{\partial \mathbf{x}}$ is nice enough, and all f_n are 0 outside a compact set in $\mathbf{v}$, then ϱ_n defined by $\varrho_n(\mathbf{x}, t) = \int f_n(\mathbf{x}, \mathbf{v}, t) \, d\mathbf{v}$ in $\mathbb{R}^N \times \mathbb{R}$ converges strongly to ϱ_∞ defined by $\varrho_\infty = \int f_\infty \, d\mathbf{v}$.

This result cannot be proven with the compensated compactness ideas that I had developed with François MURAT (because they are restricted to partial differential equations with constant coefficients), but it reminded me of a result of Lars HÖRMANDER concerning a class of hypoelliptic operators that he had introduced,[2] because an example of his general theory is that $f \in L^2_{loc}$, $\frac{\partial f}{\partial t} + v \cdot \frac{\partial f}{\partial x} \in L^2_{loc}$ and $\frac{\partial f}{\partial v_i} \in L^2_{loc}$ for $i = 1, \ldots, N$ implies that $f \in H^{1/2}_{loc}$; for the particular class of operators considered by Lars HÖRMANDER, the regularity depends upon the number of levels of commutators that one needs to compute in order to generate derivatives in all directions, and in the example one has $[\frac{\partial}{\partial v_i}, \frac{\partial}{\partial t} + v \cdot \frac{\partial}{\partial x}] = \frac{\partial}{\partial x_i}$ for $i = 1, \ldots, N$.[3]

I thought that the lack of information on the partial derivatives in v was balanced by an integration in v instead, and a precise mathematical result unifying the two types of results was obtained later by Patrick GÉRARD,[4] using

---

[1] It then appeared as a joint work of François GOLSE, Pierre-Louis LIONS, Benoît PERTHAME and Rémi SENTIS.

[2] A (linear) differential operator $P(x, D)$ is hypoelliptic if, when $P(x, D)u = f$ and f is of class C^∞ in an open set ω, then u is necessarily of class C^∞ in ω.

[3] Another example in $\mathbb{R}^2$ is that $u \in L^2_{loc}$, $\frac{\partial u}{\partial x} \in L^2_{loc}$ and $x^m \frac{\partial u}{\partial y} \in L^2_{loc}$ imply $u \in H^{1/(m+1)}_{loc}$ (if m is a nonnegative integer); here one has $[\frac{\partial}{\partial x}, x^m \frac{\partial}{\partial y}] = x^{m-1} \frac{\partial}{\partial y}, \ldots, [\frac{\partial}{\partial x}, x \frac{\partial}{\partial y}] = \frac{\partial}{\partial y}$. Using a partial Fourier transform in y, one easily proves the same result for any nonnegative real m.

[4] Patrick GÉRARD, French mathematician, born in 1961. He works at Université Paris-Sud, Orsay, France.

his *micro-local defect measures*, which are almost the same objects which I had called H-measures (but he defined them independently), the difference being that he had developed his theory for functions with values in a Hilbert space (for applying it to L^2 in the variable v), while the only examples which I had thought of were of finite dimensions. I had actually tried to use my H-measures for proving results of compactness by integration, without success, but what I had tried was different from the idea that Patrick GÉRARD used, and I checked afterward that his line of proof works with H-measures, i.e. it is not necessary to develop a theory for functions with values in infinite-dimensional Hilbert spaces. However, this approach is not good enough for finding a more precise result of Pierre-Louis LIONS, that ϱ belongs to a fractional Sobolev space.

One can avoid the general theory of Lars HÖRMANDER in some cases, like our example with information on f, $\frac{\partial f}{\partial t} + v.\frac{\partial f}{\partial x}, \frac{\partial f}{\partial v_j} \in L^2(\mathbb{R}^N \times \mathbb{R}^N \times \mathbb{R})$ for $j = 1, \ldots, N$, by using a partial Fourier transform.

Lemma 31.1. *If f, $\frac{\partial f}{\partial t} + v.\frac{\partial f}{\partial x}, \frac{\partial f}{\partial v_j} \in L^2(\mathbb{R}^N \times \mathbb{R}^N \times \mathbb{R})$ for $j = 1, \ldots, N$, then $f \in H_{loc}^{1/2}(\mathbb{R}^N \times \mathbb{R}^N \times \mathbb{R})$.*

Proof: Denoting by (ξ, τ) the dual variable of (x, t), one obtains $\mathcal{F}f$, $(\tau + v.\xi)\mathcal{F}f$, $\frac{\partial(\mathcal{F}f)}{\partial v_j} \in L^2$; then for (ξ, τ) fixed one has

$$2\Re \int_{\mathbb{R}^N} (\tau + v.\xi)\overline{\mathcal{F}f}\frac{\partial(\mathcal{F}f)}{\partial v_j}\, dv = \int_{\mathbb{R}^N} (\tau + v.\xi)\frac{\partial(|\mathcal{F}f|^2)}{\partial v_j}\, dv = -\int_{\mathbb{R}^N} \xi_j|\mathcal{F}f|^2\, dv, \tag{31.1}$$

which is true for smooth functions with compact support, and extends by a density argument in our case.[5] Multiplying (31.1) by $sign(\xi_j)$ and integrating in (ξ, τ) gives

$$\int_{\mathbb{R}^N \times \mathbb{R}^N \times \mathbb{R}} |\xi_j|\,|\mathcal{F}f|^2\, d\xi\, d\tau\, dv \leq \frac{1}{\pi}\left\|\frac{\partial f}{\partial t} + v.\frac{\partial f}{\partial x}\right\|_{L^2}\left\|\frac{\partial f}{\partial v_j}\right\|_{L^2} \text{ for } j = 1, \ldots, N. \tag{31.2}$$

Then, one notices that

$$\frac{|\tau|}{1 + |v|} \leq \frac{|\tau + v.\xi|}{1 + |v|} + \frac{|v.\xi|}{1 + |v|} \leq |\tau + v.\xi| + |\xi|, \tag{31.3}$$

so that

$$\int_{\mathbb{R}^N \times \mathbb{R}^N \times \mathbb{R}} \frac{|\tau|}{1 + |v|}|\mathcal{F}f|^2\, d\xi\, d\tau\, dv \leq \int_{\mathbb{R}^N \times \mathbb{R}^N \times \mathbb{R}} (|\tau + v.\xi| + |\xi|)\,|\mathcal{F}f|^2\, d\xi\, d\tau\, dv < \infty. \tag{31.4}$$

---

[5] The truncation step is based on the Lebesgue dominated convergence theorem, and then the regularizing step is done by convolution, noticing that $\tau + v.\xi$ has bounded partial derivatives, as one works on a compact set.

One deduces a bound in $H^{1/2}$ in all variables, if one restricts attention to a bounded set in v. □

Of course, a proof by Fourier transform is restricted to an L^2 framework, and it is useful to consider a different type of proof, valid in an L^p framework for $1 \leq p \leq \infty$; it will also show how commutators appear in a natural way in the theory.

Lemma 31.2. *One denotes* $L_0 = \frac{\partial}{\partial t} + \sum_{j=1}^{N} v_j \frac{\partial}{\partial x_j}$, *and* $L_k = \frac{\partial}{\partial v_k}$ *for* $k = 1, \ldots, N$ *(so that the commutator* $[L_k, L_0] = L_k L_0 - L_0 L_k$ *is* $\frac{\partial}{\partial x_k}$ *for* $k = 1, \ldots, N$*). For* $s \in \mathbb{R}$, *one denotes by* $S_0(s)$ *the group of operators defined by*

$$(S_0(s)f)(x, v, t) = f(x - s\,v, v, t - s) \text{ a.e. in } \mathbb{R}^N \times \mathbb{R}^N \times \mathbb{R}, \quad (31.5)$$

which is a group of isometries in $L^p(\mathbb{R}^N \times \mathbb{R}^N \times \mathbb{R})$, *with infinitesimal generator* L_0, *so that*

$$\|S_0(s)f - f\|_p \leq |s|\,\|L_0 f\|_p \text{ for all } s \in \mathbb{R}, f \in L^p(\mathbb{R}^N \times \mathbb{R}^N \times \mathbb{R}), 1 \leq p \leq \infty.$$
$$(31.6)$$

For $s \in \mathbb{R}$, *one denotes by* $V_k(s)$ *for* $k = 1, \ldots, N$ *the group of operators defined by*

$$(V_k(s)f)(x, v, t) = f(x, v - s\,e_k, t) \text{ a.e. in } \mathbb{R}^N \times \mathbb{R}^N \times \mathbb{R}, \quad (31.7)$$

where $e_1, \ldots, e_N$ *is the canonical basis of* $\mathbb{R}^N$, *which is a group of isometries in* $L^p(\mathbb{R}^N \times \mathbb{R}^N \times \mathbb{R})$, *with infinitesimal generator* L_k, *so that*

$$\|V_k(s)f - f\|_p \leq |s|\,\|L_k f\|_p \text{ for all } s \in \mathbb{R}, f \in L^p(\mathbb{R}^N \times \mathbb{R}^N \times \mathbb{R}), 1 \leq p \leq \infty.$$
$$(31.8)$$

Proof: The fact that they are isometries comes from the fact that the mappings $(x, v, t) \mapsto (x - s\,v, v, t - s)$ as well as $(x, v, t) \mapsto (x, v - s\,e_k, t)$ have Jacobian determinant 1, on $\mathbb{R}^N \times \mathbb{R}^N \times \mathbb{R}$ and for all $s \in \mathbb{R}$. The infinitesimal generator of S_0 is obtained by looking for the limit as $s \to 0$ of $\frac{f - S_0(s)f}{s}$, and for $p = +\infty$ one only asks for this limit to exist for the L^∞ weak $\star$ topology; of course, for $f \in L^1_{loc}(\mathbb{R}^N \times \mathbb{R}^N \times \mathbb{R})$ the limit exists in the sense of distributions and is $L_0 f$. The same remarks hold for the infinitesimal generator of V_k. That (31.6) and (31.8) hold follows from the characterization of the infinitesimal generators and from the fact that one deals with groups of isometries. □

There is a discrete analogue of the commutation relation $\left[\frac{\partial}{\partial v_k}, \frac{\partial}{\partial t} + v.\frac{\partial}{\partial x}\right] = \frac{\partial}{\partial x_k}$.

Lemma 31.3. *For* $s \in \mathbb{R}$, *one denotes by* $X_k(s)$ *for* $k = 1, \ldots, N$ *the group of operators defined by*

$$(X_k(s)f)(x, v, t) = f(x - s\,e_k, v, t) \text{ a.e. in } \mathbb{R}^N \times \mathbb{R}^N \times \mathbb{R}, \quad (31.9)$$

which is a group of isometries in $L^p(\mathbb{R}^N \times \mathbb{R}^N \times \mathbb{R})$, *with infinitesimal generator* $\frac{\partial}{\partial x_k}$. *For* $a, b \in \mathbb{R}$, *and* $1 \leq k \leq N$, *one has*

$$X_k(a\, b) = S_0(-b)\, V_k(-a)\, S_0(b)\, V_k(a) \tag{31.10}$$

Proof: Indeed, let g_1, g_2, g_3, g_4 be defined by $g_1 = V_k(a)f$, $g_2 = S_0(b)g_1$, $g_3 = V_k(-a)g_2$, and $g_4 = S_0(-b)g_3$. One has $g_1(x, v, t) = f(x, v - a\, e_k, t)$, $g_2(x, v, t) = g_1(x - b\, v, v, t - b) = f(x - b\, v, v - a\, e_k, t - b)$, $g_3(x, v, t) = g_2(x, v + a\, e_k, t) = f(x - b\, v - a\, b\, e_k, v, t - b)$, and $g_4(x, v, t) = g_3(x + b\, v, v, t + b) = f(x - a\, b\, e_k, v, t) = (X_k(a\, b)f)(x, v, t)$. □

Then one has the following discrete version of Lemma 31.1.

Lemma 31.4. *Let* $1 \leq p \leq \infty$, $k \in \{1, \dots, N\}$, *and let* $f \in L^p(\mathbb{R}^N \times \mathbb{R}^N \times \mathbb{R})$ *satisfy*

there exists $\alpha \in (0, 1]$ *such that* $\|S_0(s)f - f\|_p \leq A\, |s|^\alpha$ *for all* $s \in \mathbb{R}$,
$$\tag{31.11}$$
there exists $\beta_k \in (0, 1]$ *such that* $\|V_k(s)f - f\|_p \leq B_k\, |s|^{\beta_k}$ *for all* $s \in \mathbb{R}$,
$$\tag{31.12}$$

then one has

$$\|X_k(s)f - f\|_p \leq c(\alpha, \beta_k)\, A^{\beta_k/(\alpha+\beta_k)}\, B^{\alpha/(\alpha+\beta_k)}\, |s|^{\alpha\, \beta_k/(\alpha+\beta_k)} \text{ for all } s \in \mathbb{R}. \tag{31.13}$$

Proof: By (31.10), one has

$$V_k(a)S_0(b)[X_k(a\, b)f - f] = S_0(b)\, V_k(a)f - V_k(a)S_0(b)f, \tag{31.14}$$

and because $V_k(a)$ and $S_0(b)$ are isometries,

$$\|X_k(a\, b)f - f\|_p = \|S_0(b)\, V_k(a)f - V_k(a)S_0(b)f\|_p, \tag{31.15}$$

and then one observes that

$$S_0(b)\, V_k(a)f - V_k(a)S_0(b)f = S_0(b)\, (V_k(a)f - f) - V_k(a)\, (S_0(b)f - f)$$
$$+ \quad S_0(b)f - f + f - V_k(a)f, \tag{31.16}$$

and one deduces that

$$\|X_k(a\, b)f - f\|_p \leq 2\|V_k(a)f - f\|_p + 2\|S_0(b)f - f\|_p \leq 2A\, |a|^\alpha + 2B_k\, |b|^{\beta_k},$$
for all $a, b \in \mathbb{R}$.
$$\tag{31.17}$$

Finally, one minimizes the right-hand side of (31.17) for $a\, b = s$. □

If $\|L_0 f\|_p < \infty$ and $\|L_k f\|_p < \infty$, then one can take $\alpha = \beta_k = 1$, so that $\frac{\alpha\, \beta_k}{\alpha + \beta_k} = \frac{1}{2}$, and (31.13) then corresponds to f having half a derivative, but f actually belongs to an interpolation space slightly larger than $H^{1/2}$ (in x_k).

Using (31.11) and (31.13), one can estimate $||T_s f - f||_{L^p(\mathbb{R}^N \times K \times \mathbb{R})}$ for a compact $K \subset \mathbb{R}^N$, where $(T_s f)(x, v, t) = f(x, v, t - s)$.

Finally, I adapt the argument of Patrick GÉRARD to a simple situation of compactness by integration, but I refer to [18] for the definitions of the terms and properties of H-measures used in the proof.

Lemma 31.5. *Writing functions of (x, t, v), if*

$$f_n \rightharpoonup 0 \text{ in } L^2(\mathbb{R}^N \times \mathbb{R} \times \mathbb{R}^N) \text{ weak and}$$
$$\frac{\partial f_n}{\partial t} + \sum_{j=1}^N v_j \frac{\partial f_n}{\partial x_j} \to 0 \text{ in } H_{loc}^{-1}(\mathbb{R}^N \times \mathbb{R} \times \mathbb{R}^N) \text{ strong,} \qquad (31.18)$$

then defining ϱ_n by

$$\varrho_n(x, t) = \int_{\mathbb{R}^N} f_n(x, v, t) \varphi(v) \, dv \text{ for } \varphi \in L^2(\mathbb{R}^N), x \in \mathbb{R}^N, t \in \mathbb{R}, \qquad (31.19)$$

one has

$$\varrho_n \to 0 \text{ in } L_{loc}^2(\mathbb{R}^N \times \mathbb{R}) \text{ strong.} \qquad (31.20)$$

Proof: It is equivalent to show that for any sequence u_n converging weakly to 0 in $L^2(\mathbb{R}^N \times \mathbb{R})$ and keeping its support compact, the scalar product of ϱ_n and u_n converges to 0, i.e. the scalar product of f_n and $g_n = u_n \otimes \varphi$ converges to 0. If μ is the H-measure of a subsequence (f_m, g_m), it means that one must show that $\mu^{12} = 0$.

Denoting by ξ, τ, ω the dual variables of x, t, v, the localization principle transforms (31.18) into

$$(\tau + (v, \xi))\mu^{11} = (\tau + (v, \xi))\mu^{12} = 0, \qquad (31.21)$$

and because $\frac{\partial g_n}{\partial v_j} \to 0$ in $H_{loc}^{-1}(\mathbb{R}^N \times \mathbb{R} \times \mathbb{R}^N)$ strong, the localization principle implies

$$\omega_j \mu^{21} = \omega_j \mu^{22} = 0 \text{ for } j = 1, \ldots, N. \qquad (31.22)$$

On the support of μ^{12}, one then has $\tau + (v, \xi) = 0$ by (31.21) and $\omega = 0$ by (31.22), so that one cannot have $\xi = 0$, and therefore for each (x, t, ξ, τ) the set of v such that $(x, t, v, \xi, \tau, 0)$ belongs to the support of μ^{12} is included in a hyperplane and thus has Lebesgue measure 0. It remains to show that μ^{12} has an L^1 density in v to deduce by the Fubini theorem that $\mu^{12} = 0$, and this comes from the fact that

$$\mu^{22} = \nu \otimes |\varphi|^2, \qquad (31.23)$$

where ν is the H-measure corresponding to a subsequence of u_m. □

Although the proof of Patrick GÉRARD is a little more general, his method does not seem suitable to deduce some generalizations of Pierre-Louis LIONS, alone or in collaboration with Ron DiPERNA and Yves MEYER.

[Taught on Monday November 19, 2001.]

Notes on names cited in footnotes for Chapter 31, GOLSE,[6] SENTIS.[7]

---

[6] François GOLSE, French mathematician. He works at Université Paris 7 (Denis Diderot), Paris, France.
[7] Rémi SENTIS, French mathematician. He works at CEA (Commissariat à l'Énergie Atomique), France.

32

Wave Front Sets; H-Measures

In the late 1970s, I had developed the method of compensated compactness, partly with François MURAT, and I had used Young measures for explaining in more "classical" terms what it meant,[1] and I was wondering how to introduce a new object with a dual variable ξ to describe the transport of oscillations.[2]

Why was I looking for a dual variable ξ? I agree that I was short sighted, but in the late 1970s, I wanted to find if oscillations were transported in a way similar to the "propagation of singularities" in linear hyperbolic equations which Lars HÖRMANDER and his school were studying. I knew that propagation of singularities is fake physics,[3] and I understood later that it was pushed

---

[1] I thought that the parametrized measures which I had heard about in seminars on "control theory" were a classical concept, and in the summer of 1978 I had paid attention to introducing them in my Heriot–Watt course without any probabilistic language, as it is completely irrelevant to the questions of continuum mechanics and physics that I was interested in. I was the first to use Young measures for questions of partial differential equations, but I must warn the reader that a few have afterward claimed, explicitly or by omitting to mention my contributions, that it was their idea, and as they have unfortunately written a lot of nonsense corresponding to what Young measures are good for, I must say that I have had no part in their unscientific method of misleading students and researchers.

[2] Some authors insist on distinguishing between "oscillations" and "concentration effects", but their reasons are not always very good, and the basic compensated compactness result treats these two questions in a unified way, as do the H-measures, which I introduced ten years after [18].

[3] Light is described by the Maxwell–Heaviside equation, and not by the wave equation, but Lars HÖRMANDER seems to have found it too challenging to develop mathematical tools for the systems of partial differential equations that one encounters in continuum mechanics and in physics. Even if he had never studied much continuum mechanics or physics, and had not felt the difference between the Maxwell–Heaviside equation and the wave equation, he should have known that a ray of light transports energy, and could not then be related to the question

forward for that particular reason,[4] but it was possible that the oscillations for a first-order linear hyperbolic equation, satisfying

$$\sum_{j=1}^{N} b_j \frac{\partial u_n}{\partial x_j} = f_n \tag{32.1}$$

would also involve the associated bicharacteristic rays, defined by

$$\begin{aligned} \frac{dx_j}{dt} &= b_j\big(x(t)\big) = \frac{\partial P}{\partial \xi_j}, j = 1, \dots, N, \text{ with } P(x, \xi) = \sum_{j=1}^{N} b_j(x)\xi_j \\ \frac{d\xi_j}{dt} &= -\frac{\partial P}{\partial x_j}, j = 1, \dots, N, \end{aligned} \tag{32.2}$$

so my first idea was to look for a mathematical object more general than a Young measure, in that it would have a variable ξ, which would play a role in proving results of propagation. I thought of introducing functionals of the form $\int_{\Omega} F\big(x, u_n, grad(u_n)\big)$, with F being positively homogeneous of degree 0 in the last variable, but I did not find much in that direction.

After I had mentioned my idea of adding a ξ variable, George Papanico-laou had mentioned the Wigner transform,[5] which consists in associating to a function u on $\mathbb{R}^N$ the wave function W defined on $\mathbb{R}^N \times \mathbb{R}^N$ by

$$W(x, \xi) = \int_{\mathbb{R}^N} u\Big(x + \frac{y}{2}\Big) \overline{u\Big(x - \frac{y}{2}\Big)} e^{-2i\pi(y,\xi)} \, dy, \tag{32.3}$$

which makes sense for $u \in L^2(\mathbb{R}^N)$, giving $W \in C_b(\mathbb{R}^N \times \mathbb{R}^N)$. If one adds $u \in L^1(\mathbb{R}^N)$, so that W is bounded in ξ with values in $L^1(\mathbb{R}^N)$, one has

$$\int_{\mathbb{R}^N} W(x, \xi) \, dx = |\mathcal{F}u(\xi)|^2, \tag{32.4}$$

and

---

of micro-local regularity which interested him, as his wave front set is a no man's land where one does not study what happens.

[4] It was pure propaganda to call the results of propagation of micro-local regularity "propagation of singularities", and one might be surprised that Lars Hörmander had fallen hostage to that political propaganda. Others before him had fallen hostages to a political propaganda of a different kind, which consisted in brainwashing students into believing that the world is described by differential equations, as if 19th century continuum mechanics and physics had never happened, and no one had understood the difference between ordinary and partial differential equations during the whole 20th century!

[5] Jenő Pál (Eugene Paul) Wigner, Hungarian-born physicist, 1902–1995. He received the Nobel Prize in Physics in 1963, for his contributions to the theory of the atomic nucleus and the elementary particles, particularly through the discovery and application of fundamental symmetry principles, jointly with Maria Goeppert-Mayer and J. Hans D. Jensen. He had worked at Princeton University, Princeton, NJ.

$$\int_{\mathbb{R}^N} W(x,\xi)\,d\xi = |u(x)|^2. \tag{32.5}$$

I did not find a way to use this idea either.

In 1984, I had the idea of a mathematical tool for computing a correction in a problem of homogenization which had shown an unexpected quadratic effect, but I only tried to develop it in 1986 to prove results of small amplitude homogenization, where it served for computing a correction which is quadratic with respect to a small parameter, and there is not yet a general theory for computing the following terms.[6]

After having defined H-measures for that question of small amplitude homogenization, with a variable ξ in the definition, I wondered if this mathematical tool helps in proving propagation results for oscillations and concentration effects for equation (32.1), and indeed it does, with the bicharacteristic rays (32.2) playing a role.

My definition has a vague analogy with Lars HÖRMANDER's definition of the *wave front set* of a distribution T, also called the *essential singular support* of T, whose projection onto the x space is the *singular support* of T which had been defined by Laurent SCHWARTZ as the complement of the largest open set $\omega \subset \Omega$ such that the restriction of T to ω is a C^∞ function.[7] After localizing in $x \in \Omega \subset \mathbb{R}^N$ by considering φT for $\varphi \in C_c^\infty(\Omega)$, Lars HÖRMANDER declares that T is micro-locally regular at (x_0, ξ_0) if $\varphi(x_0) \neq 0$, $\xi_0 \neq 0$ and $\mathcal{F}(\varphi T)$ decays fast in a conic neighbourhood of the direction ξ_0. Then the set of points

---

[6] In a periodic framework, one can describe all the terms, and the formula for the quadratic correction suggests the general formula proven with H-measures [18], which is valid in a general case, but some people are misled by this similarity and do not understand what the mathematics says. If one invents weak convergence (which F. RIESZ did) and then for a continuous periodic function f one considers the sequence u_n defined by $u_n(x) = f(n\,x)$, it is easy to see that u_n converges in $L^\infty(\mathbb{R})$ weak $\star$ to a constant, which is the average of f in a period. If one listens to a physicist postulating a behaviour of a physical quantity to be $f\left(x, \frac{x}{\varepsilon_n}\right)$ where $f(x,y)$ is periodic in y and ε_n is a small characteristic length, one understands easily that the average $\overline{f}(x)$ of $f(x,y)$ in y may serve as a macroscopic value, but it is doubtful that one will invent weak convergence to explain that what this physicist has been doing is to say that $f\left(x, \frac{x}{\varepsilon_n}\right)$ and $\overline{f}(x)$ are very near in a weak topology, without wondering if that weak topology is adapted to the equation that this quantity satisfies (which is why homogenization was not really understood before mathematicians became interested in the question, because one needs a different topology than the classical weak topology!). No one having seen the formula for quadratic corrections in the periodic case had deduced a correct mathematical definition of H-measures [18]. Even now that I have given such a definition, no one has yet understood the definition of a mathematical object that helps calculate the following corrections in a general framework!

[7] If for a nonempty family $\omega_i, i \in I$, of open subsets of Ω the restriction of T to ω_i is a C^∞ function, then using a C^∞ partition of unity one deduces that the restriction of T to the union $\omega = \cup_{i \in I}\omega_i$ is a C^∞ function, hence there exists a largest open set where T is C^∞.

where T is micro-locally regular is open, and the complement of these points is the wave front set of T, which is then closed.

Conversely, I work with a sequence u_n converging weakly in $L^2_{loc}(\Omega)$ to u_∞, and I localize in x by considering $(u_n - u_\infty)\varphi$ for $\varphi \in C_c(\Omega)$, and then I localize in all directions $\xi \neq 0$ by extracting a subsequence $m \to \infty$ such that for every $\psi \in C(\mathbb{S}^{N-1})$ (and every $\varphi \in C_c(\Omega)$) one has

$$\int_{\mathbb{R}^N} \left|\mathcal{F}\big((u_m - u_\infty)\,\varphi\big)\right|^2 \psi\Big(\frac{\xi}{|\xi|}\Big)\, d\xi \to L(\varphi, \psi) \text{ as } m \to \infty, \qquad (32.6)$$

and it is obvious that for each $\varphi \in C_c(\Omega)$ there exists a nonnegative Radon measure $\mu_\varphi \in \mathcal{M}(\mathbb{S}^{N-1})$ such that $L(\varphi, \psi) = \langle \mu_\varphi, \psi \rangle$ for all $\psi \in C(\mathbb{S}^{N-1})$, but the interesting question is the dependence of μ_φ with respect to φ, and I proved that

> there exists $\mu \in \mathcal{M}(\Omega \times \mathbb{S}^{N-1}), \mu \geq 0$, such that $L(\varphi, \psi) = \langle \mu, |\varphi|^2 \otimes \psi \rangle$
> for all $\varphi \in C_c(\Omega), \psi \in C(\mathbb{S}^{N-1})$,

$$(32.7)$$

where μ denotes the H-measure associated to the subsequence u_m. For vector-valued functions, if $U^n \rightharpoonup U^\infty$ in $L^2_{loc}(\Omega; \mathbb{C}^p)$ weak, one can extract a subsequence $m \to \infty$ such that for every $\varphi_1, \varphi_2 \in C_c(\Omega)$ and every $\psi \in C(\mathbb{S}^{N-1})$ one has

$$\int_{\mathbb{R}^N} \mathcal{F}\big((U_j^m - U_j^\infty)\,\varphi_1\big)\, \overline{\mathcal{F}\big((U_k^m - U_k^\infty)\,\varphi_2\big)}\, \psi\Big(\frac{\xi}{|\xi|}\Big)\, d\xi \to L_{j,k}(\varphi_1, \varphi_2, \psi)$$
as $m \to \infty$, for $j, k = 1, \ldots, p$,

$$(32.8)$$

and it is obvious that for each $\varphi_1, \varphi_2 \in C_c(\Omega)$ there exists a complex Radon measure $\mu^{j,k}_{\varphi_1,\varphi_2} \in \mathcal{M}(\mathbb{S}^{N-1})$ such that $L_{j,k}(\varphi_1, \varphi_2, \psi) = \langle \mu^{j,k}_{\varphi_1,\varphi_2}, \psi \rangle$ for all $\psi \in C(\mathbb{S}^{N-1})$, but the interesting question is the dependence of $\mu^{j,k}_{\varphi_1,\varphi_2}$ with respect to φ_1, φ_2, and I proved that

> there exists an Hermitian symmetric nonnegative $\mu = (\mu^{j,k})_{j,k=1,\ldots,p}$,
> $\mu^{j,k} \in \mathcal{M}(\Omega \times \mathbb{S}^{N-1}), j, k = 1, \ldots, p$, such that
> $L_{j,k}(\varphi_1, \varphi_2, \psi) = \langle \mu^{j,k}, \varphi_1\overline{\varphi_2} \otimes \psi \rangle$
> for all $\varphi_1, \varphi_2 \in C_c(\Omega), \psi \in C(\mathbb{S}^{N-1})$,

$$(32.9)$$

where μ denotes the H-measure associated to the subsequence U^m.[8] In constructing my theory of H-measures [18], I wanted to avoid the regularity hypotheses which Joseph KOHN and Louis NIRENBERG had chosen for their theory of pseudo-differential operators,[9] and which Lars HÖRMANDER has also used for his theory of Fourier integral operators, because they are not

---

[8] Charles HERMITE, French mathematician, 1822–1901. He had worked in Paris, France.

[9] Joseph John KOHN, Czech-born mathematician, born in 1932. He works at Princeton University, Princeton, NJ.

adapted to problems from continuum mechanics or physics, where interfaces occur and discontinuous coefficients appear, and I developed a calculus of "pseudo-differential" operators with minimal regularity hypotheses.[10] An important property, from which the compensated compactness theorem follows, is what I called the *localization principle*, where for continuous coefficients $A_{j,k}, j = 1, \ldots, N, k = 1, \ldots, p$, I proved that

if $\sum_{j=1}^{N} \sum_{k=1}^{p} \frac{\partial(A_{j,k}U_k^m)}{\partial x_j}$ belongs to a compact of $H_{loc}^{-1}(\Omega)$ strong, then $\sum_{j=1}^{N} \sum_{k=1}^{p} \xi_j A_{j,k} \mu^{k,\ell} = 0$ for $\ell = 1, \ldots, p$.

$$(32.10)$$

If T satisfies $\sum_j b_j \frac{\partial T}{\partial x_j} = g$ with $b_1, \ldots, b_n, g \in C^\infty(\Omega)$, Lars HÖRMANDER proves that the wave front set of T is included in the zero set of P defined in (32.2), using an argument related to the stationary phase principle. Using a first commutation lemma (that a commutator is compact),[11] I prove that if (32.1) holds with $b_1, \ldots, b_N$ of class C^1 and f_n belonging to a compact of H_{loc}^{-1} strong, then the support of μ is included in the zero set of P.

Assuming that the b_j are real, Lars HÖRMANDER proves (using his theory of Fourier integral operators) that micro-local regularity for T is propagated along the bicharacteristic rays defined by (32.2), so that the wave front set of T is a union of bicharacteristic rays. Using a second commutation lemma (and a result of Alberto CALDERÓN for avoiding more than C^1 regularity on $b_1, \ldots, b_N$), assuming that f_n converges in L_{loc}^2 strong,[12] I prove that μ satisfies an homogeneous first-order partial differential equation in x and ξ, whose characteristic curves are related to the bicharacteristic rays defined by (32.2).[13]

---

[10] Because I deal with Radon measures, I use continuous test functions, and some care must be taken for the case of coefficients in L^∞, but one must pay attention that the transport properties use C^1 coefficients, therefore refraction effects at interfaces cannot be studied yet.

[11] In proving the existence of H-measures, I also used Laurent SCHWARTZ's kernel theorem in order to prove that a distribution kernel exists and then that it is a nonnegative measure by a positivity argument, another much simpler remark of Laurent SCHWARTZ. Jacques-Louis LIONS had told me once that he had written a simple proof of the kernel theorem with Lars GÅRDING, which I then read, so that I knew that I could avoid sending my readers to the initial proof, but I even simplified the argument a little more so that I only used classical results in functional analysis, that one teaches with Hilbert–Schmidt operators, but I did not explain that when I wrote [18].

[12] If f_n converges in L_{loc}^2 weak, one may need to extract another subsequence, and the first-order equation has a source term, related to the H-measure for the pair (u_n, f_n).

[13] The equations for bicharacteristic rays are not really defined on $\Omega \times \mathbb{S}^{N-1}$, and $\mathbb{S}^{N-1}$ should be replaced by the quotient space obtained from $\mathbb{R}^N \setminus \{0\}$ by identifying half lines. One can enforce $\xi \in \mathbb{S}^{N-1}$ by replacing the second line of (32.2)

Pseudo-differential operators were introduced (by Joseph KOHN and Louis NIRENBERG) for questions concerning elliptic operators,[14] and the mapping which to initial data associates the solution at time t of the wave equation, for example, is not given by a pseudo-differential operator, and Lars HÖRMANDER introduced the larger class of Fourier integral operators for working with questions of linear hyperbolic equations, and his approach works for the scalar wave equation (with C^∞ coefficients), but I do not think that it applies to systems with smooth coefficients if they cannot be reduced to scalar equations.[15]

It may seem a miracle then that I was able to deal with the propagation of oscillations and concentration effects for a large class of linear hyperbolic systems with C^1 coefficients (the wave equation, the Maxwell–Heaviside equation, the linearized elasticity equation), by using only H-measures [18], which mimic methods from pseudo-differential operators.

[Taught on Monday November 26, 2001 (Wednesday 21 and Friday 23 fell during Thanksgiving recess).]

Notes on names cited in footnotes for Chapter 32, GOEPPERT-MAYER,[16] J.H. JENSEN,[17] SCHMIDT.[18]

---

by $\frac{d\xi_j}{ds} = -\frac{\partial P}{\partial x_j} + \xi_j \left(\sum_{k=1}^N \xi_k \frac{\partial P}{\partial x_k} \right)$, for $j = 1, \ldots, N$. It is actually useful to distinguish ξ and $-\xi$, although for real sequences the H-measures charges in the same way ξ and $-\xi$, and this has consequences which physicists know, that one needs nonlinearity to send a beam of light in one direction without sending the same amount of energy in the opposite direction (one puts a light bulb at the focus of a parabola to send a beam in one direction, but the parabola must be a mirror to reflect forward the energy from the light bulb which is sent backward, and the nonlinearity comes from what happens inside the mirror).

[14] They are linked to singular integrals, but the specialists of harmonic analysis who had specialized on questions of singular integrals had not created a calculus where the symbols of the operators play an important role.

[15] I first heard Lars HÖRMANDER talk at a conference in Jerusalem, Israel, in the summer of 1972, and I understood that he had introduced these ideas as an attempt to characterize lacunas, i.e. describe the precise support of the elementary solution of a linear hyperbolic equation with constant coefficients, and he could at least say what the singular support of the elementary solution is.

[16] Maria GOEPPERT-MAYER, German-born physicist, 1906–1972. She received the Nobel Prize in Physics in 1963, with J. Hans D. JENSEN, for their discoveries concerning nuclear shell structure, jointly with Eugene P. WIGNER. She had worked in Chicago, IL, and at USCD (University of California San Diego), La Jolla, CA.

[17] J. Hans D. JENSEN, German physicist. He received the Nobel Prize in Physics in 1963, with Maria GOEPPERT-MAYER, for their discoveries concerning nuclear shell structure, jointly with Eugene P. WIGNER. He had worked in Hannover, and in Heidelberg, Germany.

[18] Erhard SCHMIDT, German mathematician, 1876–1959. He had worked in Bonn, Germany, in Zürich, Switzerland, in Erlangen, Germany, in Breslau (then in Germany, now Wrocław, Poland), and in Berlin, Germany.

33

H-Measures and "Idealized Particles"

For a general scalar wave equation

$$\frac{\partial}{\partial t}\left(\varrho\frac{\partial u_n}{\partial t}\right) - \sum_{i,j=1}^{N}\frac{\partial}{\partial x_i}\left(a_{i,j}\frac{\partial u_n}{\partial x_j}\right) = f_n \text{ in } \Omega \times (0,T), \qquad (33.1)$$

with continuous coefficients, one assumes that $u_n \rightharpoonup u_\infty$ in $H^1_{loc}(\Omega \times (0,T))$ weak, and one applies the theory to $U^n = grad_{t,x}(u_n)$, i.e. $p = N+1$, and one denotes $x_0 = t$. For a subsequence U^m defining an H-measure μ, the localization principle (using $curl(U^m) = 0$) gives $\mu_{j,k} = \xi_j\xi_k\nu$, for $j,k = 0,\ldots,N$ for a nonnegative Radon measure ν. Then, if f_m stays in a compact of $H^{-1}_{loc}(\Omega \times (0,T))$ strong, the localization principle (now using (33.1)) gives

$$Q\nu = 0, \text{ with } Q = \varrho\,\xi_0^2 - \sum_{i,j=1}^{N} a_{i,j}\xi_i\xi_j. \qquad (33.2)$$

Then, if one assumes that f_m converges in $L^2_{loc}(\Omega \times (0,T))$ strong, and that one really has a wave equation, i.e. the coefficients are independent of t, ϱ is real and a is real and symmetric,[1] and there exist $\alpha, \beta \in (0,\infty)$ with $a \geq \alpha\,I$ and $\varrho \geq \beta$ a.e. in Ω, one deduces that ν satisfies a partial differential equation in $(x_0,\ldots,x_N)$ and $(\xi_0,\ldots,\xi_N)$, written in weak form as

$$\langle \nu, \{Q,\Phi\}\rangle = 0 \text{ for all } \Phi \in C^1_c(\Omega \times (0,T) \times \mathbb{S}^N), \qquad (33.3)$$

where the Poisson bracket of two functions in (x,ξ) is defined by

$$\{g,h\} = \sum_{j=0}^{N}\left(\frac{\partial g}{\partial \xi_j}\frac{\partial h}{\partial x_j} - \frac{\partial g}{\partial x_j}\frac{\partial h}{\partial \xi_j}\right), \qquad (33.4)$$

---

[1] One could allow a to have complex entries and be Hermitian symmetric, but complex coefficients are not so natural for the wave equation, while they do appear naturally for the Dirac equation.

so that the characteristic curves of the first-order equation (33.3) satisfied by ν are the bicharacteristic rays associated to Q, i.e.

$$\begin{aligned} \frac{dx_j}{ds} &= \frac{\partial Q}{\partial \xi_j}, j = 0, 1, \ldots, N, \\ \frac{d\xi_j}{ds} &= -\frac{\partial Q}{\partial x_j}, j = 0, 1, \ldots, N. \end{aligned} \qquad (33.5)$$

If $(x(s), \xi(s))$ is a solution of (33.5), then for $\lambda \neq 0$, $(x(\lambda s), \lambda \xi(\lambda s))$ is also a solution, so that (33.5) induces a differential equation on the quotient space mentioned.[2] Patrick GÉRARD has pointed out to me that if ϱ and $a_{i,j}$ are only of class C^1 for $i, j = 1, \ldots, N$, then $\frac{\partial Q}{\partial x_j}$ are only continuous in x for $j = 1, \ldots, N$, and uniqueness of solutions of (33.5) might not hold.

Conservation of energy for (33.1) is

$$\frac{\partial}{\partial t}\left(\frac{\varrho}{2}\left|\frac{\partial u_n}{\partial t}\right|^2 + \sum_{i,j=1}^{N} \frac{a_{i,j}}{2}\frac{\partial u_n}{\partial x_i}\frac{\partial u_n}{\partial x_j}\right) - \sum_{i,j=1}^{N} \frac{\partial}{\partial x_i}\left(a_{i,j}\frac{\partial u_n}{\partial x_j}\frac{\partial u_n}{\partial t}\right) = f_n \frac{\partial u_n}{\partial t}$$
in $\Omega \times (0, T)$,

$$(33.6)$$

so that the difference between the limit of $\frac{\varrho}{2}\left|\frac{\partial u_m}{\partial t}\right|^2 + \sum_{i,j=1}^{N}\frac{a_{i,j}}{2}\frac{\partial u_m}{\partial x_i}\frac{\partial u_m}{\partial x_j}$ and

$\frac{\varrho}{2}\left|\frac{\partial u_\infty}{\partial t}\right|^2 + \sum_{i,j=1}^{N}\frac{a_{i,j}}{2}\frac{\partial u_\infty}{\partial x_i}\frac{\partial u_\infty}{\partial x_j}$ corresponds to a part of the energy hidden at a mesoscopic level, i.e. a form of internal energy, which is then

$$\text{internal energy} = \int_{\mathbb{S}^N}\left(\frac{\varrho\,\xi_0^2}{2} + \sum_{i,j=1}^{N}\frac{a_{i,j}\xi_i\xi_j}{2}\right)d\nu(x,\xi), \qquad (33.7)$$

and because $Q\nu = 0$ there is *equipartition of energy*,[3] i.e. half the internal energy has a kinetic origin, $\int_{\mathbb{S}^N}\frac{\varrho\,\xi_0^2}{2}d\nu(x,\xi)$, which is the limit of $\frac{\varrho}{2}\left|\frac{\partial u_n}{\partial t}\right|^2 - \frac{\varrho}{2}\left|\frac{\partial u_\infty}{\partial t}\right|^2$, and half the internal energy has a potential origin, $\int_{\mathbb{S}^N}\left(\sum_{i,j=1}^{N}\frac{a_{i,j}\xi_i\xi_j}{2}\right)d\nu(x,\xi)$, which is the limit of $\sum_{i,j=1}^{N}\frac{a_{i,j}}{2}\frac{\partial u_n}{\partial x_i}\frac{\partial u_n}{\partial x_j} - \sum_{i,j=1}^{N}\frac{a_{i,j}}{2}\frac{\partial u_\infty}{\partial x_i}\frac{\partial u_\infty}{\partial x_j}$. One should observe that it is not the internal energy that satisfies a partial differential equation in x (and it means also time, which is x_0), but another object, linked to an H-measure of the sequence, which does satisfy an equation in (x, ξ).

It is important to observe that this transport equation has not been postulated like all the equations from kinetic theory, and it has been deduced from

---

[2] It means that one can enforce $\xi \in \mathbb{S}^N$ by replacing the second line of (33.5) by $\frac{d\xi_j}{ds} = -\frac{\partial Q}{\partial x_j} + \xi_j\left(\sum_{k=0}^{N}\xi_k\frac{\partial Q}{\partial x_k}\right)$, for $j = 0, 1, \ldots, N$.

[3] Oscillating solutions of the Maxwell–Heaviside equation show another form of *equipartition of energy*, because $(D^n.E^n) - (B^n.H^n) \rightharpoonup (D^\infty.E^\infty) - (B^\infty.H^\infty)$ (i.e. the *action* is a robust quantity), but the density of electromagnetic energy is $\frac{1}{2}((D^n.E^n) + (B^n.H^n))$, so that for the electromagnetic energy which is hidden at a mesoscopic level, half has electric origin, the limit of $\frac{1}{2}((D^n.E^n) - (D^\infty.E^\infty))$, and half has magnetic origin, the limit of $\frac{1}{2}((B^n.H^n) - (B^\infty.H^\infty))$.

a balance law. The explanation is that some part of a conserved quantity may hide itself at a mesoscopic level, and because of the linearity of the equation a complete analysis was possible, and all kinds of ways of hiding energy at a mesoscopic level have been automatically taken into account. Of course, it was important that only oscillating solutions compatible with (33.1) were considered: if a guess or probabilities had been used, it would have made the result doubtful.

In the case $\varrho = 1$ and $a_{i,j} = c^2 \delta_{i,j}$, one has $Q = \xi_0^2 - c^2 \left(\sum_{i,j=1}^N \xi_j^2 \right)$, so that ν lives on $\xi_0^2 = c^2 \left(\sum_{i,j=1}^N \xi_j^2 \right)$, and one can parametrize those points on $\mathbb{S}^N$ by choosing $\eta \in \mathbb{S}^{N-1}$ and then having $\xi_0 = \frac{\pm c}{\sqrt{1+c^2}}$ and $\xi_j = \frac{\eta_j}{\sqrt{1+c^2}}$ for $j = 1, \ldots, N$ (because Q is independent of x one has $\frac{d\xi}{ds} = 0$, so that ξ stays on $\mathbb{S}^N$). One has $\{Q, \Phi\} = 2\xi_0 \frac{\partial \Phi}{\partial t} - 2c^2 \sum_{j=1}^N \xi_j \frac{\partial \Phi}{\partial x_j} = \frac{2c}{\sqrt{1+c^2}} \left(\pm \frac{\partial \Phi}{\partial t} - c \sum_{j=1}^N \eta_j \frac{\partial \Phi}{\partial x_j} \right)$, so that the equation for ν corresponds to a transport with velocity c in the direction $\mp \eta$. One could say that the energy hidden at a mesoscopic level is transported by "idealized particles", moving in all directions with velocity c, and because the equation is linear, these "idealized particles" do not interact when they go through the same point with different directions.

One might be tempted to call these "idealized particles" photons, but there is of course no possible quantification $h\nu$, and because H-measures do not use any characteristic length they cannot distinguish between different frequencies, so that if there were photons, the H-measure would only see the total energy of all the photons moving in a given direction, for all frequencies (supposed to be very large). Actually, it is still not clear to me what are these photons that physicists mention,[4] but they cannot be properties of the wave equation or of the Maxwell–Heaviside equation, which are linear and do not contain the Planck constant h in their coefficients, and my conjecture is that they are related to the coupling of the Maxwell–Heaviside equation and the Dirac equation, in the way DIRAC had proposed but without the zero-order term containing the mass of the electron; in this coupled equation the density of charge ϱ and the density of current j are expressed in terms of $\psi \in \mathbb{C}^4$ which describes matter, and the equation for ψ has a coupling term with a coefficient in $\frac{1}{h}$, which is linear in ψ, and linear in the scalar potential V and in the vector potential A.[5]

---

[4] I find appealing the proposition of BOSTICK concerning electrons, but I cannot guess what his proposition concerning photons means.

[5] Photons seem to result from interaction of light and matter, but as I have not been able yet to develop a theory valid for semi-linear hyperbolic systems, I conjecture that for oscillating solutions with large frequency ν of the Maxwell–Heaviside/Dirac coupled equation, the only possible transfers of energy between the electromagnetic field and the matter field described by ψ are (almost) multiples of $h\nu$. In some way, I think about photons in the way gusts of wind are just a particular type of solution of the equations of hydrodynamics, and no one thinks of explaining laminar flows as a superposition of small gusts of wind.

I find a sign that EINSTEIN had not understood the ideas of POINCARÉ about relativity, is that he seemed to believe in forces at a distance in having proposed a quite impossible scenario where light rays are bent near the sun because of the mass of the sun, and if he had understood that the Maxwell–Heaviside equation is hyperbolic (as is the wave equation, which does not really describe light!), he would have known that a light ray only feels the local properties of matter along its way (although he could not know much about what is going on near the surface of the sun, as it is not so clear that we know enough about that now), but he should certainly have thought about two other phenomena concerning light. The first one is about mirages, which correspond to objects hidden behind the horizon, where it is not the mass of the earth which plays a role, but the Brillouin effect,[6] that the index of refraction of air depends upon its temperature. The second one is about a computation by AIRY,[7] who had wondered why the solution of the wave equation is not zero in the shadow of an obstacle, i.e. the shadow only exists in the approximation of geometrical optics, but there what happens is not so clear, because it was only in the 1950s that Joseph KELLER developed his *geometric theory of diffraction* (GTD),[8] where in guessing how grazing rays follow the geodesics of the boundary, he had taken into account some explicit computations made in the same spirit as AIRY. Although Joseph KELLER had mentioned early on that his expansions are not good near caustics, it is still not really understood why they give good results away from caustics, but he had mentioned something else to me more recently (in the fall of 1990 in Stanford, I think), that the phenomenon of grazing rays which he had studied is similar to the tunnelling effect in quantum mechanics, and I consider that a good possibility to avoid the probabilistic ideas that physicists use for this question, but I have not been able to find a way to explain his computations; however, after discussing this question with Michael VOGELIUS (in the summer of 2005, in Grenoble, France), I had the feeling that one might explain his computation by the existence of a boundary layer with a width of order $\nu^{-1/3}$ in places with a finite radius of curvature (for large frequencies ν).

As there is not yet a generalization of my theory of H-measures to semi-linear hyperbolic systems, there are still guesses of quantum mechanics that cannot be explained in a rational way, but certainly the proof of transport theorems for H-measures has already shown a crucial error of quantum

---

[6] Léon BRILLOUIN, French physicist, 1889–1969. He had worked in Paris, France.

[7] George Biddell AIRY, English mathematician, 1801–1892. He had worked in Greenwich, England, as the seventh Astronomer Royal.

[8] Joseph Bishop KELLER, American mathematician, born in 1923. He received the Wolf Prize for 1996/97, for his innovative contributions, in particular to electromagnetic, optical, acoustic wave propagation and to fluid, solid, quantum and statistical mechanics, jointly with Yakov G. SINAI. He worked at NYU (New York University), New York, NY, and at Stanford University, Stanford, CA.

mechanics: there are no particles playing esoteric games, there are only waves, but waves may hide conserved quantities like energy and momentum at various mesoscopic levels, and one needs effective equations for describing the macroscopic effects of these hidden quantities. Of course, thermodynamics was only a first guess for that question, and this theory should be generalized in view of the new understanding which came out of the new mathematical tools which I developed in the last quarter of the 20th century for understanding this question which I call *beyond partial differential equations*, which is for me the key to understanding the continuum mechanics and the physics of the 20th century, plasticity, turbulence, and atomic physics.

It was a mistake to start from ordinary differential equations (with Hamiltonian structure) and to deduce partial differential equations of Schrödinger type, and before H-measures it was already clear that Schrödinger equations are simplified models where one has let the velocity of light c tend to ∞,[9] but after H-measures it is clear that one should start from partial differential equations, preferably of hyperbolic type, or of an intermediate type that one would have proven to be natural (i.e. without postulating it), and one should derive effective equations, without postulating either that they correspond to an ordinary differential system of Hamiltonian type!

[Taught on Wednesday November 28, 2001.]

Notes on names cited in footnotes for Chapter 33, SINAI.[10]

---

[9] In the spring of 1983, while I visited MSRI (Mathematical Sciences Research Institute) in Berkeley, CA, I had stumbled upon an article from a physics journal where one started from the Dirac equation and one deduced the Schrödinger equation by letting c tend to ∞, but I do not know if everything was proven there, because I was not able to find that article again when I looked for its reference a few years after.

[10] Yakov Grigor'evich SINAI, Russian-born mathematician, born in 1935. He received the Wolf Prize for 1996/97, for his fundamental contributions to mathematically rigorous methods in statistical mechanics and the ergodic theory of dynamical systems and their applications in physics, jointly with Joseph B. KELLER. He works at Princeton University, Princeton, NJ.

34

Variants of H-Measures

Homogenization, in the way I had developed it with François MURAT in the early 1970s, has no small characteristic length in it (as the periodically modulated framework is only a particular example, which too many only consider, as if they could not understand the general case), and no probabilities (which is one of the diseases which has plagued 20th century sciences, for which I am trying to find a cure), and it was natural that I should first develop H-measures, which use no characteristic length, because the test functions ψ are homogeneous of degree zero.

From my proof of a transport theorem for the H-measures associated to solutions of (32.1), I saw how to generalize it to a large class of systems, those admitting a sesqui-linear balance law for their complex solutions,[1] and the first example of the wave equation (33.1) led to the transport equation (33.3), and what it says is that in the limit of infinite frequencies the rules of geometrical optics apply to all solutions of (33.1), and it is worth pointing out that this is not what the usual understanding of geometrical optics is about.

The classical geometrical optics approach to the wave equation[2]

$$\frac{\partial^2 u}{\partial t^2} - c^2 \Delta u = 0, \tag{34.1}$$

is to construct asymptotic solutions of the form

$$u(x,t) = A\, e^{i\nu\varphi}, \tag{34.2}$$

---

[1] Sesqui is a prefix meaning one and a half, as the antilinearity is counted as half.

[2] The transport result for H-measures applies to many hyperbolic equations or systems (like the Maxwell–Heaviside equation, the equation of linearized elasticity, or the Dirac equation), but geometrical optics only seems to apply to the wave equation, apart from homogeneous isotropic media, where the components of solutions of the Maxwell–Heaviside equation or of the Lamé equation satisfy scalar wave equations.

where the amplitude $A(x, t)$ and the phase $\varphi(x, t)$ have an expansion in terms of the large frequency ν,

$$A = A^0 + \frac{A^1}{\nu} + \ldots; \quad \varphi = \varphi^0 + \frac{\varphi^1}{\nu} + \ldots. \tag{34.3}$$

Putting (34.2) in (34.1), and identifying the coefficients of ν^2 gives an Hamilton–Jacobi equation for φ^0 (called the *eikonal equation*)

$$|\varphi_t^0|^2 = c^2 |grad_x(\varphi^0)|^2, \tag{34.4}$$

whose solution stops being smooth at caustics, and then identifying the coefficients of ν gives a linear transport equation for A^0:

$$A_t^0 \varphi_t^0 - c^2 \big(grad_x(A^0), grad_x(\varphi^0)\big) + \frac{\varphi_{tt}^0 - c^2 \Delta \varphi^0}{2} A^0 = 0. \tag{34.5}$$

Based on similar considerations, it has been guessed that energy is transported along bicharacteristic rays, but if one looks at what has been done, one sees that at best, i.e. if one estimates all the coefficients A_j and φ_j and one proves convergence of the series (34.3) in some domain, one has only proven that it is true for a particular type of solutions, and away from caustics. For what happens at caustics, MASLOV has proposed a formal expansion,[3] which I think predicts a jump of $\frac{\pi}{2}$ for the phase when one crosses caustics.[4]

   Conversely, what my theorem with H-measures says is that for all solutions of the wave equation, in the limit of infinite frequencies where some energy may be hidden at a mesoscopic level, this energy is transported along bicharacteristic rays and the amounts moving in various directions are taken into account by a new variable ξ, which is related to the direction of the gradient of the phase in the particular case considered by geometrical optics, and there is no difficulty in having waves moving in infinitely many directions at the same time because the Radon measure ν takes care of recording how much energy moves in each direction. Geometrical optics gets in trouble at caustics because it is designed to follow just one distorted plane wave, and caustics are precisely the points where one needs to consider plane waves arriving with slightly different directions. Actually, a difficulty appears at caustics for the H-measures, if one wants to study the regularity of its density in ξ, because

---

[3] Victor P. MASLOV, Russian mathematician. He works in Moscow, Russia.

[4] Because Jean LERAY had written a one page preface to the French translation (by LASCOUX) of a book by MASLOV, I asked him (in the early 1990s, I think) if my interpretation was right, and I was surprised by his answer. Although he had written in his preface that Lars HÖRMANDER's theory of Fourier integral operators is the wrong thing and that MASLOV was looking at the right question, he only answered to me that what MASLOV had done is formal, i.e. it is not mathematics. Of course, I knew this, but I had asked that question to Jean LERAY because I thought that he had read the book, and that he could then tell me what MASLOV was conjecturing.

it is precisely at caustics that a limitation of the regularity occurs, and one then sees the advantage of the weak formulation (33.3).

Having learnt more continuum mechanics or physics than most mathematicians, I have difficulty in being interested in oversimplified physical situations, and if I have to consider an oversimplified model, for example one which uses only one characteristic length, I usually warn about the limitations of such questions. However, for a simple question of showing one limitation of H-measures and how to overcome it, I had proposed a way to introduce one characteristic length ε_n (tending to 0), by adding one variable x_{N+1} for introducing the sequence

$$V^n(x_1, \ldots, x_N, x_{N+1}) = U^n(x_1, \ldots, x_N) \cos \frac{x_{N+1}}{\varepsilon_n}, \qquad (34.6)$$

and then by considering the H-measure $\mu \in \mathcal{M}(\Omega \times \mathbb{R} \times \mathbb{S}^N)$ for the sequence V^n; actually, because μ is independent of the variable x_{N+1}, it really corresponds to an element of $\mathcal{M}(\Omega \times \mathbb{S}^N)$. Shortly after, I learnt that Patrick GÉRARD had already made a more elaborate proposal where, assuming u^n scalar as a simplification, he considered a subsequence for which

$$\lim_{m \to \infty} \int_{\mathbb{R}^N} |\mathcal{F}(\varphi\, u^m)|^2 \psi(\varepsilon_m \xi)\, d\xi = \langle \mu_{sc}, |\varphi|^2 \otimes \psi \rangle,$$
$$\text{for all } \varphi \in C_c^\infty(\mathbb{R}^N), \psi \in \mathcal{S}(\mathbb{R}^N), \qquad (34.7)$$

and he called $\mu_{sc} \in \mathcal{M}(\Omega \times \mathbb{R}^N)$ the *semi-classical measure* associated to the subsequence (because some examples that he had in mind are related to questions that physicists call semi-classical). For technical reasons he used $\mathcal{F}(\varphi\, u^m)$ and not $\mathcal{F}(\varphi\, (u^m - u^\infty))$ in his definition, and his regularity hypothesis on ψ has the reason that he had in mind a more general localization principle, using higher-order derivatives (multiplied by the correct power of ε_n). Although rather different, our two definitions are actually quite related, and if my choice of $\mathbb{S}^N$ as a quotient of $\mathbb{R}^{N+1}$ is not so good, his definition consists in using $\{x \in \mathbb{R}^{N+1} \mid x_{N+1} = 1\}$, which misses what my definition puts on the equator $\{x \in \mathbb{S}^N \mid x_{N+1} = 0\}$, and this defect is related to having chosen test functions ψ which vanish at ∞, but another defect is to have chosen test functions ψ which are continuous at 0, and my approach has this defect too. The motivation for the use of $\psi(\varepsilon_m \xi)$ in (34.7) comes from the fact that if $u^n(x) = v(x)w\left(\frac{x}{\varepsilon_n}\right)$ with v smooth with compact support and w periodic, then $\mathcal{F}(\varphi\, u^n)$ is mostly localized at distances $O\left(\frac{1}{\varepsilon_n}\right)$ from the origin.[5] As a consequence, if $\frac{\delta_n}{\varepsilon_n} \to 0$, and $u^n(x) = v(x)w\left(\frac{x}{\delta_n}\right)$, then the semi-classical measure computed with ε_n is 0, because ψ is 0 at ∞; if $\eta_n \to 0$ and $\frac{\eta_n}{\varepsilon_n} \to \infty$, and $u^n(x) = v(x)w\left(\frac{x}{\eta_n}\right)$, then the semi-classical measure computed with ε_n is concentrated at 0 but it mixes the information corresponding to various directions, because ψ is continuous at 0. This second defect can be corrected by

---

[5] Because ξ is the dual variable of x, and the use of $e^{\pm 2i\pi(\xi, x)}$ forces (ξ, x) to have no dimension, if ε_n is used to scale x, then $\frac{1}{\varepsilon_n}$ is used to scale ξ.

using test functions ψ which behave like $\psi_0\left(\frac{\xi}{|\xi|}\right)$ near 0 (with $\psi_0 \in C(\mathbb{S}^{N-1})$), and the first defect can be corrected by using test functions ψ which behave like $\psi_\infty\left(\frac{\xi}{|\xi|}\right)$ near ∞ (with $\psi_\infty \in C(\mathbb{S}^{N-1})$), for example.[6] Using such test functions ψ corresponds to considering $\mathbb{R}^N \setminus \{0\}$ and compactifying it with a sphere $\mathbb{S}^{N-1}$ at 0 and a sphere $\mathbb{S}^{N-1}$ at ∞, and a generalized measure (for which I prefer not to use a term like semi-classical, because it is not wise to give too many different names to variants of H-measures) may charge these two spheres and by a natural projection of this compactified space on $\mathbb{S}^{N-1}$ one recovers the H-measure.

Without taking such precautions near 0 and at ∞, it is false that the knowledge of a semi-classical measure for a sequence gives the H-measure of that sequence,[7] as was written by Pierre-Louis LIONS and Thierry PAUL,[8] when they later found a different way to define the same objects that Patrick GÉRARD had introduced, which they wanted to call a different name, *Wigner measures*, because they had discovered a way to introduce semi-classical measures by using the Wigner transform. After George PAPANICOLAOU had told me about the Wigner transform (32.3), I could not have thought of doing what Pierre-Louis LIONS and Thierry PAUL did, i.e. to look at

$$W_n(x,\xi) = \int_{\mathbb{R}^N} u\left(x + \frac{\varepsilon_n y}{2}\right) \overline{u\left(x - \frac{\varepsilon_n y}{2}\right)} e^{-2i\pi(y,\xi)} \, dy, \qquad (34.8)$$

and to show that

$$W_m \rightharpoonup \mu_{sc} \text{ as } m \to \infty \qquad (34.9)$$

because I did not want to use any characteristic length in my general construction. I understood later that WIGNER had observed that (32.3) implies that

$$\text{if } i\, u_t - \Delta u = 0, \text{ then } \frac{\partial W}{\partial t} - \sum_{j=1}^{N} \xi_j \frac{\partial W}{\partial x_j} = 0, \qquad (34.10)$$

and he would have liked to interpret $W(x,\xi)$ as a density of particles moving with velocity ξ, if W had been nonnegative. Marc FEIX told me afterward that WIGNER had proven that the convolution in ξ with $e^{-\alpha |\xi|^2}$ is nonnegative,[9] and he had characterized the best $\alpha > 0$, and he told me that he had

---

[6] A careful analysis concerning a commutation lemma shows that at ∞ it is enough to ask that ψ belong to the space $BUC(\mathbb{R}^N)$ of bounded uniformly continuous functions. However, one must pay attention that this space is not separable.

[7] Except, of course, if the spheres at 0 and at ∞ are not charged by the generalized measure, and Patrick GÉRARD had coined two words to express this fact. In other words, it is only true in a dull physical world with only one characteristic length, and it is worth pointing out that there are people who know the statement to be wrong but nevertheless repeat it, probably because they like to advocate fake continuum mechanics or physics.

[8] Thierry PAUL, French mathematician. He works at Université Paris IX-Dauphine, Paris, France.

[9] Marc R. FEIX, French physicist, 1928–2005. He had worked in Orléans, France.

mentioned this fact to Pierre-Louis LIONS, who with his coauthor nevertheless attributed the idea to a Japanese. I had not tried to read the detail of what they had written, as I thought that they only wanted to show that they had read about physics, while they had also shown a complete lack of physical intuition in thinking that H-measures could be deduced from semi-classical measures, but after Patrick GÉRARD had explained to me that they wanted to show that the limit of W_n is nonnegative, I immediately suggested a simpler proof using correlations, which actually opens the road to a new kind of generalization.

Before talking about correlations, I find it useful to mention another important observation learnt from H-measures, which shows a new kind of defect of the classical equations of kinetic theory; I had mentioned it once to Pierre-Louis LIONS in the early 1990s, but I may not have mentioned it in print before. It is that the density of particles $f(\mathbf{x}, \mathbf{v}, t)$, that one uses in the Boltzmann equation, or other equations in kinetic theory, looks pretty much like the density of a variant of H-measures, i.e. one should think of it already as a quadratic micro-local object with respect to the waves, which are the only real thing behind all that. It is then not so logical to introduce quadratic quantities in f, and it would be more natural to have a (micro-local) cubic quantity in the waves appear, and although such a general object has not been constructed yet, one can have a guess about that by using three-point correlations.

I have mentioned before the Percus–Yevick equation for correlations, which I think was postulated, so that I would not attach too much faith to it, but I suggest that it should be understood as a hint that the ideas used in kinetic theory have been terribly simplistic, and that new ideas like correlations of positions should be thought about. When I discussed Kepler's laws I pointed out that in the Boltzmann equation one mostly thinks in terms of two-body problems with hyperbolas as trajectories, forgetting completely the case of trajectories looking like ellipses, which must occur more and more if the gas is less and less rarefied, and in describing a gas with plenty of trajectories like that, one might find that correlations of positions play an important role.

I should also recall the delays which take place during the close encounters, and mention the importance of considering equations with nonlocal terms, in time, but also in space.

However, after making a list of other ideas to use in classical descriptions, one must recall that in the end, the main problem is that real gases are not classical at all, and they are made of waves![10]

---

[10] One may be interested in what happens to a "gas" of small metallic spheres rolling on a smooth plane surface and colliding, and one may compare theoretical results to experiments, and there is no doubt that one may improve on the Boltzmann equation for that, but in the end one will know more about a hard-sphere model, which no real gas follows!

Apart from probabilistic frameworks, which I do not recommend for explaining what happens in the real world, one needs a characteristic length (or time) for defining correlations. If u is periodic with period T, one defines k-point correlations by computing $C_k(h_1, \ldots, h_k) = \frac{1}{T} \int_0^T u(t + h_1) \cdots u(t + h_k)\, dt$, and by applying this idea to the fast variable in a periodically modulated framework $u\left(x, \frac{x}{\varepsilon_n}\right)$ one is led to the following natural definition.

Definition 34.1. *If for $k \geq 2$ one has $u_n \rightharpoonup u_\infty$ in $L^k_{loc}(\Omega)$ weak, one defines the k-point correlation measure $C_k(h_1, \ldots, h_k)$ using the characteristic length ε_n tending to 0 by*

$$\langle C_k(h_1, \ldots, h_k), \varphi \rangle = \lim_{n \to \infty} \int_\Omega u_n(x + \varepsilon_n h_1) \cdots u_n(x + \varepsilon_n h_k)\, \varphi(x)\, dx,$$
for $\varphi \in C_c(\Omega), h_1, \ldots, h_k \in \mathbb{R}^N$,

$$(34.11)$$

for real functions, but also

$$\langle C_2(h_1, h_2), \varphi \rangle = \lim_{n \to \infty} \int_\Omega u_n(x + \varepsilon_n h_1)\overline{u_n(x + \varepsilon_n h_2)}\, \varphi(x)\, dx, \quad (34.12)$$
for $\varphi \in C_c(\Omega), h_1, h_2 \in \mathbb{R}^N$,

for complex functions, in the particular case $k = 2$.

The definition makes sense because $\varepsilon_n \to 0$ and for $x \in support(\varphi)$, one has $x + \varepsilon_n h_1, \ldots, x + \varepsilon_n h_k \in \Omega$ for n large enough. For given $h_1, \ldots, h_k$, the sequence $u_n(\cdot + \varepsilon_n h_1) \cdots u_n(\cdot + \varepsilon_n h_k)$ is bounded in $L^1_{loc}(\Omega)$, so that there exists a subsequence which converges in $\mathcal{M}(\Omega)$ weak $\star$, and using the Cantor diagonal argument one can extract a subsequence such that (34.11) holds for all $h_1, \ldots, h_k \in \mathbb{Q}^N$, and then using the uniform continuity of φ it also holds for all $h_1, \ldots, h_k \in \mathbb{R}^N$. One should notice that although a local bound in L^3 seems natural for defining three-point correlations, such an hypothesis is not really adapted to hyperbolic equations, because of an observation of Walter LITTMAN concerning the lack of L^p estimates for the wave equation if $p \neq 2$,[11] and it might be that either new functional spaces must be invented, or that one must use ideas of compensated regularity (which is not the same thing as compensated compactness!) for defining some special parts of the three-point correlation measures.

Lemma 34.2. *One has*

$$C_k(h_1 + z, \ldots, h_k + z) = C_k(h_1, \ldots, h_k) \text{ for all } h_1, \ldots, h_k, z \in \mathbb{R}^N, \quad (34.13)$$

and

$$\sum_{i,j=1}^m C_2(h_i, h_j)\lambda_i\overline{\lambda_j} \geq 0 \text{ for all } m \geq 1, h_1, \ldots, h_m \in \mathbb{R}^N, \lambda_1, \ldots, \lambda_m \in \mathbb{C}^N,$$

$$(34.14)$$

---

[11] Walter LITTMAN, American mathematician. He worked at University of Minnesota, Minneapolis, MN.

so that $C_2(h,0)$ is the Fourier transform (in its second argument $\xi \in \mathbb{R}^N$) of a nonnegative measure $\in \mathcal{M}(\Omega \times \mathbb{R}^N)$.

Proof: Translating all the h_j by z corresponds to evaluating φ at $x - \varepsilon_n z$ in (34.11), and the uniform continuity of φ gives (34.13), while (34.14) is just saying that the weak $\star$ limit of $\left|\sum_{j=1}^m \lambda_j u_n(x + \varepsilon_n h_j)\right|^2$ is ≥ 0. Denoting $\Gamma_2(h) = C_2(h + z, z)$ for all $h, z \in \mathbb{R}^N$, (34.14) says that $\sum_{i,j=1}^m \Gamma_2(h_i - h_j)\lambda_i \overline{\lambda_j} \geq 0$, so that a theorem of BOCHNER on functions of positive type,[12] extended by Laurent SCHARTZ to (tempered) distributions of positive type, tells us that Γ_2 is the Fourier transform of a nonnegative measure.   □

It is not too difficult to check that it is the semi-classical measure μ_{sc} which is behind this formula, but the interest of this lemma is more that it helps understand what is behind the definition of the Wigner transform, that it is like the Fourier transform of a two-point correlation function, and that for questions of symmetry it is better to use $\Gamma_2(h) = C_2\left(\frac{h}{2}, \frac{-h}{2}\right)$.

Although I know of no analogous result that would play the role of the Bochner theorem for what concerns k-point correlations with $k \geq 3$, one may nevertheless obtain partial differential equations satisfied by C_k directly,[13] but it might be important to investigate "natural formulations" for the cases $k \geq 3$, in parallel with the search for cubic and higher-order corrections in small amplitude homogenization, or a question which I consider of a greater importance, extending the theory of H-measures to semi-linear hyperbolic systems. My approach is not to try to read what physicists have done, as they often use what I call pseudo-logic,[14] or put in their hypotheses what they want to find in the conclusion, but I would not be surprised that a mathematical answer might explain some of the strangely efficient formal methods introduced by FEYNMAN, although by using completely different ideas.[15]

---

[12] Salomon BOCHNER, Polish-born mathematician, 1899–1982. He had worked in München (Munich), Germany, and at Princeton University, Princeton, NJ.

[13] For $k = 2$, I had made such an observation for an equation of the form $u_t^n - \varepsilon_n^2 \Delta u^n = f_n$, where I used natural bounds in the L^2 norm of $\varepsilon_n grad(u^n)$, but Patrick GÉRARD then taught me a simpler derivation which uses only bounds in the L^2 norm of u^n, and we then checked that this method gives partial differential equations for C_k for $k \geq 3$.

[14] This is how I qualify an "argument" that an hypothesis A seems to imply a conclusion like B, and as one observes something that looks like B it must be that A is true! Apart from showing a strange lack of imagination, it suggests that whoever uses that kind of "reasoning" has never heard of basic logic.

[15] One attributes all kinds of statements to FEYNMAN, and it might be true that after having shown a formal argument, and then heard a mathematician mention that he could prove that in a mathematical way, he had wondered why anyone would bother to do such a (useless) thing. In the presence of a formal argument presented by a physicist, I think that the question for a mathematician is not to

The first question which I had overlooked concerning H-measures concerned taking into account initial conditions for the transport equations that I had obtained for H-measures, because the scalar case that I had solved was not general enough. For the wave equation with smooth coefficients, it was done by Gilles FRANCFORT and François MURAT,[16] with the technical advice of Patrick GÉRARD, using classical pseudo-differential operators, but more remains to be done for general systems.

In the same way, there is not much understood about the boundary conditions to impose for H-measures.

The second question which I had overlooked concerned the Dirac equation, and I had asked my student Nenad ANTONIC to look at it,[17] but I was surprised to see that there was nothing in the answer that suggested a question about electrons. The reason was that the zero-order term containing m_0 (the mass of the electron) plays no role if one considers it independent of n.

What Patrick GÉRARD observed is that the zero-order term is large and has a behaviour in $\frac{1}{\varepsilon_n^2}$ for a small characteristic length, and that one should then consider the semi-classical measure associated with $\varepsilon_n \to 0$. To do this analysis, he had to freeze the coefficients involving the potentials V and A, assumed to be smooth and given, so that the Dirac equation is then linear in ψ, with the velocity of light c, the charge of the electron e and the mass of the electron m_0 appearing in its coefficients. He observed that the equation that he obtains for the semi-classical measure can be interpreted as describing two types of particles, of charge $\pm e$ and relativistic mass $\frac{m_0}{\sqrt{1-v^2/c^2}}$ evolving under the Lorentz force $\pm e(E + v \times B)$, with E and B related to V and A as usual, but E and B have not been asked to solve the Maxwell–Heaviside equation. It shows that DIRAC had really done a superb job in creating his equation, and the work of Patrick GÉRARD explains in a mathematical way what the physicists had meant by saying that the Dirac equation both describes "electrons" and "positrons".

It is important to notice that Patrick GÉRARD's computation shows that the Lorentz force does not exist at the level of the "particles", here called "electrons" and "positrons" because of the values of their mass and their electric charge, but that it is dependent upon m_0 appearing explicitly in the equation, and on V and A being smooth enough on a scale much larger than ε_n.

In 1984, I had suggested that the term containing m_0 should appear as a homogenization correction and would correspond to the mass being entirely made of electromagnetic energy stored inside the particles (with "Einstein

---

make sense of the path that he/she has followed, but usually to understand what he/she was really looking for.

[16] Gilles FRANCFORT, French mathematician, born in 1957. He works at Université Paris-Nord, Villetaneuse, France.

[17] Nenad ANTONIC, Croatian mathematician. He works in Zagreb, Croatia. He was my PhD student (1992) at CMU (Carnegie Mellon University), Pittsburgh, PA.

equation" $e = m\,c^2$), and shortly afterwards I read of a similar proposition by BOSTICK (but not involving the Dirac equation), and this should make V and A change enough on a scale of the order of ε_n.

These considerations seem to imply that once again one is thrown into the semi-linear world, which is not understood yet.

Conclusion: It is important to observe then which mathematical results are proven, for what equations, and which are the hypotheses used. It is important to understand enough about how the accepted "laws of physics" should be modified when the rules used by physicists seem illogical, because physicists have assumed some macroscopic equations to be valid at a mesoscopic level or a microscopic level, while it seems clear that the equations have a different form at these levels, even though this form might not be understood yet.

It might be useful to recall how ideas about chemistry have evolved. First one observed reactions in given proportions, and one assumed that the same proportions were used at a microscopic level. Then one invoked time for the reactions to take place, so that one entered the realm of ordinary differential equations. Then one invoked also space for the constituents to be moved to the place where reactions took place, so that one entered the realm of partial differential equations. Then one observed the appearance of small scales created by turbulent mixing, and one invented all kind of variants of thermodynamics, in order to avoid having to think about what was really happening at a microscopic level or at a mesoscopic level: one had then moved from early chemistry to chemical engineering.

At some point one started changing the equations at a microscopic level, because the physicists had invented quantum mechanics,[18] without questioning some of the strange rules that had been invented, and one started computing orbitals and requiring larger and larger computers for playing a game which should have been criticized from the start, for example because the rules of quantum mechanics had been invented in order to fit what one had observed for electromagnetism in the vacuum, and that is hardly the kind of environment that one finds in chemistry!

The art of the engineers makes it possible to tame some phenomena for which one does not have the right equations, but it is the role of the scientists to discover what these missing equations are, as part of their duty to find the real laws of nature.

It is my feeling that one has not really found the laws of nature, because one has made the mistake to continue thinking in terms of the classical mechanics of the 18th century and the continuum mechanics of the 19th century, with the

---

[18] Many chemists think that they should mimic physicists, and many physicists think that they should mimic mathematicians, and choose to do astrophysics, probably because of an irrational tendency of believing a questionable classification of a philosopher of sciences, COMTE, who had put mathematics above astronomy, physics and chemistry, in that order.

mathematics created in the 19th and the 20th century, instead of observing that the continuum mechanics and the physics of the 20th century require mathematical tools that are *beyond partial differential equations*, which should be developed in the 21st century, maybe along the line of what I have started doing since the 1970s.

[Taught on Friday November 30, 2001.]

Notes on names cited in footnotes for Chapter 34, LASCOUX,[19] COMTE.[20]

---

[19] Jean LASCOUX, French physicist. He worked at École Polytechnique, Palaiseau, France.
[20] Auguste COMTE, French philosopher, 1798–1857. He had worked in Paris, France.

35
Biographical Information

[In a reference a-*b*, a is the lecture number, 0 referring to the Preface, and *b* the footnote number in that lecture.]

ABEL, 0-*52*
ADAMS J.C., 27-*9*
AIRY, 33-*7*
ALAOGLU, 10-*16*
D'ALEMBERT, 24-*2*
ALEXANDER the G., 2-*32*
ALFVÉN, 1-*1*
AL KHWARIZMI, 9-*23*
AL MAMUN, 9-*26*
AMPÈRE, 1-*4*
ANTONIČ, 34-*17*
D'ARC, 0-*58*
AVOGADRO, 1-*38*
— — —
BABUŠKA, 0-*29*
BACHELIER, 8-*5*
BALL W.R., 16-*7*
BANACH, 9-*2*
BATEMAN, 4-*1*
BEALE, 17-*2*
BECQUEREL, 0-*60*
BELLMAN, 4-*7*
BENILAN, 4-*11*
BERKELEY, 0-*57*
BERNOULLI D., 10-*3*
BESSEL, 2-*35*
BIOT, 1-*6*

BOCHNER, 34-*12*
BOLTZMANN, 0-*12*
BONAPARTE N., 1-*68*
BOREL E., 1-*73*
BOSE, 11-*5*
BOSTICK, 23-*24*
BOYLE, 1-*19*
BRAHÉ, 27-*5*
BRENIER, 5-*5*
BRILLOUIN, 33-*6*
BROADWELL, 14-*7*
DE BROGLIE L., 0-*15*
BROUWER, 2-*6*
BROWN N., 0-*53*
BROWN R., 1-*41*
BRUN, 7-*2*
BUNYAKOVSKY, 10-*11*
BURGERS, 0-*19*
— — —
CABANNES, 14-*13*
CAFLISCH, 16-*6*
CALDERÓN A., 13-*4*
CALVIN, 0-*6*
CANTOR, 20-*2*
CARATHÉODORY, 4-*6*
CARLEMAN, 13-*2*
CARLESON, 14-*27*

CARNEGIE, 2-*25*
CARNOT S., 1-*70*
CARTAN E., 8-*24*
CARTAN H., 8-*23*
CAUCHY, 1-*12*
CAVENDISH, 0-*47*
CELSIUS, 1-*28*
CHALLIS, 1-*50*
CHAPMAN S., 20-*4*
CHARLES IV, 0-*56*
CHARLES X, 1-*67*
CHARLES J., 1-*37*
CHÉRET, 7-*1*
CHISHOLM-YOUNG, 19-*10*
CIORANESCU, 23-*14*
CLAUSIUS, 1-*69*
CLEBSCH, 2-*20*
COIFMAN, 18-*6*
COLE, 4-*4*
COMTE A., 34-*20*
CONLEY, 6-*5*
CORIOLIS, 23-*22*
CORNELL, 0-*55*
COURANT, 1-*71*
CRAFOORD, 13-*16*
CRANDALL, 4-*12*
CRISTIN, 1-*29*

CURIE, 0-*44*

– – –

DAFERMOS C., 0-*25*

D'ALEMBERT, 24-*2*

D'ARC, 0-*58*

DAUTRAY, 0-*4*

DA VINCI, 0-*33*

DE BROGLIE L., 0-*15*

DEBYE, 23-*15*

DE GIORGI, 18-*1*

DE KLEIN, 8-*10*

DE PRETTO, 23-*25*

DE VRIES, 0-*21*

DIDEROT, 0-*46*

DI PERNA, 0-*24*

DIRAC, 0-*17*

DIRICHLET, 2-*33*

DUHEM, 1-*56*

DUKE, 0-*54*

DUNFORD, 15-*2*

– – –

EARNSHAW, 3-*2*

EINSTEIN, 0-*14*

EKSTRÖM, 1-*30*

ENSKOG, 20-*5*

EÖTVÖS L., 14-*32*

EUCLID, 2-*13*

EULER, 0-*41*

– – –

FAHRENHEIT, 1-*26*

FARADAY, 1-*8*

FEFFERMAN C., 14-*22*

FEIX, 34-*9*

FERDINAND II, 1-*23*

FERMAT, 10-*1*

FEYNMAN, 1-*10*

FICK, 4-*18*

FIELDS, 0-*45*

FOIAS, 2-*8*

FOKKER, 8-*10*

FORSYTH, 4-*20*

FOURIER J.-B., 2-*30*

FRANCFORT, 34-*16*

FRÉCHET, 9-*22*

FRIEDRICHS, 1-*72*

FROBENIUS, 2-*21*

FROSTMAN, 14-*28*

FUBINI, 9-*4*

FULLER, 1-*64*

– – –

GAGLIARDO, 18-*10*

GALILEI, 1-*16*

GALLE, 27-*11*

GANTMAKHER, 12-*12*

GÅRDING, 2-*10*

GATIGNOL, 14-*1*

GAUSS, 1-*5*

GAY LUSSAC, 1-*35*

GEL'FAND, 6-*4*

GEORGE II, 6-*12*

GÉRARD P., 31-*4*

GERMAIN, 8-*1*

GERSHGORIN, 12-*5*

GLIMM, 0-*61*

GODUNOV, 2-*2*

GOEPPERT MAYER, 32-*16*

GOLSE, 31-*6*

GOUDSMIT, 8-*20*

GRAD, 14-*20*

GREEN, 8-*15*

GRÖNWALL, 28-*2*

GUIRAUD, 9-*21*

– – –

HAAR, 10-*8*

HADAMARD, 9-*24*

HAMDACHE, 17-*6*

HAMILTON, 1-*18*

HARDINGE, 19-*11*

HARDY, 13-*7*

HARVARD, 1-*76*

HEATH, 8-*6*

HEAVISIDE, 1-*3*

HEDBERG, 16-*4*

HERIOT, 21-*8*

HERMITE, 32-*8*

HERSCHEL, 27-*7*

HESSE, 6-*2*

HILBERT D., 0-*40*

HIRZEBRUCH, 2-*31*

HODGE, 16-*1*

HÖLDER O., 20-*1*

HOPF E., 0-*36*

HOPKINS, 4-*16*

HÖRMANDER, 2-*11*

Hugo of St Victor, 0-*42*

HUGONIOT, 1-*54*

– – –

ILLNER, 16-*8*

ITO, 0-*51*

– – –

JACOBI, 3-*6*

JENSEN J.H., 32-*17*

JOHN, 13-*8*

JORDAN C., 12-*8*

JOST, 8-*2*

JOULE, 1-*34*

– – –

KANIEL, 30-*9*

KAWASHIMA, 17-*5*

KELLER J.B., 33-*8*

Kelvin, 1-*33*

KEPLER, 1-*11*

KEYFITZ, 4-*10*

KIRCHGÄSSNER, 27-*6*

KIRCHHOFF, 1-*48*

KNOPS, 23-*12*

KNUDSEN, 30-*1*

KODAIRA, 4-*21*

KOHN J., 32-*9*

KOLMOGOROV, 8-*16*

KOLODNER, 13-*3*

KORTEWEG, 0-*20*

KRONECKER, 2-*12*

KRUZHKOV, 0-*37*

KURTZ, 20-*8*

– – –

Abbreviations and Mathematical Notation

Abbreviations for states: For those not familiar with geography, I have mentioned England, Scotland, and Wales, without mentioning that they are part of UK (United Kingdom), I have mentioned British Columbia and Ontario, without mentioning that they are part of Canada, and I have mentioned a few of the fifty states in the United States of America: AZ = Arizona, CA = California, CO = Colorado, CT = Connecticut, IL = Illinois, IN = Indiana, KY = Kentucky, LA = Louisiana, MA = Massachusetts, MD = Maryland, MI = Michigan, MN = Minnesota, MO = Missouri, NC = North Carolina, NJ = New Jersey, NM = New Mexico, NY = New York, OH = Ohio, PA = Pennsylvania, RI = Rhode Island, TX = Texas, UT = Utah, VA = Virginia, WI = Wisconsin.

- a.e.: almost everywhere.
- $B(x,r)$: open ball centred at x and radius $r > 0$, i.e. $\{y \in E \mid ||x - y||_E < r\}$ (in a normed space E).
- $BMO(\mathbb{R}^N)$: space of functions of bounded mean oscillation on $\mathbb{R}^N$, i.e. semi-norm $||u||_{BMO} < \infty$, with $||u||_{BMO} = \sup_{cubes\ Q} \frac{\int_Q |u - u_Q|\,dx}{|Q|} < \infty$ ($u_Q = \frac{\int_Q u\,dx}{|Q|}$, $|Q| = meas(Q)$).
- $BV(\Omega)$: space of functions of bounded variation in Ω, whose partial derivatives (in the sense of distributions) belong to $\mathcal{M}_b(\Omega)$, i.e. have finite total mass.
- $C(\Omega)$: space of scalar continuous functions in an open set $\Omega \subset \mathbb{R}^N$ ($\mathcal{E}_0(\Omega)$ in the notation of L. SCHWARTZ).
- $C(\Omega; \mathbb{R}^m)$: space of continuous functions from an open set $\Omega \subset \mathbb{R}^N$ into $\mathbb{R}^m$.
- $C(\overline{\Omega})$: space of scalar continuous and bounded functions on $\overline{\Omega}$, for an open set $\Omega \subset \mathbb{R}^N$.
- $C_0(\Omega)$: space of scalar continuous bounded functions tending to 0 at the boundary of an open set $\Omega \subset \mathbb{R}^N$, equipped with the sup norm.

- $C_c(\Omega)$: space of scalar continuous functions with compact support in an open set $\Omega \subset \mathbb{R}^N$.
- $C_c^k(\Omega)$: space of scalar functions of class C^k with compact support in an open set $\Omega \subset \mathbb{R}^N$.
- $C^k(\Omega)$: space of scalar continuous functions with continuous derivatives up to order k in an open set $\Omega \subset \mathbb{R}^N$.
- $C^k(\overline{\Omega})$: restrictions to $\overline{\Omega}$ of functions in $C^k(\mathbb{R}^N)$, for an open set $\Omega \subset \mathbb{R}^N$.
- $C^{0,\alpha}(\Omega)$: space of scalar Hölder continuous functions of order $\alpha \in (0,1)$ (Lipschitz continuous functions if $\alpha = 1$), i.e. bounded functions for which there exist M such that $|u(x) - u(y)| \leq M\,|x - y|^\alpha$ for all $x, y \in \Omega \subset \mathbb{R}^N$; it is included in $C(\overline{\Omega})$.
- $C^{k,\alpha}(\Omega)$: space of functions of $C^k(\Omega)$ whose derivatives of order k belong to $C^{0,\alpha}(\Omega) \subset C(\overline{\Omega})$, for an open set $\Omega \subset \mathbb{R}^N$.
- curl: rotational operator $(curl(u))_i = \sum_{jk} \varepsilon_{ijk} \frac{\partial u_j}{\partial x_k}$, used for open sets $\Omega \subset \mathbb{R}^3$.
- D^α: $\frac{\partial^{\alpha_1}}{\partial x_1^{\alpha_1}} \cdots \frac{\partial^{\alpha_N}}{\partial x_N^{\alpha_N}}$ (for a multi-index α with α_j nonnegative integers, $j = 1, \ldots, N$).
- $\mathcal{D}'(\Omega)$: space of distributions T in Ω, dual of $C_c^\infty(\Omega)$ ($\mathcal{D}(\Omega)$ in the notation of L. SCHWARTZ, equipped with its natural topology), i.e. for every compact $K \subset \Omega$ there exists $C(K)$ and an integer $m(K) \geq 0$ with $|\langle T, \varphi \rangle| \leq C(K) \sup_{|\alpha| \leq m(K)} ||D^\alpha \varphi||_\infty$ for all $\varphi \in C_c^\infty(\Omega)$ with support in K.
- div: divergence operator $div(u) = \sum_i \frac{\partial u_i}{\partial x_i}$.
- $\mathcal{F}$: Fourier transform, $\mathcal{F}f(\xi) = \int_{\mathbb{R}^N} f(x) e^{-2i\pi(x,\xi)}\,dx$.
- $\overline{\mathcal{F}}$: inverse Fourier transform, $\overline{\mathcal{F}}f(\xi) = \int_{\mathbb{R}^N} f(x) e^{+2i\pi(x,\xi)}\,dx$.
- $grad(u)$: gradient operator, $grad(u) = \left(\frac{\partial u}{\partial x_1}, \ldots, \frac{\partial u}{\partial x_N} \right)$.
- $H^s(\mathbb{R}^N)$: Sobolev space of temperate distributions ($\in \mathcal{S}'(\mathbb{R}^N)$), or functions in $L^2(\mathbb{R}^N)$ if $s \geq 0$, such that $(1 + |\xi|^2)^{s/2} \mathcal{F}u \in L^2(\mathbb{R}^N)$ ($L^2(\mathbb{R}^N)$ for $s = 0$, $W^{s,2}(\mathbb{R}^N)$ for s a positive integer).
- $H^s(\Omega)$: space of restrictions to Ω of functions from $H^s(\mathbb{R}^N)$ (for $s \geq 0$), for an open set $\Omega \subset \mathbb{R}^N$.
- $H_0^s(\Omega)$: for $s \geq 0$, closure of $C_c^\infty(\Omega)$ in $H^s(\Omega)$, for an open set $\Omega \subset \mathbb{R}^N$.
- $H^{-s}(\Omega)$: for $s \geq 0$, dual of $H_0^s(\Omega)$, for an open set $\Omega \subset \mathbb{R}^N$.
- $H(div; \Omega)$: space of functions $u \in L^2(\Omega; \mathbb{R}^N)$ with $div(u) \in L^2(\Omega)$, for an open set $\Omega \subset \mathbb{R}^N$.
- $H(curl; \Omega)$: space of functions $u \in L^2(\Omega; \mathbb{R}^3)$ with $curl(u) \in L^2(\Omega; \mathbb{R}^3)$, for an open set $\Omega \subset \mathbb{R}^3$.
- $\mathcal{H}^1(\mathbb{R}^N)$: Hardy space of functions $f \in L^1(\mathbb{R}^N)$ with $R_j f \in L^1(\mathbb{R}^N)$, $j = 1, \ldots, N$, where R_j, $j = 1, \ldots, N$ are the (M.) Riesz operators.
- $\mathcal{H}(\theta)$: class of Banach spaces satisfying $(E_0, E_1)_{\theta,1;J} \subset E \subset (E_0, E_1)_{\theta,\infty;K}$.
- $ker(A)$: kernel of a linear operator $A \in \mathcal{L}(E; F)$, i.e. $\{e \in E \mid A\,e = 0\}$.
- $\mathcal{L}(E; F)$: space of linear continuous operators M from the normed space E into the normed space F, i.e. with $||M||_{\mathcal{L}(E;F)} = \sup_{e \neq 0} \frac{||M\,e||_F}{||e||_E} < \infty$.

- $L^p(A)$, $L^\infty(A)$: Lebesgue space of (equivalence classes of a.e. equal) measurable functions u with $||u||_p = \left(\int_A |u(x)|^p \, dx\right)^{1/p} < \infty$ if $1 \leq p < \infty$, with $||u||_\infty = \inf\{M \mid |u(x)| \leq M \text{ a.e. in } A\} < \infty$, for a Lebesgue measurable set $A \subset \mathbb{R}^N$ (spaces also considered for the induced $(N-1)$-dimensional Hausdorff measure if $A = \partial\Omega$ for an open set $\Omega \subset \mathbb{R}^N$ with a smooth boundary).
- $L^p_{loc}(A)$: (equivalence classes of) measurable functions whose restriction to every compact $K \subset A$ belongs to $L^p(K)$ (for $1 \leq p \leq \infty$), for a Lebesgue measurable set $A \subset \mathbb{R}^N$.
- $L^p\big((0,T); E\big)$: (weakly or strongly) measurable functions u from $(0,T)$ into a separable Banach space E, such that $t \mapsto ||u(t)||_E$ belongs to $L^p(0,T)$ (for $1 \leq p \leq \infty$).
- $|\alpha|$: length of a multi-index $\alpha = (\alpha_1, \ldots, \alpha_N)$, $|\alpha| = |\alpha_1| + \ldots + |\alpha_N|$.
- $Lip(\Omega)$: space of scalar Lipschitz continuous functions, also denoted $C^{0,1}(\Omega)$, i.e. bounded functions for which there exists M such that $|u(x) - u(y)| \leq M |x - y|$ for all $x, y \in \Omega \subset \mathbb{R}^N$; it is included in $C(\overline{\Omega})$.
- loc: for any space Z of functions in an open set $\Omega \subset \mathbb{R}^N$, Z_{loc} is the space of functions u such that $\varphi u \in Z$ for all $\varphi \in C^\infty_c(\Omega)$.

- $M f$: maximal function of f, i.e. $M f(x) = \sup_{r>0} \frac{\int_{B(x,r)} |f(y)| \, dy}{|B(x,r)|}$.
- $\mathcal{M}(\Omega)$: space of Radon measures μ in an open set $\Omega \subset R^N$, dual of $C_c(\Omega)$ (equipped with its natural topology), i.e. for every compact $K \subset \Omega$ there exists $C(K)$ with $|\langle \mu, \varphi \rangle| \leq C(K)||\varphi||_\infty$ for all $\varphi \in C_c(\Omega)$ with support in K.
- $\mathcal{M}_b(\Omega)$: space of Radon measures μ with finite total mass in an open set $\Omega \subset \mathbb{R}^N$, dual of $C_0(\Omega)$, the space of continuous bounded functions tending to 0 at the boundary of Ω (equipped with the sup norm), i.e. there exists C with $|\langle \mu, \varphi \rangle| \leq C ||\varphi||_\infty$ for all $\varphi \in C_c(\Omega)$.
- meas(A): Lebesgue measure of A, sometimes denoted $|A|$.
- $|\cdot|$: norm in H, or sometimes the Lebesgue measure of a set.
- $||\cdot||$: norm in V.
- $||\cdot||_*$: dual norm in V'.
- p': conjugate exponent of $p \in [1, \infty]$, i.e. $\frac{1}{p} + \frac{1}{p'} = 1$.
- p^*: Sobolev exponent of $p \in [1, N)$, i.e. $\frac{1}{p^*} = \frac{1}{p} - \frac{1}{N}$ for $\Omega \subset R^N$ and $N \geq 2$.
- $\mathbb{R}_+$: $(0, \infty)$.
- $\mathbb{R}^N_+$: $\{x \in \mathbb{R}^N \mid x_N > 0\}$.
- $R(A)$: range of a linear operator $A \in \mathcal{L}(E; F)$, i.e. $\{f \in F \mid f = A e \text{ for some } e \in E\}$.
- R_j: Riesz operators, $j = 1, \ldots, N$, defined by $\mathcal{F}(R_j u)(\xi) = \frac{i\xi_j \mathcal{F}u(\xi)}{|\xi|}$ on $L^2(\mathbb{R}^N)$; natural extensions to $\mathbb{R}^N$ of the Hilbert transform, they map $L^p(\mathbb{R}^N)$ into itself for $1 < p < \infty$, and $L^\infty(\mathbb{R}^N)$ into $BMO(\mathbb{R}^N)$.
- $\mathcal{S}(\mathbb{R}^N)$: Schwartz space of functions $u \in C^\infty(\mathbb{R}^N)$ with $x^\alpha D^\beta u$ bounded for all multi-indices α, β with α_j, β_j nonnegative integers for $j = 1, \ldots, N$.
- $\mathcal{S}'(\mathbb{R}^N)$: temperate distributions, dual of $\mathcal{S}(\mathbb{R}^N)$, i.e. $T \in \mathcal{D}'(\mathbb{R}^N)$ and there exists C and an integer $m \geq 0$ with $|\langle T, \psi \rangle| \leq C \sup_{|\alpha|,|\beta| \leq m} ||x^\alpha D^\beta \psi||_\infty$ for all $\psi \in \mathcal{S}(\mathbb{R}^N)$.

- $\star$: convolution product $(f \star g)(x) = \int_{\mathbb{R}^N} f(x-y)g(y)\,dy$.
- $supp(\cdot)$: support; for a continuous function u from a topological space into a vector space, it is the closure of $\{x \mid u(x) \neq 0\}$, but for a locally integrable function f, a Radon measure μ, or a distribution T defined on an open set $\Omega \subset \mathbb{R}^N$, it is the complement of the largest open set ω where f, μ, or T is 0, i.e. where $\int_\omega \varphi f\,dx = 0$, or $\langle \mu, \varphi \rangle = 0$ for all $\varphi \in C_c(\Omega)$, or $\langle T, \varphi \rangle = 0$ for all $\varphi \in C_c^\infty(\Omega)$.
- $W^{m,p}(\Omega)$: Sobolev space of functions in $L^p(\Omega)$ whose derivatives (in the sense of distributions) of length $\leq m$ belong to $L^p(\Omega)$, for an open set $\Omega \subset \mathbb{R}^N$.
- $W^{m,p}(\Omega; \mathbb{R}^m)$: Sobolev space of functions from Ω into $\mathbb{R}^m$ whose components belong to $W^{m,p}(\Omega)$, for an open set $\Omega \subset \mathbb{R}^N$.
- x': in $\mathbb{R}^N$, $x = (x', x_N)$, i.e. $x' = (x_1, \dots, x_{N-1})$.
- x^α: $x_1^{\alpha_1} \dots x_N^{\alpha_N}$ for a multi-index α with α_j nonnegative integers for $j = 1, \dots, N$, for $x \in \mathbb{R}^N$.

- Δ: Laplacian $\sum_{j=1}^N \frac{\partial^2}{\partial x_j^2}$, defined on any open set $\Omega \subset \mathbb{R}^N$.
- δ_{ij}: Kronecker symbol, equal to 1 if $i = j$ and equal to 0 if $i \neq j$ (for $i, j = 1, \dots, N$).
- ε_{ijk}: for $i, j, k \in \{1, 2, 3\}$, completely antisymmetric tensor, equal to 0 if two indices are equal, and equal to the signature of the permutation $123 \mapsto ijk$ if indices are distinct (i.e. $\varepsilon_{123} = \varepsilon_{231} = \varepsilon_{312} = +1$ and $\varepsilon_{132} = \varepsilon_{321} = \varepsilon_{213} = -1$).
- γ_0: trace operator, defined for smooth functions by restriction to the boundary $\partial\Omega$, for an open set $\Omega \subset \mathbb{R}^N$ with a smooth boundary, and extended by density to functional spaces in which smooth functions are dense.
- Λ_1: Zygmund space, $|u(x+h) + u(x-h) - 2u(x)| \leq M\,|h|$ for all $x, h \in \mathbb{R}^N$.
- ν: exterior normal to $\Omega \subset \mathbb{R}^N$, open set with Lipschitz boundary.
- ρ_ε: smoothing sequence, with $\rho_\varepsilon(x) = \frac{1}{\varepsilon^N}\rho_1\left(\frac{x}{\varepsilon}\right)$ with $\varepsilon > 0$ and $\rho_1 \in C_c^\infty(\mathbb{R}^N)$ with $\int_{x \in \mathbb{R}^N} \rho_1(x)\,dx = 1$, and usually $\rho_1 \geq 0$.
- τ_h: translation operator of $h \in \mathbb{R}^N$, acting on a function $f \in L^1_{loc}(\mathbb{R}^N)$ by $\tau_h f(x) = f(x-h)$ a.e. $x \in \mathbb{R}^N$.
- Ω_F: $\{x \in \mathbb{R}^N \mid x_N \geq F(x')\}$, for a continuous function F, where $x' = (x_1, \dots, x_{N-1})$.

References

[1] CARLEMAN T., *Problèmes mathématiques dans la théorie cinétique des gaz*, Publ. Sci. Inst. Mittag-Leffler. 2, Almqvist & Wiksells Boktryckeri Ab, Uppsala 1957, 112 pp.

[2] CHÉRET R., *Detonation of Condensed Explosives*, Springer-Verlag, New York, 1993.

[3] COURANT R. & FRIEDRICHS K.O., *Supersonic Flow and Shock Waves*, Interscience Publishers, Inc., New York, 1948, xvi+464 pp. Reprinting of the 1948 original, Applied Mathematical Sciences, Vol. 21. Springer-Verlag, New York-Heidelberg, 1976, xvi+464 pp.

[4] DAFERMOS C., *Hyperbolic Conservation Laws in Continuum Physics* (Grundlehren der Mathematischen Wissenschaften, 325. Springer-Verlag, Berlin, 2000, xvi+443 pp.

[5] DAUTRAY Robert & LIONS Jacques-Louis, *Mathematical Analysis and Numerical Methods for Science and Technology*, Vol. 1. *Physical Origins and Classical Methods*, xviii+695 pp., Springer-Verlag, Berlin-New York, 1990.

[6] DAUTRAY Robert & LIONS Jacques-Louis, *Mathematical Analysis and Numerical Methods for Science and Technology*, Vol. 2. *Functional and Variational Methods*, xvi+561 pp., Springer-Verlag, Berlin-New York, 1988.

[7] DAUTRAY Robert & LIONS Jacques-Louis, *Mathematical Analysis and Numerical Methods for Science and Technology*, Vol. 3. *Spectral Theory and Applications*, x+515 pp., Springer-Verlag, Berlin, 1990.

[8] DAUTRAY Robert & LIONS Jacques-Louis, *Mathematical Analysis and Numerical Methods for Science and Technology*, Vol. 4. *Integral Equations and Numerical Methods*, x+465 pp., Springer-Verlag, Berlin, 1990.

[9] DAUTRAY Robert & LIONS Jacques-Louis, *Mathematical Analysis and Numerical Methods for Science and Technology*, Vol. 5. *Evolution Problems. I*, xiv+709 pp., Springer-Verlag, Berlin, 1992.

[10] DAUTRAY Robert & LIONS Jacques-Louis, *Mathematical Analysis and Numerical Methods for Science and Technology*, Vol. 6. *Evolution Problems. II*, xii+485 pp., Springer-Verlag, Berlin, 1993.

[11] DAUTRAY Robert & LIONS Jacques-Louis, *Mathematical Analysis and Numerical Methods for Science and Technology*, Vol. 7. *Évolution: Fourier, Laplace*, xliv+344+xix pp., INSTN: Collection Enseignement. Masson, Paris, 1988 (reprint of the 1985 edition).

[12]   DAUTRAY Robert & LIONS Jacques-Louis, *Mathematical Analysis and Numerical Methods for Science and Technology*, Vol. 8. *Évolution: semi-groupe, variationnel*, xliv+345–854+xix pp., INSTN: Collection Enseignement. Masson, Paris, 1988 (reprint of the 1985 edition).

[13]   DAUTRAY Robert & LIONS Jacques-Louis, *Mathematical Analysis and Numerical Methods for Science and Technology*, Vol. 9. *Évolution: numérique, transport*, xliv+855–1303 pp., INSTN: Collection Enseignement. Masson, Paris, 1988 (reprint of the 1985 edition).

[14]   FEYNMAN R., LEIGHTON R.B. & SANDS M., *The Feynman Lectures on Physics: The Definitive and Extended Edition*, 3 vol, Addison-Wesley, 2005.

[15]   FEYNMAN R., *Surely You're Joking, Mr. Feynman!*, Vintage, UK, 1992.

[16]   GATIGNOL R., *Théorie cinétique des gaz à répartition discrète de vitesses*, Lecture Notes in Physics 36, Springer, Berlin, 1975.

[17]   TARTAR L., *Une introduction à la théorie mathématique des systèmes hyperboliques de lois de conservation*, Publicazioni 682, Istituto di Analisi Numerica, Pavia, 1989.

[18]   TARTAR L., H-measures, a new approach for studying homogenisation, oscillations and concentration effects in partial differential equations. *Proc. Roy. Soc. Edinburgh Sect. A* **115**, 1990, no. 3-4, 193–230.

[19]   TARTAR L., Compensation effects in partial differential equations. *Memorie di Matematica e Applicazioni, Rendiconti della Accademia Nazionale delle Scienze detta dei XL*, Ser. V, vol. XXIX, 2005, 395–454.

[20]   TARTAR L., *An Introduction to Navier–Stokes Equation and Oceanography*, 271 pp., Lecture Notes of Unione Matematica Italiana, Vol. 1, Springer, Berlin-Heidelberg-New York, 2006.

[21]   TARTAR L., *An Introduction to Sobolev Spaces and Interpolation Spaces*, 248 pp., Lecture Notes of Unione Matematica Italiana, Vol. 3, Springer, Berlin-Heidelberg-New York, 2007.

[22]   TRUESDELL C. & MUNCASTER R., *Fundamentals of Maxwell's Kinetic Theory of a Simple Monatomic Gas. Treated as a Branch of Rational Mechanics.* Pure and Applied Mathematics, 83. Academic Press, Inc. [Harcourt Brace Jovanovich, Publishers], New York-London, 1980, xxvii+593 pp.

Index

Editor in Chief: Franco Brezzi

Editorial Policy

1. The UMI Lecture Notes aim to report new developments in all areas of mathematics and their applications - quickly, informally and at a high level. Mathematical texts analysing new developments in modelling and numerical simulation are also welcome.

2. Manuscripts should be submitted (preferably in duplicate) to
   Redazione Lecture Notes U.M.I.
   Dipartimento di Matematica
   Piazza Porta S. Donato 5
   I – 40126 Bologna
   and possibly to one of the editors of the Board informing, in this case, the Redazione about the submission. In general, manuscripts will be sent out to external referees for evaluation. If a decision cannot yet be reached on the basis of the first 2 reports, further referees may be contacted. The author will be informed of this. A final decision to publish can be made only on the basis of the complete manuscript, however a refereeing process leading to a preliminary decision can be based on a pre-final or incomplete manuscript. The strict minimum amount of material that will be considered should include a detailed outline describing the planned contents of each chapter, a bibliography and several sample chapters.

3. Manuscripts should in general be submitted in English. Final manuscripts should contain at least 100 pages of mathematical text and should always include
   – a table of contents;
   – an informative introduction, with adequate motivation and perhaps some historical remarks: it should be accessible to a reader not intimately familiar with the topic treated;
   – a subject index: as a rule this is genuinely helpful for the reader.

4. For evaluation purposes, manuscripts may be submitted in print or electronic form (print form is still preferred by most referees), in the latter case preferably as pdf- or zipped ps- files. Authors are asked, if their manuscript is accepted for publication, to use the LaTeX2e style files available from Springer's web-server at
   ftp://ftp.springer.de/pub/tex/latex/svmonot1/   for monographs
   and at
   ftp://ftp.springer.de/pub/tex/latex/svmultt1/   for multi-authored volumes

5. Authors receive a total of 50 free copies of their volume, but no royalties. They are entitled to a discount of 33.3% on the price of Springer books purchased for their personal use, if ordering directly from Springer.

6. Commitment to publish is made by letter of intent rather than by signing a formal contract. Springer-Verlag secures the copyright for each volume. Authors are free to reuse material contained in their LNM volumes in later publications: A brief written (or e-mail) request for formal permission is sufficient.